STABILITY OF INFINITE DIMENSIONAL STOCHASTIC DIFFERENTIAL EQUATIONS WITH APPLICATIONS

T0174313

CHAPMAN & HALL/CRC
Monographs and Surveys in
Pure and Applied Mathematics   **135**

STABILITY OF INFINITE

DIMENSIONAL STOCHASTIC

DIFFERENTIAL EQUATIONS

WITH APPLICATIONS

Kai Liu

CRC Press
Taylor & Francis Group
Boca Raton London New York

CRC Press is an imprint of the
Taylor & Francis Group, an **informa** business

A CHAPMAN & HALL BOOK

CRC Press
Taylor & Francis Group
6000 Broken Sound Parkway NW, Suite 300
Boca Raton, FL 33487-2742

First issued in paperback 2019

© 2006 by Taylor & Francis Group, LLC
CRC Press is an imprint of Taylor & Francis Group, an Informa business

No claim to original U.S. Government works

ISBN-13: 978-1-58488-598-6 (hbk)
ISBN-13: 978-0-367-39225-3 (pbk)

Library of Congress Cataloging-in-Publication Data

Catalog record is available from the Library of Congress

Visit the Taylor & Francis Web site at
http://www.taylorandfrancis.com

and the CRC Press Web site at
http://www.crcpress.com

To My Parents: Zhiyun Liu and Fengdi Cai

Contents

Preface

In most dynamical systems which describe processes in engineering, physics and economics, stochastic components and random noise are included. The stochastic aspects of the models are used to capture the uncertainty about the environment in which the system is operating and the structure and parameters of the models of physical processes being studied. Stochastic differential equations in infinite dimensional spaces are motivated by the development of analysis and the theory of stochastic processes itself such as stochastic partial differential equations and stochastic delay differential equations on the one hand, and by such topics as stochastic control, population biology and turbulence in applications on the other. The analysis and control of such systems then involves investigating their stability, which is a qualitative property and often regarded as the first characteristic of the dynamical systems (or models) studied.

Although the theory of stochastic differential equations in infinite dimensional spaces is already an established area of research, the corresponding study of stability properties only saw a rapid growth in the last two decades. In particular, most of the existing results are scattered throughout research journals and conference proceedings. These results have been obtained by using various methods, concepts and theorems from functional analysis, stochastic partial differential equations and functional differential equations. This makes it difficult for a newcomer to enter this interesting and important field.

The purpose of this book is to provide a systematic presentation of the contemporary theory of stability of stochastic differential equations in infinite dimensional spaces, mainly Hilbert spaces. The applications of this theory to various concrete problems are also shown. Hopefully, the style of presentation will be helpful in making the up to date material in this field accessible and meanwhile lay the foundation for future researches. The fundamental prerequisite for an intelligent reading of most material in this book is a knowledge of stochastic differential equations in infinite dimensions, for instance, at the level of the initial seven chapters in Da Prato and Zabczyk's book [1]. An acquaintance with stability theory of finite dimensional stochastic differential equations such as those in Khas'minskii's classic book [1] will prove useful, although not essential.

A brief description of the organization of this work follows. In Chapter 1, we recall basic concepts of the theory of stochastic differential equations in infinite dimensional spaces, mainly Hilbert spaces. In this way, such notions as Q-Wiener processes, stochastic integrals, strong and mild solutions will be appropriately reviewed. We will also present some proofs of the results which are not available in existing books and are to be found scattered in the literature.

Chapter 2 of the book is devoted to an investigation of stability for the essential classes of linear stochastic evolution equations. The central result is a formulation of the characteristic conditions of mean square exponential stability in terms of stationary Lyapunov type equations. We also deal in this chapter with the topic of almost sure pathwise stability of solution processes (strong and mild) of stochastic linear evolution systems.

In Chapter 3, we proceed to the study of nonlinear stochastic differential equations. This chapter contains fundamental theory and interesting examples concerning a number of topics involved with the stability behavior of nonlinear systems. In particular, we generalize linear characteristic results in Chapter 2 in a suitable way to obtain a version for a class of nonlinear stochastic evolution systems. Motivated by the idea of reducing the stability of a nonlinear stochastic system to problems about a linear system, we explore the so-called Lyapunov function characterization method and the associated first order approximation technique. Various interesting topics such as stabilization, Lyapunov exponents and stochastic decay rates of stochastic systems are also investigated.

Chapter 4 is an extensive study of the stability theory of stochastic functional differential equations in infinite dimensions. The reduction of a stochastic neutral equation to a stochastic retarded equation with infinite delay and the notion of the L^2-stability in mean are developed. As a consequence, the established result that L^2-stability in mean implies asymptotic stability allows one to apply the Lyapunov function method to obtain stability criteria. For nonlinear cases, the methods of Lyapunov functionals and Razumikhin types are emphasized and contrasted, and applications are illustrated by interesting examples.

In Chapter 5, selected topics and applications are presented in which the choice of the material reflects my own personal preference. The treatment here is somewhat sketchy and by no means the only approach or even the usual one. The purpose of our formulation here is to show some of the topics, especially applications such as stochastic optimal control and feedback stabilization, stochastic reaction-diffusion, Navier-Stokes equations and stochastic population dynamics, beyond the main scheme of the book, but associated with the stability theory of stochastic evolution equations. I hope also to stimulate further work in these and relevant fields.

The appendix contains one proof of existence and uniqueness of strong solutions for a certain class of stochastic functional differential equations. As far as I know, there seems not to be a similar presentation in the existing literature. At this stage, I feel it appropriate to present this as a background for the arguments in the previous chapters.

Notes and comments at the end of each chapter contain historical and related background material as well as references to the results discussed in that chapter. The pervading influence of a variety of authors' work on this book is obvious. I have drawn freely on their work and hopefully I can acknowledge my scientific debt to them by some remarks shown there.

The book is mainly devoted to the stability theory of stochastic differential equations in Hilbert spaces. It should be emphasized that my choice of material is highly subjective. This subject appears to be important and has recently been receiving increasing attention. As a consequence, the volume of relevant literature is rapidly expanding. In particular, the lengthy list of references at the end of the book is somewhat incomplete and only includes those titles which pertain directly to the contents. The author wishes to apologize to those people whose research might have been overlooked.

It should be pointed out that this work would not have been possible without the inspiration and wisdom of my colleagues. I am indebted to many good friends who read the first draft on which this book is based, corrected errors, and suggested improvements. In particular, I would like to express my sincere thanks to T. Caraballo, P. L. Chow, J. A. Langa, V. Mandrekar, B. Maslowski and Y. F. Shi for their valuable advice. Gratitude goes to Professor A. Truman for his constructive suggestions on the initial two chapters and Professor P. Giblin for his careful check of language errors. Thanks also go to Professor Jiezhong Zou, Professor Zhiyuan Huang and Dr. Xiaogu Zheng for their lasting concern and encouragement.

The author acknowledges the various financial supports from EPSRC, Royal Society, LMS and the University of Liverpool in UK during the preparation of this book.

Kai Liu

Chapter 1

Stochastic Differential Equations in Infinite Dimensions

We begin by recalling some basic definitions and preliminaries, especially those on stochastic integration and stochastic differential equations in infinite dimensional spaces. We recall important inequalities for stochastic integrals with respect to Wiener processes which are essential for the subsequent developments. We also establish two notions of solutions, strong and mild, and investigate the existence and uniqueness of these two kinds of solutions under suitable assumptions. To present the proofs of all of these results here would require preparatory background material which would considerably increase both the size and scope of this book. Therefore, we would like to adopt the approach of omitting the proofs of those results which are treated in detail in well-known standard books, such as Da Prato and Zabczyk [1], Rozovskii [1] and Métivier [1]. However, those proofs will be presented which are not available in existing books and are to be found scattered in the literature, or which discuss ideas specially relevant to our purposes.

1.1 Notations, Definitions and Preliminaries

A *measurable space* is a pair $(\Omega, \mathcal{F})$ where Ω is a set and $\mathcal{F}$ is a σ-field, also called a σ-algebra, of subsets of Ω. This means that the family $\mathcal{F}$ contains the set Ω and is closed under the operation of taking complements and countable unions of its elements. If $(\Omega, \mathcal{F})$ and $(S, \mathcal{S})$ are two measurable spaces, then a mapping ξ from Ω into S such that the set $\{\omega \in \Omega : \xi(\omega) \in A\} = \{\xi \in A\}$ belongs to $\mathcal{F}$ for arbitrary $A \in \mathcal{S}$ is called a *measurable mapping* or a *random variable* from $(\Omega, \mathcal{F})$ into $(S, \mathcal{S})$. A random variable is called *simple* if it takes on only a finite number of values. In this book, we shall only be concerned with the case when S is a complete, separable, metric space, then $\mathcal{S} = \mathcal{B}(S)$, the Borel σ-field of S which is the smallest σ-field containing all closed (or open) subsets of S. If S is a separable Banach or Hilbert space, we shall denote its norm by $\| \cdot \|_S$ and its topological dual by S^*.

A *probability measure* on a measurable space $(\Omega, \mathcal{F})$ is a σ-additive function P from $\mathcal{F}$ into $[0, 1]$ such that $P(\Omega) = 1$. The triplet $(\Omega, \mathcal{F}, P)$ is called a

1

probability space. If $(\Omega, \mathcal{F}, P)$ is a probability space, we set

$$\bar{\mathcal{F}} = \{A \subset \Omega : \exists B, C \in \mathcal{F}; \ B \subset A \subset C, \ P(B) = P(C)\}.$$

Then $\bar{\mathcal{F}}$ is a σ-field, called the *completion* of $\mathcal{F}$. If $\mathcal{F} = \bar{\mathcal{F}}$, the probability space $(\Omega, \mathcal{F}, P)$ is said to be *complete*.

Let $(\Omega, \mathcal{F}, P)$ denote a complete probability space. A family $\{\mathcal{F}_t\}$, $t \geq 0$, for which the $\mathcal{F}_t$ are sub-σ-fields of $\mathcal{F}$ and form an increasing family of σ-fields, is called a *filtration* if $\mathcal{F}_s \subset \mathcal{F}_t \subset \mathcal{F}$ for $s \leq t$. With the $\{\mathcal{F}_t\}_{t \geq 0}$, one can associate two other filtrations by setting σ-fields: $\mathcal{F}_{t-} = \bigvee_{s<t} \mathcal{F}_s$ if $t > 0$, $\mathcal{F}_{t+} = \bigcap_{s>t} \mathcal{F}_s$ if $t \geq 0$, where $\bigvee_{s<t} \mathcal{F}_s$ is the smallest σ-field containing $\bigcup_{s<t} \mathcal{F}_s$. The σ-field $\mathcal{F}_{0-}$ is not defined and, by convention, we put $\mathcal{F}_{0-} = \mathcal{F}_0$, and also $\mathcal{F}_\infty = \bigvee_{t \geq 0} \mathcal{F}_t$. An increasing family $\{\mathcal{F}_t\}_{t \geq 0}$ is *right-continuous* if

(i). $\mathcal{F}_{t+} = \mathcal{F}_t$ for each $t \geq 0$.

For many purposes we need the following assumption:

(ii). $\mathcal{F}_0$ contains all P-null sets in $\mathcal{F}$.

Unless otherwise stated, completeness of $(\Omega, \mathcal{F}, P)$ and Conditions (i), (ii) will always be assumed to hold in this book. Sometimes, we also call a filtration satisfying (i), (ii) *normal* or say that it satisfies *the usual conditions*.

If ξ is a random variable from $(\Omega, \mathcal{F})$ into $(S, \mathcal{S})$ and P a probability measure on Ω, then by $\mathcal{Q}(\xi)(\cdot)$ we will denote the image of P under the mapping ξ:

$$\mathcal{Q}(\xi)(A) = P\{\omega \in \Omega : \xi(\omega) \in A\}, \quad \forall A \in \mathcal{S}.$$

The measure is called the *distribution* or the *law* of ξ.

Definition 1.1.1   With respect to an increasing family $\{\mathcal{F}_t\}_{t \geq 0}$ of sub-σ-fields of the type discussed above, we call a function $\tau : \Omega \to \bar{\bar{\mathbf{R}}}_+ = [0, \infty]$ a *stopping time* for $\{\mathcal{F}_t\}$, $t \geq 0$, if $\{\omega : \tau(\omega) \leq t\} \in \mathcal{F}_t$ for every $t \geq 0$. A similar notion is defined for a family $\{\mathcal{F}_t\}$, $t \in [0, T]$, $0 \leq T \leq \infty$, with the assumption that the range of τ is now in $[0, T]$.

Define the σ-field of events prior to τ, denoted by $\mathcal{F}_\tau$, as

$$\mathcal{F}_\tau = \{A \in \mathcal{F}_\infty : A \cap \{\tau \leq t\} \in \mathcal{F}_t \text{ for every } t\}.$$

Theorem 1.1.1   *For σ and τ stopping times relative to $\{\mathcal{F}_t\}_{t \geq 0}$,*

(a). *$\mathcal{F}_\tau$ is a σ-field;*
(b). *τ is $\mathcal{F}_\tau$-measurable;*
(c). *$\sigma \leq \tau$ almost surely implies $\mathcal{F}_\sigma \subset \mathcal{F}_\tau$;*
(d). *$\sigma \wedge \tau$ and $\sigma \vee \tau$ are also stopping times with respect to $\{\mathcal{F}_t\}_{t \geq 0}$.*

Assume now that S is a separable Banach space with norm $\|\cdot\|_S$ and ξ is an S-valued random variable on $(\Omega, \mathcal{F}, P)$. By a standard limit argument, we can define the integral $\int_\Omega \xi dP$ of ξ with respect to the probability measure P, often denoted by $E(\xi)$. The integral defined in this way is called *Bochner's integral*. We denote by $L^1(\Omega, \mathcal{F}, P; S)$ the set of all equivalence classes of S-valued random variables with respect to the equivalence relation of almost sure equality. In the same way as for real random variables, one can check that $L^1(\Omega, \mathcal{F}, P; S)$, equipped with the norm

$$\|\xi\|_1 = E\|\xi\|_S,$$

is a Banach space. In a similar manner, one can define $L^p(\Omega, \mathcal{F}, P; S)$ for arbitrary $p > 1$ with norms

$$\|\xi\|_p = (E\|\xi\|_S^p)^{1/p}, \qquad p \in (1, \infty),$$

and

$$\|\xi\|_\infty = \operatorname*{ess.\,sup}_{w \in \Omega} \|\xi(\omega)\|_S.$$

If Ω is an interval $[0, T]$, $\mathcal{F} = \mathcal{B}([0, T])$, $0 \leq T < \infty$, and P is the Lebesgue measure on $[0, T]$, we also write $L^p(0, T; S)$, or more simply $L^p(0, T)$, for the spaces defined above when no confusion arises.

Of great interest to us will be operator-valued random variables and their integrals. Let K and H be two separable Hilbert spaces with norms $\|\cdot\|_K$, $\|\cdot\|_H$ and inner products $\langle\cdot, \cdot\rangle_K$, $\langle\cdot, \cdot\rangle_H$, respectively. We denote by $\mathcal{L}(K, H)$ the set of all linear bounded operators from K into H, equipped with the usual operator norm $\|\cdot\|$. From now on, without further specification we always use the same symbol $\|\cdot\|$ to denote norms of operators regardless of the spaces involved when no confusion is possible. The set $\mathcal{L}(K, H)$ is a linear space and, equipped with the operator norm, becomes a Banach space. Unfortunately, if both the spaces K and H are infinite dimensional, the space $\mathcal{L}(K, H)$ is not generally separable. A direct consequence of this inseparability is that Bochner's integral definition cannot be applied to $\mathcal{L}(K, H)$-valued random variables. One of the methods of overcoming these difficulties is to introduce a weaker concept of measurability.

A mapping $\Phi(\cdot)$ from Ω into $\mathcal{L}(K, H)$ is said to be *strongly measurable* if for arbitrary $k \in K$, $\Phi(\cdot)k$ is measurable as a mapping from $(\Omega, \mathcal{F})$ into $(H, \mathcal{B}(H))$. Let $\mathcal{F}(\mathcal{L}(K, H))$ be the smallest σ-field of subsets of $\mathcal{L}(K, H)$ containing all sets of the form

$$\{\Phi \in \mathcal{L}(K, H) : \Phi k \in A\}, \quad k \in K, \quad A \in \mathcal{B}(H),$$

then $\Phi : \Omega \to \mathcal{L}(K, H)$ is a strongly measurable mapping from $(\Omega, \mathcal{F})$ into the space $(\mathcal{L}(K, H), \mathcal{F}(\mathcal{L}(K, H)))$. Elements of $\mathcal{F}(\mathcal{L}(K, H))$ are called *strongly measurable*. Mapping Φ is said to be *Bochner integrable* with respect to the measure P if for arbitrary k, the mapping $\Phi(\cdot)k$ is Bochner integrable and there exists a bounded linear operator $\Psi \in \mathcal{L}(K, H)$ such that

$$\int_\Omega \Phi(\omega)kP(d\omega) = \Psi k, \qquad k \in K.$$

The operator Ψ is then denoted as

$$\Psi = \int_\Omega \Phi(\omega)P(d\omega)$$

and called the *strong Bochner integral* of Φ. This integral has many of the properties of the Lebesgue integral. For instance, it is easy to show that if K and H are both separable, then $\|\Phi(\cdot)\|$ is a measurable function and

$$\|\Psi\| \leq \int_\Omega \|\Phi(\omega)\|P(d\omega).$$

The following operator spaces are of fundamental importance. Let K and H be two separable Hilbert spaces. We denote by $\mathcal{L}_1(K, H)$ the space of all nuclear operators and by $\mathcal{L}_2(K, H)$ the space of all Hilbert-Schmidt operators from K into H. It can be proved that the spaces $\mathcal{L}_1(K, H)$ and $\mathcal{L}_2(K, H)$ are both separable Banach spaces ($\mathcal{L}_2(K, H)$ is actually a Hilbert space) under the usual norms and are strongly measurable subsets of $\mathcal{L}(K, H)$.

Assume that S is a separable Banach space with norm $\|\cdot\|_S$ and let $\mathcal{B}(S)$ be the σ-field of its Borel subsets. Let $(\Omega, \mathcal{F}, P)$ be a probability space. An arbitrary family $X = \{X_t\}_{t\geq 0}$ of S-valued random variables X_t, $t \geq 0$, defined on Ω is called a *stochastic process*. Sometimes, we also write $X(t, \omega) = X(t) = X_t(\omega)$ for all $t \geq 0$ and $\omega \in \Omega$. The functions $X.(\omega)$ are called the *trajectories* of X. In our study of stochastic processes, we need some additional regularities to proceed with our programme. To this end, we now introduce several definitions of regularity for a process X on $I = [0, T)$, where T could be finite or infinite.

(a). X is *measurable* if the mapping $X(\cdot, \cdot) : I \times \Omega \to S$ is $\mathcal{B}(I) \times \mathcal{F}$-measurable (all stochastic processes considered in this book will be assumed to be measurable);

(b). Let $\{\mathcal{F}_t\}$, $t \in I$, be an increasing family of σ-fields. The process X is $\{\mathcal{F}_t\}_{t\in I}$-*adapted* if each X_t is measurable with respect to $\mathcal{F}_t$, $t \in I$;

(c). X is *stochastically continuous* at $t_0 \in I$ if, $\forall \varepsilon > 0$, $\forall \delta > 0$, $\exists \rho > 0$ such that

$$P\{\|X_t - X_{t_0}\|_S \geq \varepsilon\} \leq \delta, \qquad \forall t \in [t_0 - \rho, t_0 + \rho] \cap [0, T);$$

(d). X is *stochastically continuous in I* if it is stochastically continuous at every point of I;

(e). X is *continuous with probability one (or continuous)* if its trajectories $X(\cdot, \omega)$ are continuous almost surely.

Definition 1.1.2

(i). The S-valued processes $X = \{X_t\}$ and $Y = \{Y_t\}$, $t \in I$, defined on the probability space $(\Omega, \mathcal{F}, P)$ are called *equivalent* if for every $\{t_1, \cdots, t_n\} \subset I$ and set $B_i \in \mathcal{B}(S)$, $i = 1, 2, \cdots, n$,

$$P\{\omega : X_{t_1} \in B_1, \cdots, X_{t_n} \in B_n\} = P\{\omega : Y_{t_1} \in B_1, \cdots, Y_{t_n} \in B_n\};$$

(ii). A stochastic process $Y = \{Y_t\}$, $t \in I$, is called a *modification* or a *version* of $X = \{X_t\}$ if

$$P\{\omega \in \Omega : X_t(\omega) \neq Y_t(\omega)\} = 0, \qquad \forall t \in I;$$

(iii). The processes $X = \{X_t\}$ and $Y = \{Y_t\}$, $t \in I$, are called *indistinguishable* if

$$P\{\omega \in \Omega : X_t(\omega) = Y_t(\omega); \ \forall t \in I\} = 1.$$

An immediate consequence of Definition 1.1.2 is that if Y is a modification of X, and supposing that both of the processes have almost surely continuous sample paths, then X and Y are indistinguishable.

Note that if X is a stochastic process on $I = [0, T)$, the function $X(\cdot, \cdot)$ need not be measurable in the product space $\mathcal{B}([0, T)) \times \mathcal{F}$. However, we have the following result:

Theorem 1.1.2 *Let X_t, $t \in I$, be a stochastically continuous process with values in the separable Banach space S. Then X has a measurable modification.*

Definition 1.1.3 Suppose $X = \{X_t\}$, $t \in I$, is an S-valued process and $\{\mathcal{F}_t\}_{t \in I}$ is a normal increasing family of sub-σ-fields of $\mathcal{F}$. X is said to be *progressively measurable* with respect to $\{\mathcal{F}_t\}_{t \in I}$ if for every $t \in I$, the mapping

$$[0, t] \times \Omega \to S, \qquad (s, \omega) \to X(s, \omega),$$

is $\mathcal{B}([0, t]) \times \mathcal{F}_t$-measurable.

It is obvious that if X is progressively measurable with respect to $\{\mathcal{F}_t\}_{t \in I}$, then it must be both measurable and $\{\mathcal{F}_t\}_{t \in I}$-adapted. The converse of this statement is known to be false. However, if we are concerned only with versions of a given process, this converse can hold in the following sense:

Theorem 1.1.3 *Let X_t, $t \in I$, be a stochastically continuous and adapted process with values in the separable Banach space S. Then X has a progressively measurable modification.*

Given an S-valued process $X = \{X_t\}$, $t \in I$, and supposing that $\sigma : \Omega \to I$ is a real random variable, it is desirable for many applications that the mapping $X_\sigma : \Omega \to S$ defined by $X_\sigma(\omega) = X_{\sigma(\omega)}(\omega)$ is also measurable. This will probably be true if X is a measurable process, although it is not generally true otherwise. More information associated with this is provided by the following result:

Theorem 1.1.4   *Let $X = \{X_t\}$, $t \in I$, be an S-valued progressively measurable process with respect to $\{\mathcal{F}_t\}_{t \in I}$, $t \in I$, and let $\tau \in I$ be a finite stopping time. Then the random variable X_τ is $\mathcal{F}_\tau$-measurable.*

Using some of the notions defined above for stochastic processes, we can construct interesting examples of stopping times as follows. Let X_t, $t \in I$, be a process with state space S which is continuous and $\{\mathcal{F}_t\}_{t \in I}$-adapted. For any Borel set B in $\mathcal{B}(S)$, define τ_B, called the *hitting time* for the set B, by

$$\tau_B = \begin{cases} \inf\{t \geq 0 : X_t(\omega) \in B\}, \\ \infty \quad \text{if the set is empty.} \end{cases}$$

Then τ_B is in fact a stopping time relative to the family $\{\mathcal{F}_t^X\} := \sigma\{X_s, \, s \leq t \in I\}$, the σ-fields generated by the process X up to time t.

The following σ-field $\mathcal{P}_\infty$ of subsets of $[0, \infty) \times \Omega$ plays an essential role in the construction of stochastic integrals. That is, $\mathcal{P}_\infty$ is defined as the σ-field generated by sets of the form:

$$(s, t] \times F, \quad 0 \leq s < t < \infty, \quad F \in \mathcal{F}_s \text{ and } \{0\} \times F, \quad F \in \mathcal{F}_0.$$

This σ-field is called *predictable* and its elements are called *predictable sets*. The restriction of the σ-field $\mathcal{P}_\infty$ to $[0, T] \times \Omega$, $0 \leq T < \infty$, will be denoted by $\mathcal{P}_T$. An arbitrary measurable mapping from $([0, \infty) \times \Omega, \mathcal{P}_\infty)$ or $([0, T] \times \Omega, \mathcal{P}_T)$ into $(S, \mathcal{B}(S))$ is called a *predictable process*. A predictable process is necessarily an adapted one.

Theorem 1.1.5

(i). *An adapted process X_t, $t \in I$, which takes its values in $(\mathcal{L}(K, H), \mathcal{F}(\mathcal{L}(K, H)))$ such that for arbitrary $k \in K$ and $h \in H$ the process $\langle X_t k, h \rangle_H$, $t \in I$, has left continuous trajectories, is predictable;*

(ii). *Assume that X_t, $t \in I$, is an adapted and stochastically continuous process on the interval I. Then the process X has a predictable version on I.*

Let $\mathcal{E}$ be an arbitrary sub-σ-field of $\mathcal{F}$ and $E(\cdot \mid \mathcal{E})$ denote the conditional expectation given $\mathcal{E}$. An S-valued process $X = X_t$, $t \in I$, defined on $(\Omega, \mathcal{F}, P)$

and adapted to the family $\{\mathcal{F}_t\}_{t\in I}$ is said to be a *Markov process* with respect to $\{\mathcal{F}_t\}_{t\in I}$ if the following property is satisfied: for all s and t $(T \geq t \geq s)$,

$$E(f(X_t) \mid \mathcal{F}_s) = E(f(X_t) \mid X_s)$$

almost surely for every bounded real-valued Borel function $f(\cdot)$ on S. We say that X_t, $t \in I$, is a Markov process if it is Markov process with respect to $\{\mathcal{F}_t^X\}_{t\in I}$. Let X be a Markov process with state space S. Then it can be shown that there exists a function $P(s, x, t, \Gamma)$ $(s < t, x \in S, \Gamma \in \mathcal{B}(S))$ with the following properties:

(a). For all (s, x, t), $P(s, x, t, \cdot)$ is a probability measure on $\mathcal{B}(S)$;
(b). For each (s, t, Γ), $P(s, \cdot, t, \Gamma)$ is $\mathcal{B}(S)$-measurable;
(c). $P(X_t \in \Gamma \mid \mathcal{F}_s^X) = P(s, x, t, \Gamma)$ almost surely where $P(X_t \in \Gamma \mid \mathcal{F}_s^X) = E(\chi_{\{X_t\in\Gamma\}} \mid \mathcal{F}_s^X)$ and $\chi_{\{X_t\in\Gamma\}}$ is the indicator function of $\{X_t \in \Gamma\}$;
(d). The function $P(s, x, t, \Gamma)$ satisfying the properties (a), (b) and (c) is called the *transition probability function* of the Markov process if it further satisfies the following Chapman-Kolmogorov equation

$$P(s, x, t, \Gamma) = \int_S P(s, x, u, dy) P(u, y, t, \Gamma)$$

for all $x \in S$, $\Gamma \in \mathcal{B}(S)$ and (s, u, t) such that $s < u < t$.

For each s and t $(0 \leq s \leq t < T)$, define $P_s^t f$ for Borel measurable function $f : S \to \mathbf{R}^1$ by

$$P_s^s = I \quad \text{(the identity operator)},$$
$$P_s^t f(x) = \int_S f(y) P(s, x, t, dy),$$

provided the integral on the right exists. On the class of bounded Borel-measurable functions, P_s^t defines a family of operators satisfying the semigroup property

$$P_s^u f = P_s^t P_t^u f, \quad 0 \leq s \leq t \leq u < T.$$

The equation is essentially a restatement of the Chapman-Kolmogorov equation. The family $\{P_s^t\}$ is called the *semigroup of the Markov process* X.

The Markov process X_t, $t \geq 0$, is said to have *homogeneous* transition probability functions if

$$P(s, x, t, \Gamma) = P(0, x, t - s, \Gamma).$$

It is clear that in this case, we have $P_s^t = P_0^{t-s}$. In particular, write

$$P(t - s, x, \Gamma) = P(0, x, t - s, \Gamma) \quad \text{and} \quad P_{t-s} = P_0^{t-s},$$

the semigroup is now given by $\{P_t\}$, $t \geq 0$,

$$P_t f(x) = \int_S f(y) P(t, x, dy).$$

Let $C_b = C_b(S)$ be the class of real-valued, bounded continuous functions on S.

Definition 1.1.4   The semigroup $\{P_s^t\}$ has the *Feller property*, or is called a *Feller semigroup* if, for arbitrary $f \in C_b$ and $s \geq 0$, $u > 0$, the function

$$x \to P_s^{s+u} f(x)$$

is continuous.

Let S be a separable Banach space with norm $\| \cdot \|_S$ and $M = M_t$, $t \in I = [0, T)$, an S-valued stochastic process defined on $(\Omega, \mathcal{F}, \{\mathcal{F}_t\}_{t \in I}, P)$. If $E\|M_t\|_S < \infty$ for all $t \in I$, then the process is called *integrable*. An integrable and adapted S-valued process M_t, $t \in I$, is said to be a *martingale* with respect to $\{\mathcal{F}_t\}_{t \in I}$ if

$$E(M_t \mid \mathcal{F}_s) = M_s \qquad P - a.s. \tag{1.1.1}$$

for arbitrary $t \geq s$, t, $s \in I$. By the definition of conditional expectations, the identity (1.1.1) is equivalent to the following statement

$$\int_F M_t dP = \int_F M_s dP, \quad \forall F \in \mathcal{F}_s, \ s \leq t, \ s, t \in I.$$

We also recall that a real-valued integrable and adapted process M_t, $t \in I$, is said to be a *submartingale* (resp. *supermartingale*) with respect to $\{\mathcal{F}_t\}_{t \in I}$ if

$$E(M_t \mid \mathcal{F}_s) \geq M_s, \ (\text{resp. } E(M_t \mid \mathcal{F}_s) \leq M_s), \ P - a.s.$$

for any $s \leq t$, $s, t \in I$. The following is a straightforward consequence of classical martingale theory.

Theorem 1.1.6   *The following statements hold:*

(i). If M_t is an S-valued martingale, then $\|M_t\|_S$, $t \in I$, is a submartingale;
(ii). If $g(\cdot)$ is an increasing convex function from $[0, \infty)$ into $[0, \infty)$ and $Eg(\|M_t\|_S) < \infty$ for all $t \in I$, then $g(\|M_t\|_S)$, $t \in [0, T)$, is a submartingale.

As an immediate consequence of Theorem 1.1.6 and the maximal inequalities for real-valued submartingales, we obtain the following inequalities due to Doob.

Theorem 1.1.7   *The following statements hold:*

(i). If M_t, $t \in L \subset [0, \infty)$, is an S-valued martingale, L a countable set and $p \geq 1$, for arbitrary $\lambda > 0$,

$$P\left\{ \sup_{t \in L} \|M_t\|_S \geq \lambda \right\} \leq \lambda^{-p} \sup_{t \in L} E(\|M_t\|_S^p); \qquad (1.1.2)$$

(ii). If, in addition, $p > 1$, then

$$E\left(\sup_{t \in L} \|M_t\|_S^p \right) \leq \left(\frac{p}{p-1} \right)^p \sup_{t \in L} E(\|M_t\|_S^p); \qquad (1.1.3)$$

(iii). The above estimates remain true if the set L is uncountable and the martingale M_t, $t \in L$, is continuous.

Let $[0, T]$, $0 \leq T < \infty$, be a subinterval of $[0, \infty)$. A continuous S-valued stochastic process M_t, $t \in [0, T]$, defined on $(\Omega, \mathcal{F}, \{\mathcal{F}_t\}_{t \in [0,T]}, P)$, is a *continuous square integrable martingale* with respect to $\{\mathcal{F}_t\}_{t \in [0,T]}$ if it is a martingale with almost surely continuous trajectories and satisfies, in addition, $\sup_{t \in [0,T]} E\|M_t\|_S^2 < \infty$. Let us denote by $\mathcal{M}_T^2(S)$ the space of all S-valued continuous, square integrable martingales M. By a standard argument, we can prove, using Theorem 1.1.7, that:

Theorem 1.1.8   *The space $\mathcal{M}_T^2(S)$, equipped with the norm*

$$\|M\|_{\mathcal{M}_T^2(S)} = \left(E \sup_{t \in [0,T]} \|M_t\|_S^2 \right)^{1/2},$$

is a Banach space.

Denote by $\mathcal{L}_1 = \mathcal{L}_1(K) = \mathcal{L}_1(K, K)$ the space of all nuclear operators from the separable Hilbert space K into itself, equipped with the usual nuclear norm. Then $\mathcal{L}_1$ is a separable Banach space. An $\mathcal{L}_1$-valued process V. is said to be *increasing* if the operators V_t, $t \in [0, T]$, are nonnegative, denoted by $V_t \geq 0$, i.e., for any $k \in K$, $\langle V_t k, k \rangle_K \geq 0$, $t \in [0, T]$, and $0 \leq V_s - V_t$ if $0 \leq t \leq s \leq T$. An $\mathcal{L}_1$-valued continuous, adapted and increasing process V_t such that $V_0 = 0$ is said to be a *tensor quadratic variation process* of the martingale $M_t \in \mathcal{M}_T^2(K)$ if and only if for arbitrary $a, b \in K$, the process

$$\langle M_t, a \rangle_K \langle M_t, b \rangle_K - \langle V_t a, b \rangle_K, \quad t \in [0, T],$$

is a continuous $\mathcal{F}_t$-martingale, or equivalently, if and only if the process

$$M_t \otimes M_t - V_t, \qquad t \in [0, T],$$

is a continuous $\mathcal{F}_t$-martingale, where $(a \otimes b)k := a\langle b, k \rangle_K$ for any $k \in K$ and $a, b \in K$. One can show that the process V_t is uniquely determined and can be denoted therefore by $\ll M_t \gg$, $t \in [0, T]$.

On the other hand, one can also show that there exists a real-valued, increasing, continuous process which is uniquely determined up to probability one, denoted by $[M_t]$ with $[M_0] = 0$, called the *quadratic variation* of M_t, such that

$$\|M_t\|_K^2 - [M_t]$$

is an $\mathcal{F}_t$-martingale.

With regard to the relation between $\ll M_t \gg$ and $[M_t]$ of M_t, we have the following:

Theorem 1.1.9   *For arbitrary $M_t \in \mathcal{M}_T^2(K)$, there exists a unique predictable, positive symmetric element $Q_M(\omega, t)$, or simply $Q(\omega, t)$, of $\mathcal{L}_1(K)$ such that*

$$\ll M_t \gg = \int_0^t Q_M(\omega, s)d[M_s]$$

for all $t \in [0, T]$. In particular, we also call the K-valued stochastic process M_t, $t \geq 0$, a $Q_M(\omega, t)$-martingale process.

In a similar manner, one can define the so-called *cross quadratic variation* for any $M_t \in \mathcal{M}_T^2(K)$, $N_t \in \mathcal{M}_T^2(K)$ as a unique continuous process, denoted by $\ll M_t, N_t \gg$, such that

$$M_t \otimes N_t - \ll M_t, N_t \gg, \qquad t \in [0, T],$$

is a continuous $\mathcal{F}_t$-martingale (cf. Da Prato and Zabczyk [1]).

As a special case, we can consider a symmetric non-negative operator $Q \in \mathcal{L}(K)$ with $trQ < \infty$, where trA denotes the trace of operator A. Then there exists a complete orthonormal system $\{e_k\}_{k \geq 1}$ in K, and a bounded sequence of nonnegative real numbers λ_k such that

$$Qe_k = \lambda_k e_k, \qquad k = 1, 2, \cdots.$$

A K-valued stochastic process W_t, $t \geq 0$, is called a *Q-Wiener process* if

(1). $W_0 = 0$;
(2). W_t has continuous trajectories;
(3). W_t has independent increments;
(4). $E(W_t) = 0$ and $\text{Cov}(W_t - W_s) = (t - s)Q$ for all $t \geq s \geq 0$, where $\text{Cov}(X)$ denotes the *covariance operator* of $X \in H$ (cf. Da Prato and Zabczyk [1]).

It is straightforward to see that the tensor quadratic variation of a Q-Wiener process in K, with $trQ < \infty$, is given by the formula $\ll W_t \gg = tQ$, $t \geq 0$.

1.2   Wiener Processes and Stochastic Integration

Let K be a real separable Hilbert space with norm $\|\cdot\|_K$ and inner product $\langle\cdot,\cdot\rangle_K$, respectively. A probability measure $\mathcal{N}$ on $(K,\mathcal{B}(K))$ is called *Gaussian* if for arbitrary $u \in K$, there exist numbers $\mu \in \mathbf{R}^1$, $\sigma \geq 0$, such that

$$\mathcal{N}\{x \in K : \langle u,x\rangle_K \in A\} = N(\mu,\sigma^2)(A), \qquad A \in \mathcal{B}(\mathbf{R}^1),$$

where $N(\mu,\sigma^2)$ is the usual one dimensional normal distribution with mean μ and standard deviation σ. It can be proved that if $\mathcal{N}$ is Gaussian, then there exist an element $m \in K$ and a symmetric nonnegative trace class operator $Q \in \mathcal{L}(K)$ such that

$$\int_K \langle k,x\rangle_K \mathcal{N}(dx) = \langle m,k\rangle_K, \qquad \forall k \in K,$$

$$\int_K \langle k_1,x\rangle_K \langle k_2,x\rangle_K \mathcal{N}(dx) - \langle m,k_1\rangle_K \langle m,k_2\rangle_K$$
$$= \langle Qk_1,k_2\rangle_K, \qquad \forall k_1,\ k_2 \in K,$$

and, furthermore the characteristic function

$$\hat{\mathcal{N}}(\lambda) = \int_K e^{i\langle\lambda,x\rangle_K}\mathcal{N}(dx) = e^{i\langle\lambda,m\rangle_K - \frac{1}{2}\langle Q\lambda,\lambda\rangle_K}, \qquad \lambda \in K.$$

Therefore, the measure $\mathcal{N}$ is uniquely determined by m and Q and denoted also by $\mathcal{N}(m,Q)$. In particular, in this case we call m the *mean* and Q the *covariance operator* of $\mathcal{N}$.

Recall that we always assume the probability space $(\Omega,\mathcal{F},P)$ is equipped with a right continuous filtration $\{\mathcal{F}_t\}_{t\geq 0}$ such that $\mathcal{F}_0$ contains all sets of P-measure zero. Let $W(t)$ or W_t, $t \geq 0$, be a K-valued, Q-Wiener process which is assumed to be adapted to $\{\mathcal{F}_t\}_{t\geq 0}$ and for every $t > s$ the increments $W_t - W_s$ are independent of $\mathcal{F}_s$. Hence, W_t, $t \geq 0$, is a continuous martingale relative to $\{\mathcal{F}_t\}_{t\geq 0}$ and we have the following representation of W_t:

$$W_t = \sum_{i=1}^{\infty} B_t^i e_i, \tag{1.2.1}$$

where $\{e_i\}$ is an orthonormal set of eigenvectors of Q, $\{B_t^i\}$, $t \geq 0$, is a family of mutually independent real Wiener processes with incremental covariance $\lambda_i > 0$, $Qe_i = \lambda_i e_i$ and $trQ = \sum_{i=1}^{\infty}\lambda_i < \infty$. It is also clear that $EW_t = 0$ and for all $t \geq s \geq 0$, the distribution $\mathcal{Q}(W_t - W_s) = \mathcal{N}(0,(t-s)Q)$. Let $K_0 = \text{Ran}\,Q^{1/2}$, the image of K under the operator $Q^{1/2}$, which is a Hilbert space equipped with the inner product $\langle\cdot,\cdot\rangle_{K_0}$,

$$\langle u,v\rangle_{K_0} = \langle Q^{-1/2}u, Q^{-1/2}v\rangle_K, \qquad u,\ v \in K_0,$$

where $Q^{-1/2}$ is the pseudo-inverse of $Q^{1/2}$, then for arbitrary g, $h \in K_0$ and t, $s \geq 0$,

$$E\langle g, W_t\rangle_{K_0}\langle h, W_s\rangle_{K_0} = \langle h, g\rangle_{K_0} t \wedge s.$$

Therefore, the process W_t, $t \geq 0$, is also called a *cylindrical* Wiener process on K_0. Its covariance operator with respect to K_0 is the identity operator.

Theorem 1.2.1   *For an arbitrary symmetric nonnegative trace class operator Q on the real separable Hilbert space K, there exists a Q-Wiener process W_t, $t \geq 0$. Moreover, the series (1.2.1) is uniformly convergent on $[0,T]$ almost surely for arbitrary $T \geq 0$.*

We may also derive the following direct generalization of Lévy's celebrated characterization result.

Theorem 1.2.2   *A continuous martingale $M_t \in \mathcal{M}_T^2(K)$, $M_0 = 0$, is a Q-Wiener process on $[0,T]$ adapted to the filtration $\{\mathcal{F}_t\}_{t\geq0}$ and with increments $M_t - M_s$, $0 \leq s \leq t \leq T$, independent of $\mathcal{F}_s$, $s \in [0,T]$, if and only if $\ll M_t \gg = tQ$, $t \in [0,T]$.*

Roughly speaking, the stochastic integral $\int_0^t \Phi(s,\omega)dW_s$ may be defined in the following way. Define first of all a proper space of operators

$$\mathcal{L}_2^0(K_0, H) = \left\{ G \in \mathcal{L}(K_0, H) : tr\left[\left(G \cdot Q^{\frac{1}{2}}\right)\left(G \cdot Q^{\frac{1}{2}}\right)^*\right] < \infty \right\},$$

i.e., $\mathcal{L}_2^0(K_0, H)$ is the family of all Hilbert-Schmidt operators from K_0 into H, equipped with the usual Hilbert-Schmidt norm topology. For arbitrarily given $T \geq 0$, let $\Phi(t,\omega)$, $t \in [0,T]$, be an $\mathcal{F}_t$-adapted, $\mathcal{L}_2^0(K_0, H)$-valued process. We define the following norm for arbitrary $t \in [0,T]$,

$$|\Phi|_t := \left\{ E \int_0^t \|\Phi(s,\omega)\|_{\mathcal{L}_2^0}^2 ds \right\}^{\frac{1}{2}}$$

$$= \left\{ E \int_0^t tr\left[\left(\Phi(s,\omega) \cdot Q^{\frac{1}{2}}\right)\left(\Phi(s,\omega) \cdot Q^{\frac{1}{2}}\right)^*\right]ds \right\}^{\frac{1}{2}}. \quad (1.2.2)$$

In general, we denote all $\mathcal{L}_2^0(K_0, H)$-valued predictable processes Φ such that $|\Phi|_T < \infty$ by $\mathcal{W}^2([0,T]; \mathcal{L}_2^0)$. In particular, if $\Phi(t,\omega) \in \mathcal{L}_2^0(K_0, H)$, $t \in [0,T]$, is an $\mathcal{F}_t$-adapted, $\mathcal{L}(K, H)$-valued process, (1.2.2) turns out to be

$$|\Phi|_t = \left\{ E \int_0^t tr\left(\Phi(s,\omega)Q\Phi(s,\omega)^*\right)ds \right\}^{\frac{1}{2}}$$

and on this occasion, we still write $\Phi \in \mathcal{W}^2([0,T]; \mathcal{L}_2^0)$.

The stochastic integral $\int_0^t \Phi(s,\omega)dW_s \in H$ may be well defined for all $\Phi(t,\omega) \in \mathcal{W}^2([0,T]; \mathcal{L}_2^0)$ by

$$\int_0^t \Phi(s,\omega)dW_s = L^2 - \lim_{n\to\infty} \sum_{i=1}^n \int_0^t \Phi(s,\omega)e_i dB_s^i, \qquad t \in [0,T],$$

where $W_t = \sum_{i=1}^\infty B_t^i e_i$ and the limit is taken in mean square sense with respect to the probability P. For a detailed description and relevant properties of stochastic integrations, the reader is referred to Da Prato and Zabczyk [1] or Métivier [1].

For future purposes, we review briefly the following results which are directly deduced by standard arguments from the definition of stochastic integration and carrying out a standard argument.

Proposition 1.2.1 *For arbitrary $T \geq 0$, let $\Phi(\cdot) \in \mathcal{W}^2([0,T]; \mathcal{L}_2^0)$, then the stochastic integral $\int_0^t \Phi(s,\omega)dW_s$ is a continuous, square integrable H-valued martingale on $[0,T]$ and*

$$E\left\| \int_0^t \Phi(s,\omega)dW_s \right\|_H^2 = |\Phi|_t^2, \qquad t \in [0,T]. \tag{1.2.3}$$

As a matter of fact, the stochastic integral

$$\int_0^t \Phi(s,\omega)dW_s, \qquad t \geq 0,$$

may be generalized for any $\mathcal{L}_2^0(K_0, H)$-valued adapted process $\Phi(\cdot,\omega)$ satisfying

$$P\left\{ \int_0^t \|\Phi(s,\omega)\|_{\mathcal{L}_2^0}^2 ds < \infty, \ 0 \leq t \leq T \right\} = 1.$$

Moreover, we may deduce the following generalized relation of (1.2.3)

$$E\left\| \int_0^t \Phi(s,\omega)dW_s \right\|_H^2 \leq E\int_0^t \|\Phi(s,\omega)\|_{\mathcal{L}_2^0}^2 ds, \quad 0 \leq t \leq T, \tag{1.2.4}$$

with the equality holding in (1.2.4) if the right hand side is finite.

By virtue of the definitions of stochastic integrals and appropriate limiting arguments, we immediately obtain some important properties.

Proposition 1.2.2 *Let $\Phi \in \mathcal{W}^2([0,T]; \mathcal{L}_2^0)$, then $\int_0^t \Phi(s,\omega)dW_s$ is a continuous square integrable martingale, and its tensor quadratic variation is of the form*

$$\ll \int_0^t \Phi(s,\omega)dW_s \gg = \int_0^t Q_\Phi(s,\omega)ds,$$

where

$$Q_\Phi(t,\omega) = \left(\Phi(t,\omega)Q^{1/2}\right)\left(\Phi(t,\omega)Q^{1/2}\right)^*, \quad t \in [0,T].$$

Proposition 1.2.3   *Assume that both* Φ_1, $\Phi_2 \in \mathcal{W}^2([0,T]; \mathcal{L}_2^0)$. *Then*

$$E\int_0^t \Phi_i(s,\omega)dW_s = 0, \quad E\left\|\int_0^t \Phi_i(s,\omega)dW_s\right\|_H^2 < \infty, \quad t \in [0,T], \; i = 1, 2.$$

Moreover, the correlation operators

$$V(s,t) = Cor\left(\int_0^s \Phi_1(u,\omega)dW_u, \int_0^t \Phi_2(u,\omega)dW_u\right), \quad s, \; t \in [0,T],$$

are given by the formula

$$V(s,t) = \int_0^{s \wedge t} \left(\Phi_1(u,\omega)Q^{1/2}\right)\left(\Phi_2(u,\omega)Q^{1/2}\right)^* du.$$

Corollary 1.2.1   *Under the hypotheses of Proposition 1.2.3, we have*

$$E\left\langle \int_0^s \Phi_1(u,\omega)dW_u, \int_0^t \Phi_2(u,\omega)dW_u \right\rangle_H$$

$$= E\int_0^{s \wedge t} tr\left[\left(\Phi_1(u,\omega)Q^{1/2}\right)\left(\Phi_2(u,\omega)Q^{1/2}\right)^*\right]du. \qquad (1.2.5)$$

In particular, if the processes Φ_1 *and* Φ_2 *both are* $\mathcal{L} = \mathcal{L}(K,H)$-*valued, then we can rewrite formula (1.2.5) in a slightly more compact way:*

$$E\left\langle \int_0^s \Phi_1(u,\omega)dW_u, \int_0^t \Phi_2(u,\omega)dW_u \right\rangle_H = E\int_0^{s \wedge t} tr\left[\Phi_1(u,\omega)Q\Phi_2(u,\omega)^*\right]du.$$

$$(1.2.6)$$

We will quite often use the so-called maximal inequalities formulated in the following theorems.

Theorem 1.2.3 (Doob's inequalities)   *Assume* $T \geq 0$ *and*

$$E\int_0^T \|\Phi(s,\omega)\|_{\mathcal{L}_2^0}^2 \, ds < \infty.$$

(i). For arbitrary $p \geq 1$ *and* $\lambda > 0$,

$$P\left\{\sup_{0 \leq t \leq T} \left\|\int_0^t \Phi(s,\omega)dW_s\right\|_H \geq \lambda\right\} \leq \frac{1}{\lambda^p}E\left\|\int_0^T \Phi(s,\omega)dW_s\right\|_H^p.$$

(ii). For arbitrary $p > 1$,

$$E\left(\sup_{0 \leq t \leq T} \left\| \int_0^t \Phi(s,\omega)dW_s \right\|_H^p \right) \leq \frac{p}{p-1} E \left\| \int_0^T \Phi(s,\omega)dW_s \right\|_H^p.$$

Theorem 1.2.4 (Burkholder-Davis-Gundy) *For arbitrary $p > 0$, there exists a constant $C = C_p > 0$, dependent only on p, such that for any $T \geq 0$,*

$$E\left\{ \sup_{0 \leq t \leq T} \left\| \int_0^t \Phi(s,\omega)dW_s \right\|_H^p \right\} \leq C_p \cdot E\left\{ \int_0^T \|\Phi(s,\omega)\|_{\mathcal{L}_2^0}^2 ds \right\}^{p/2}.$$

Assume that A is a linear operator, generally unbounded, on H and $T(t)$, $t \geq 0$, a strongly continuous semigroup of bounded linear operators with infinitesimal generator A. Suppose $\Phi(t,\omega) \in \mathcal{W}^2([0,T]; \mathcal{L}_2^0)$, $t \in [0,T]$, is an $\mathcal{L}_2^0(K_0, H)$-valued process such that the stochastic integral

$$\int_0^t T(t-s)\Phi(s,\omega)dW_s = W_A^\Phi(t,\omega), \qquad t \in [0,T],$$

is well defined, then the process $W_A^\Phi(t,\omega)$ is called the *stochastic convolution* of Φ. In general, the stochastic convolution is no longer a martingale. However, we have the following result which could be regarded as an infinite dimensional version of Burkholder-Davis-Gundy type of inequality for stochastic convolutions.

Theorem 1.2.5 *Let $p > 2$, $T \geq 0$ and assume $\Phi(s,\omega) \in \mathcal{W}^2([0,T]; \mathcal{L}_2^0)$ is an $\mathcal{L}_2^0(K_0, H)$-valued process such that $E\left(\int_0^T \|\Phi(s,\omega)\|_{\mathcal{L}_2^0}^p ds \right) < \infty$. Then there exists a constant $C = C_{p,T} > 0$, dependent on p and T, such that*

$$E \sup_{t \in [0,T]} \left\| \int_0^t T(t-s)\Phi(s,\omega)dW_s \right\|_H^p \leq C_{p,T} \cdot E\left(\int_0^T \|\Phi(s,\omega)\|_{\mathcal{L}_2^0}^p ds \right).$$

$$(1.2.7)$$

In Theorem 1.2.5, there are some weak points from the practical viewpoint due to the assumption $p > 2$. A version including the case $p = 2$ is possible if A is assumed to generate a pseudo contraction C_0-semigroup $T(t)$, i.e., $\|T(t)\| \leq e^{at}$, $t \geq 0$, for some $a \in \mathbf{R}^1$. Based on this version, the continuity of a suitable modification of the stochastic convolution can be also derived.

Theorem 1.2.6 *Let $p \geq 2$. Assume that A generates a contraction semigroup $T(t)$, $t \geq 0$, and $\Phi(t,\omega) \in \mathcal{W}^2([0,T]; \mathcal{L}_2^0)$ is an $\mathcal{L}_2^0(K_0, H)$-valued process. Then the stochastic convolution $W_A^\Phi(t)$ has a continuous modification* *and there exists a constant $C = C_{p,T} > 0$, dependent of p and T, such that*

$$E\left(\sup_{t \in [0,T]} \left\| \int_0^t T(t-s)\Phi(s,\omega)dW_s \right\|_H^p \right) \leq C_{p,T} \cdot E\left(\int_0^T \|\Phi(s,\omega)\|_{\mathcal{L}_2^0}^2 ds \right)^{p/2}.$$

$$(1.2.8)$$

As another important tool, we mention the following infinite dimensional version of the classic Itô's formula which will play an essential role in the remainder of this book. Suppose that $V(t,x) : I \times H \to \mathbf{R}^1$ is a continuous function with properties:

(i). $V(t,x)$ is differentiable in t and $V_t'(t,x)$ is continuous on $I \times H$;
(ii). $V(t,x)$ is twice Fréchet differentiable in x, $V_x'(t,x) \in H$ and $V_{xx}''(t,x) \in \mathcal{L}(H)$ are continuous on $I \times H$, where $I = [0,T]$, $T > 0$.

Assume that $\Phi(t,\omega) \in \mathcal{W}^2([0,T]; \mathcal{L}_2^0)$ is an $\mathcal{L}_2^0(K_0,H)$-valued process, $\phi(t,\omega)$ is an H-valued continuous, Bochner integrable process on $[0,T]$, and x_0 is an $\mathcal{F}_0$-measurable, H-valued random variable. Then the following H-valued process

$$X_t = x_0 + \int_0^t \phi(s,\omega)ds + \int_0^t \Phi(s,\omega)dW_s, \qquad t \in [0,T], \qquad (1.2.9)$$

is well defined.

Theorem 1.2.7 (Itô's formula)   *Suppose the above conditions (i) and (ii) hold, then for all $t \in [0,T]$, $Z(t) = V(t,X_t)$ has the stochastic differential*

$$dZ(t) = \left\{ V_t'(t,X_t) + \langle V_x'(t,X_t), \phi(t) \rangle_H \right.$$

$$+ \frac{1}{2} tr \left[V_{xx}''(t,X_t) \left(\Phi(t)Q^{1/2} \right) \left(\Phi(t)Q^{1/2} \right)^* \right] \right\} dt$$

$$+ \langle V_x'(t,X_t), \Phi(t)dW_t \rangle_H.$$

1.3   Stochastic Evolution Equations

The theory of stochastic differential equations in infinite dimensional spaces, mainly Hilbert and Banach spaces, is a natural generalization of the classical Itô's stochastic differential equations introduced by Itô and in a slightly different form by Gihman in the 1940s. The theory is motivated by the internal development of analysis and the theory of stochastic differential equations such as stochastic partial differential equations, stochastic flows and stochastic delay differential equations on the one hand, and by a need to analyse certain random phenomena studied in the natural sciences and engineering like stochastic Navier-Stokes equations in turbulence, stochastic models of population genetics in biology and stochastic optimal control in control theory on the other. The reader is also referred to Da Prato and Zabczyk [1] for some more motivation. On this occasion, we content ourselves by explaining briefly this formulation in the case of a deterministic heat equation with external forces.

Roughly speaking, a stochastic partial differential equation, like a partial differential equation, can be viewed in two different ways. Firstly, we can consider its solution as a real-valued function of t and x, where t is the time parameter, and x (which varies typically in a domain $\mathcal{O}$ of $\mathbf{R}^n$) is a space parameter. Denoting as usual by Δ the Laplace operator $\sum_{i=1}^n \frac{\partial^2}{\partial x_i^2}$, the heat equation with external forces and Dirichlet boundary conditions can be formulated as

$$\frac{\partial u}{\partial t}(t, x) = \Delta u(t, x) + f(t, x), \qquad t > 0, \qquad x \in \mathcal{O},$$
$$u(0, x) = u_0(x), \quad x \in \mathcal{O}; \quad u(t, x) = 0, \quad t > 0, \quad x \in \partial\mathcal{O}.$$

For a nonnegative integer m, we denote by $C^m(\mathcal{O})$ the set of all m-times continuously differentiable real-valued functions in $\mathcal{O}$, and by $C_0^m(\mathcal{O})$ the subspace of $C^m(\mathcal{O})$ consisting of those functions which have compact supports in $\mathcal{O}$. For $u \in C^m(\mathcal{O})$ and $1 \leq p < \infty$, we define

$$\|u\|_{m,p} = \left(\int_{\mathcal{O}} \sum_{|\alpha| \leq m} |D^\alpha u(x)|^p dx \right)^{1/p},$$

and for $p = 2$, $u, v \in C^m(\mathcal{O})$,

$$\langle u, v \rangle_{m,2} = \int_{\mathcal{O}} \sum_{|\alpha| \leq m} D^\alpha u(x) \cdot D^\alpha v(x) dx.$$

Let $\tilde{C}_p^m(\mathcal{O})$ be the subset of $C^m(\mathcal{O})$ consisting of those functions u for which $\|u\|_{m,p} < \infty$. We define $W^{m,p}(\mathcal{O})$ and $W_0^{m,p}(\mathcal{O})$ to be the completions in the norm $\| \cdot \|_{m,p}$ of $\tilde{C}_p^m(\mathcal{O})$ and $C_0^m(\mathcal{O})$. It is well known that $W^{m,p}$ and $W_0^{m,p}$ are Banach spaces and $W_0^{m,p}(\mathcal{O}) \subset W^{m,p}(\mathcal{O})$. We will also let

$$H^m(\mathcal{O}) = W^{m,2}(\mathcal{O}), \qquad H_0^m(\mathcal{O}) = W_0^{m,2}(\mathcal{O}).$$

The spaces $H^m(\mathcal{O})$ and $H_0^m(\mathcal{O})$ are Hilbert spaces under the scalar product $\langle \cdot, \cdot \rangle_{m,2}$ defined above. The spaces $W^{m,p}(\mathcal{O})$ consist of functions $u \in L^p(\mathcal{O})$ whose derivatives $D^\alpha u$, in the sense of distributions, of order $|\alpha| \leq m$ are in $L^p(\mathcal{O})$. The space $W_0^{m,p}$ is the subspace of elements of $W^{m,p}(\mathcal{O})$ which vanish in some generalized sense on $\partial\mathcal{O}$.

The point is that one can also consider a solution of the equation above as a function of t with values in a space of functions of x, say $L^2(\mathcal{O})$. We can write Δ as an abstract operator, say A, from the Sobolev space $H^2(\mathcal{O}) \cap H_0^1(\mathcal{O})$ into $L^2(\mathcal{O})$, or from $H_0^1(\mathcal{O})$ into $H^{-1}(\mathcal{O})$, the dual of $H^1(\mathcal{O})$. Note that the Dirichlet boundary condition is implicit in the fact that we look for the solution in $H_0^1(\mathcal{O})$. Assuming that the initial condition u_0 is in $L^2(\mathcal{O})$ and that $t \to f(t)$ is in $L^2(0, \infty; L^2(\mathcal{O}))$, we have the following formulation:

$$\frac{du(t)}{dt} = Au(t) + f(t), \qquad t > 0,$$

$$u(0) = u_0 \in L^2(\mathcal{O}).$$

In the same way, a solution of a stochastic partial differential equation can be considered either as a real-valued random field indexed by t and x, or as a stochastic process indexed by t with values in an infinite dimensional space, e.g., a Hilbert space, of functions of x. Similar formulation of other types of stochastic systems which are different from the heat equation shown above can be found, for instance, in Da Prato and Zabczyk [1] and Rozovskii [1]. In this book, we shall mainly adopt the second viewpoint as the basic setting for our analysis.

Generally, we are concerned with two ways of giving a rigorous meaning to solutions of stochastic differential equations in infinite dimensional spaces, that is, the variational one (cf. Rozovskii [1] and Pardoux [1]) and the semi-group one (cf. Da Prato and Zabczyk [1]). Correspondingly, as in the case of deterministic evolution equations, we have two notions of *strong* and *mild* solutions.

1.3.1 Variational Approach and Strong Solutions

Let V be a real reflexive Banach space (i.e., $V^{**} = (V^*)^* = V$), which is continuously embedded in the real separable Hilbert space H, which we identify with its dual. Hence, we have the following relation:

$$V \hookrightarrow H \equiv H^* \hookrightarrow V^*$$

where " $\hookrightarrow$ " denotes an injection. Unless otherwise specified, we always denote by $\|\cdot\|_V$, $\|\cdot\|_H$ and $\|\cdot\|_{V^*}$ the norms in V, H and V^*, respectively; by $\langle \cdot, \cdot \rangle_{V,V^*}$ the dual product between V and V^*, and by $\langle \cdot, \cdot \rangle_H$ the inner product in H. Recall that K is a separable Hilbert space with norm $\|\cdot\|_K$ and we assume W_t, $t \geq 0$, is a K-valued Wiener process with covariance operator $Q \in \mathcal{L}_1(K)$. Here W_t, $t \geq 0$, is supposed to be defined on some complete probability space $(\Omega, \mathcal{F}, P)$ equipped with a normal filtration $\{\mathcal{F}_t\}_{t \geq 0}$ with respect to which $\{W_t\}_{t \geq 0}$ is a continuous martingale.

Consider the following nonlinear stochastic differential equation in V^*:

$$X_t = x_0 + \int_0^t A(s, X_s)ds + \int_0^t B(s, X_s)dW_s, \qquad (1.3.1)$$

where $A(t,\cdot) : V \to V^*$ and $B(t,\cdot) : V \to \mathcal{L}(K,H)$, are two families of measurable nonlinear operators satisfying that $t \in [0,T] \to A(t,x) \in V^*$, $t \in [0,T] \to B(t,x) \in \mathcal{L}(K,H)$ are Lebesgue measurable for any $x \in V$, $T \geq 0$.

Definition 1.3.1   For arbitrarily given numbers $T \geq 0$, $p > 1$ and $x_0 \in H$, a stochastic process X_t, $0 \leq t \leq T$, is said to be a *strong solution* of Equation (1.3.1) if the following conditions are satisfied:

(a). For any $0 \leq t \leq T$, X_t is a V-valued $\mathcal{F}_t$-measurable random variable;

(b). $X_t \in M^p(0,T; V)$, where $M^p(0,T; V)$ denotes the space of all V-valued processes $(X_t)_{t\in[0,T]}$ which are measurable from $[0,T] \times \Omega$ into V and satisfy

$$\int_0^T E\|X_t\|_V^p \, dt < \infty;$$

(c). Equation (1.3.1) in V^* is satisfied for every $t \in [0,T]$ with probability one.

In particular, if T is replaced by ∞ above, X_t, $t \geq 0$, is called a *global strong solution*, or *strong solution on* $[0,\infty)$, of (1.3.1).

In order to obtain the existence and uniqueness of Equation (1.3.1), we shall impose the following conditions on $A(\cdot,\cdot)$ and $B(\cdot,\cdot)$: there exist constants $\alpha > 0$, $p > 1$ and $\lambda, \gamma \in \mathbf{R}^1$ such that

(a) (Coercivity).

$$2\langle v, A(t,v)\rangle_{V,V^*} + \|B(t,v)\|_{\mathcal{L}_2^0}^2$$
$$\leq -\alpha\|v\|_V^p + \lambda\|v\|_H^2 + \gamma, \quad \forall v \in V, \quad 0 \leq t \leq T, \tag{1.3.2}$$

where $\|\cdot\|_{\mathcal{L}_2^0}$ denotes the Hilbert-Schmidt norm

$$\|B(t,v)\|_{\mathcal{L}_2^0}^2 = tr(B(t,v)QB(t,v)^*);$$

(b) (Growth). There exists a constant $c > 0$ such that

$$\|A(t,v)\|_{V^*} \leq c(1 + \|v\|_V^{p-1}), \quad \forall v \in V, \quad 0 \leq t \leq T; \tag{1.3.3}$$

(c) (Monotonicity).

$$-2\langle u - v, A(t,u) - A(t,v)\rangle_{V,V^*} + \lambda\|u-v\|_H^2$$
$$\geq \|B(t,u) - B(t,v)\|_{\mathcal{L}_2^0}^2 \quad \text{for all} \quad u, v \in V, \quad 0 \leq t \leq T; \tag{1.3.4}$$

(d) (Continuity). The map $\theta \in \mathbf{R}^1 \rightarrow \langle w, A(t, u + \theta v) \rangle_{V, V^*} \in \mathbf{R}^1$ is continuous for arbitrary u, v, $w \in V$ and $0 \le t \le T$;

(e) (Lipschitz). There exists constant $L > 0$ such that

$$\|B(t, u) - B(t, v)\|_{\mathcal{L}_2^0} \le L\|u - v\|_V \quad \text{for all} \quad u, \ v \in V, \quad 0 \le t \le T.$$

$$(1.3.5)$$

Theorem 1.3.1 (Pardoux [1])   *Under the assumptions (a)–(e) above, Equation (1.3.1) has a unique $\{\mathcal{F}_t\}$-progressively measurable strong solution X_t, $0 \le t \le T$, which satisfies:*

$$X.(\omega) \in M^p(0, T; V) \quad \text{for all} \quad T \ge 0,$$

and $X.(\omega) \in C([0, T]; H)$ almost surely where $C(0, T; H)$ denotes the space of all continuous functions from $[0, T]$ into H.

We shall omit the proof of this theorem and refer the reader to Pardoux [1]. Also, see the Appendix for a statement of uniqueness and existence of strong solutions of stochastic delay differential equations which contains as a special case Theorem 1.3.1.

It can also be shown (cf. Rozovskii [1]) that under the same conditions as above, the strong solution of (1.3.1) possesses the Markov property.

Theorem 1.3.2   *Within the same assumptions as Theorem 1.3.1, then for each Borel set Γ in H and arbitrary s, $t \in [0, T]$ with $s \le t$,*

$$P\{X_t \in \Gamma \mid \mathcal{F}_s^X\} = P\{X_t \in \Gamma \mid X_s\}, \quad (1.3.6)$$

where $\{\mathcal{F}_s^X\}$ is the σ-field generated by the solution X_r of (1.3.1) for $r \in [0, s]$. That is, the strong solution is Markovian and moreover the corresponding semigroup has the Feller property in the sense of Definition 1.1.4.

We now present an example of strong solutions to illustrate the conditions imposed above.

Proposition 1.3.1   *Let $T(t)$, $t \ge 0$, be a strongly continuous semigroup on the real Hilbert space H with infinitesimal generator A and $S(t)$, $t \in \mathbf{R}^1$, a strongly continuous group on H with infinitesimal generator G. Let $\mathcal{D}(A)$ and $\mathcal{D}(G^2)$ denote the domains of A and G^2, respectively, and assume $\mathcal{D}(A) \subset \mathcal{D}(G^2)$ and that $T(\cdot)$ and $S(\cdot)$ commute. Then $X_t = T(t)S(B_t)x_0$, $x_0 \in \mathcal{D}(A)$, is a strong solution of*

$$dX_t = \left(A + \frac{1}{2}G^2\right)X_t dt + GX_t dB_t, \qquad X_0 = x_0, \qquad (1.3.7)$$

where B_t, $t \geq 0$, is a real standard one-dimensional Brownian motion.

Proof Applying Theorem 1.2.7 to the function $v(t,x) = T(t)S(x)x_0$ and the process B_t, $t \geq 0$, then we get $v_t(t, B_t) = AT(t)S(B_t)x_0$, $v_x(t, B_t) = GT(t)S(B_t)x_0$ and $v_{xx}(t, B_t) = G^2 T(t)S(B_t)x_0$ from which the desired result follows. □

Remark As to the coercive condition (1.3.2), let us consider now a linear version of Equation (1.3.7):

$$dX_t = \frac{1}{2}\frac{\partial^2 X_t}{\partial x^2}dt + \theta\frac{\partial X_t}{\partial x}dB_t, \qquad t \geq 0,$$

where θ is some real number. Note that the condition (1.3.2) of Theorem 1.3.1 requires $\theta^2 < 1$. In the case $\theta = 1$ (or -1), if the initial condition $x_0(\cdot)$ is smooth enough, the equation has an explicit solution. Indeed, by a similar argument to that of Proposition 1.3.1, it is easy to deduce $X_t(x) = x_0(x+B_t)$, $t \geq 0$. Hence, the case $\theta^2 = 1$ is qualitatively different from the case $\theta^2 < 1$. In the latter case, the equation has a regularizing effect: the solution is more regular at time $t > 0$ than at time 0. Instead, in the first case, the regularity in x of the solution is constant in time. The second case corresponds to a parabolic behavior, while the first one corresponds to a hyperbolic behavior. Indeed, if we rewrite the equation with $\theta = 1$ in Stratonovich form, we obtain

$$dX_t = \frac{\partial X_t}{\partial x} \circ dB_t, \qquad t \geq 0,$$

which is a first order equation. Note that the equation becomes ill-posed for $\theta^2 > 1$. (Also see, for instance, Ikeda and Watanabe [1] for the exact definition and relevant properties of the Stratonovich integral.)

The proofs of Theorems 1.3.1 and 1.3.2 rely in an essential way on a version of Itô's formula for strong solutions which we now present in what follows. The reader is referred to Pardoux [1] for more details of its proof.

Theorem 1.3.3 (Itô's formula) Let $X_t \in M^p(0,T; V)$, $p > 1$, be a continuous process with values in V^*. Suppose there exist $x_0 \in H$, $\phi(\cdot) \in M^q(0,T; V^*)$, $1/p+1/q = 1$, and $\Phi(\cdot) \in \mathcal{W}^2([0,T]; \mathcal{L}_2^0)$ such that

$$X_t = x_0 + \int_0^t \phi(s)ds + \int_0^t \Phi(s)dW_s, \qquad t \in [0,T].$$

Then $X. \in C([0,T]; H)$ almost surely and moreover

$$\|X_t\|_H^2 = \|x_0\|_H^2 + 2\int_0^t \langle X_s, \phi(s)\rangle_{V,V^*}ds + 2\int_0^t \langle X_s, \Phi(s)dW_s\rangle_H$$

$$+ \int_0^t tr[\Phi(s)Q\Phi(s)^*]ds \qquad\qquad (1.3.8)$$

for arbitrary $0 \le t \le T$.

Remark   We specialise Itô's formula to the function $x \to \|x\|_H^2$. As a matter of fact, a similar result holds for a large class of functions defined on H. Precisely, suppose

$$V \quad \text{and} \quad V^* \quad \text{are uniformly convex} \qquad\qquad (1.3.9)$$

(cf. Dunford and Schwartz [1]). Let $\Psi(t,x) : \mathbf{R}_+ \times H \to \mathbf{R}^1$, $\mathbf{R}_+ = [0,\infty)$, be a function satisfying:

(i). $\Psi(\cdot,x)$, $x \in V$, is the first order differentiable and $\Psi(t,\cdot)$, $t \ge 0$, twice (Fréchet) differentiable with $\Psi'_t(\cdot,\cdot)$, $\Psi'_x(\cdot,\cdot)$ and $\Psi''_{xx}(\cdot,\cdot)$ locally bounded on $\mathbf{R}_+ \times H$;

(ii). $\Psi(\cdot,\cdot)$, $\Psi'_x(\cdot,\cdot)$ are continuous on $\mathbf{R}_+ \times H$;

(iii). For all trace class operators S, $tr(S\Psi''_{xx}(\cdot,\cdot))$ is continuous on $\mathbf{R}_+ \times H \to \mathbf{R}^1$;

(iv). If $x \in V$, then $\Psi'_x(t,x) \in V$, $t \ge 0$, and $\langle \Psi'_x(\cdot,\cdot), v^* \rangle_{V,V^*}$ is continuous on $\mathbf{R}_+ \times H \to \mathbf{R}^1$ for each $v^* \in V^*$;

(v). $\|\Psi'_x(t,x)\|_V \le C \cdot (1 + \|x\|_V)$, $t \ge 0$, for some constant $C > 0$, $\forall x \in V$.

It may be shown that under the conditions (i)–(v) above, Itô's formula remains true for the function $\Psi(\cdot,\cdot)$. When one is only interested in the case $\Psi(x) = \|x\|_H^2$, the condition (1.3.9) may be relaxed by the assumption that V^* is strictly convex. This is always true however after an equivalent change of norms because V^* is reflexive (see Pardoux [1], [2] for more details).

1.3.2   Semigroup Approach and Mild Solutions

In most situations, one finds that the concept of strong solution is too limited to include important examples. There is a weaker concept, mild solution, which is found to be more appropriate for practical purposes.

Consider the following semilinear stochastic differential equation on $I = [0,T]$, $T \ge 0$,

$$\begin{cases} dX_t = (AX_t + F(t,X_t))dt + G(t,X_t)dW_t, \\ X_0 = x_0 \in H, \end{cases} \qquad (1.3.10)$$

where A, generally unbounded, is the infinitesimal generator of a C_0-semigroup $T(t)$, $t \ge 0$, of bounded linear operators on the Hilbert space H. The coefficients $F(\cdot,\cdot)$ and $G(\cdot,\cdot)$ are two nonlinear measurable mappings from $[0,T] \times H$ to H and $[0,T] \times H$ to $\mathcal{L}(K,H)$, respectively, satisfying the following Lipschitz continuity conditions:

$$\|F(t,y) - F(t,z)\|_H \leq \alpha(T)\|y - z\|_H, \quad \alpha(T) > 0, \quad y, \ z \in H, \quad t \in [0,T],$$
$$\|G(t,y) - G(t,z)\| \leq \beta(T)\|y - z\|_H, \quad \beta(T) > 0, \quad y, \ z \in H, \quad t \in [0,T].$$

$$(1.3.11)$$

Definition 1.3.2 A stochastic process $X_t, t \in I$, defined on $(\Omega, \mathcal{F}, \{\mathcal{F}_t\}_{t \geq 0}, P)$ is called a *mild solution* of (1.3.10) if

(i). X_t is adapted to $\mathcal{F}_t$, $t \geq 0$;

(ii). For arbitrary $0 \leq t \leq T$, $P\left\{\omega : \int_0^t \|X_s(\omega)\|_H^2 ds < \infty\right\} = 1$ and

$$X_t = T(t)x_0 + \int_0^t T(t-s)F(s, X_s)ds + \int_0^t T(t-s)G(s, X_s)dW_s,$$

$$(1.3.12)$$

for any $x_0 \in H$ almost surely.

As a direct application of the properties of semigroup theory, it may be proved that:

Proposition 1.3.2 *For arbitrary $x_0 \in \mathcal{D}(A)$, the domain of A, assume $X_t \in \mathcal{D}(A)$, $t \in I$, is a solution of (1.3.10) in the sense of satisfying*

$$X_t = x_0 + \int_0^t (AX_s + F(s, X_s))ds + \int_0^t G(s, X_s)dW_s, \qquad (1.3.13)$$

then it is also a mild solution.

Bearing Proposition 1.3.2 in mind and also for purposes of future use, we would like to introduce the following concept which is actually an appropriate version of Definition 1.3.1 for the equation (1.3.10).

Definition 1.3.3 A stochastic process $X_t, t \in I$, defined on $(\Omega, \mathcal{F}, \{\mathcal{F}_t\}_{t \geq 0}, P)$ is called a *strong solution* of (1.3.10) if

(i). $X_t \in \mathcal{D}(A)$, $0 \leq t \leq T$, almost surely and is adapted to $\mathcal{F}_t$, $t \in I$;

(ii). X_t is continuous in $t \in I$ almost surely. For arbitrary $0 \leq t \leq T$, $P\left\{\omega : \int_0^t \|X_s(\omega)\|_H^2 ds < \infty\right\} = 1$ and

$$X_t = x_0 + \int_0^t (AX_s + F(s, X_s))ds + \int_0^t G(s, X_s)dW_s, \qquad (1.3.14)$$

for any $x_0 \in \mathcal{D}(A)$ almost surely.

By a straightforward argument, it is possible to establish the following result.

Proposition 1.3.3   *Assume the condition (1.3.11) holds, then there exists at most one mild solution of (1.3.10). In other words, under the condition (1.3.11) the mild solution of (1.3.10) is unique.*

The following stochastic version of the classic Fubini theorem will be frequently used in the book and its proof can be found in Da Prato and Zabczyk [1].

Proposition 1.3.4   *Let $I = [0, T]$, $T \geq 0$, and*

$$G : I \times I \times \Omega \to (\mathcal{L}(K, H), \mathcal{F}(\mathcal{L}(K, H)))$$

be strongly measurable in the sense of Section 1.1 such that $G(s, t)$ is $\{\mathcal{F}_t\}$-measurable for each $s \geq 0$ with

$$\int_0^T \int_0^T \|G(s, t)\|^2 ds dt < \infty \qquad a.s.$$

Then

$$\int_0^T \int_0^T G(s, t) dW_t ds = \int_0^T \int_0^T G(s, t) ds dW_t \qquad a.s. \qquad (1.3.15)$$

The following result gives sufficient conditions for a mild solution to be also a strong solution.

Proposition 1.3.5   *Suppose that the following conditions hold:*

(1). $x_0 \in \mathcal{D}(A)$, $T(t - s)F(s, x) \in \mathcal{D}(A)$, $T(t - s)G(s, x)k \in \mathcal{D}(A)$ *for each* $x \in H$, $k \in K$, *and* $t \geq s$;

(2). $\left\| AT(t - s)F(s, x) \right\|_H \leq f(t - s)\|x\|_H$, $f(\cdot) \in L^1(0, T; \mathbf{R}_+)$;

(3). $\left\| AT(t - s)G(s, x) \right\| \leq g(t - s)\|x\|_H$, $g(\cdot) \in L^2(0, T; \mathbf{R}_+)$.

Then a mild solution X_t, $t \in I$, of (1.3.10) is also a strong solution with $X_t \in \mathcal{D}(A)$, $t \in I$, in the sense of Definition 1.3.3.

Proof   It suffices to prove that the mild solution X_t, $t \in I$, satisfies (1.3.14). By the above conditions, we have almost surely

$$\int_0^T \int_0^t \left\| AT(t - r)F(r, X_r) \right\|_H dr dt < \infty,$$

$$\int_0^T \int_0^t \mathrm{tr}\big((AT(t-r)G(r,X_r))Q(AT(t-r)G(r,X_r))^*\big)drdt < \infty.$$

Thus by the classic Fubini's theorem, we have

$$\int_0^t \int_0^s AT(s-r)F(r,X_r)drds = \int_0^t \int_r^t AT(s-r)F(r,X_r)dsdr$$
$$= \int_0^t T(t-r)F(r,X_r)dr - \int_0^t F(r,X_r)dr.$$

On the other hand, in view of Proposition 1.3.4, we also have

$$\int_0^t \int_0^s AT(s-r)G(r,X_r)dW_r ds = \int_0^t \int_r^t AT(s-r)G(r,X_r)dsdW_r$$
$$= \int_0^t T(t-r)G(r,X_r)dW_r$$
$$- \int_0^t G(r,X_r)dW_r.$$

Hence, AX_t is integrable almost surely and

$$\int_0^t AX_s ds = T(t)x_0 - x_0 + \int_0^t T(t-r)F(r,X_r)dr - \int_0^t F(r,X_r)dr$$
$$+ \int_0^t T(t-r)G(r,X_r)dW_r - \int_0^t G(r,X_r)dW_r$$
$$= X_t - x_0 - \int_0^t F(r,X_r)dr - \int_0^t G(r,X_r)dW_r.$$

In other words, $X_t \in \mathcal{D}(A)$, $t \in I$, is a strong solution of (1.3.10). $\qquad\square$

By the standard Picard iteration procedure or a probabilistic fixed-point theorem type of argument, we can establish an existence theorem for mild solutions of (1.3.10) in the following form.

Theorem 1.3.4 *Assume the condition (1.3.11) holds. Let $x_0 \in H$ be an arbitrarily given $\mathcal{F}_0$-measurable random variable with $E\|x_0\|_H^p < \infty$ for some integer $p \geq 2$. Then there exists a unique mild solution of (1.3.10) in the space $C\big(0,T; L^p(\Omega, \mathcal{F}, P; H)\big)$.*

As we pointed out in Section 1.2, the stochastic convolution in (1.3.12) is no longer a martingale. A remarkable consequence of this fact is that we cannot employ Itô's formula for mild solutions directly on most occasions of our arguments. We can handle this problem, however, by introducing approximating systems with strong solutions to which Itô's formula can be well applied and by using a limiting argument. In particular, by virtue of Proposition 1.3.5, we may obtain an approximation result of mild solutions, which will play an important role in the subsequent stability analysis. To this end, we introduce an approximating system of (1.3.10) as follows:

$$dX_t = AX_t dt + R(l)F(t, X_t)dt + R(l)G(t, X_t)dW_t,$$
$$X_0 = R(l)x_0, \quad x_0 \in H, \tag{1.3.16}$$

where $l \in \rho(A)$, the resolvent set of A and $R(l) := lR(l, A)$, $R(l, A)$ is the resolvent of A.

Proposition 1.3.6   *Let x_0 be an arbitrarily given random variable in H with $E\|x_0\|_H^p < \infty$ for some integer $p > 2$. Suppose the nonlinear terms $F(\cdot, \cdot)$, $G(\cdot, \cdot)$ in (1.3.10) satisfy the Lipschitz condition (1.3.11). Then, for each $l \in \rho(A)$, the stochastic differential equation (1.3.16) has a unique strong solution $X_t(l) \in \mathcal{D}(A)$, which lies in $L^p(\Omega, \mathcal{F}, P; C(0, T; H))$ for all $T > 0$ and $p > 2$. Moreover, there exists a subsequence, denoted by X_t^n, such that for arbitrary $T > 0$, $X_t^n \to X_t$ almost surely as $n \to \infty$, uniformly with respect to $[0, T]$.*

Proof   The existence of unique strong solution $X_t(l)$ is an immediate consequence of Theorem 1.3.4 and Proposition 1.3.5 by noting the fact that $AR(l) = AlR(l, A) = l - l^2 R(l, A)$ are bounded operators. To prove the remainder of the proposition, let us consider for any $t \geq 0$,

$$X_t - X_t(l) = T(t)(x_0 - R(l)x_0)$$
$$+ \int_0^t T(t - s)[F(s, X_s) - R(l)F(s, X_s(l))]ds$$
$$+ \int_0^t T(t - s)[G(s, X_s) - R(l)G(s, X_s(l))]dW_s.$$

Since $\|R(l)\| \leq 2$ for $l > 0$ large enough, we have for any $T \geq 0$, $p > 2$,

$$E \sup_{0 \leq t \leq T} \|X_t - X_t(l)\|_H^p$$

$$\leq 3^p E \sup_{0 \leq t \leq T} \left\| \int_0^t T(t - s)R(l)[F(s, X_s) - F(s, X_s(l))]ds \right\|_H^p$$

$$+ 3^p E \sup_{0 \leq t \leq T} \left\| \int_0^t T(t - s)R(l)[G(s, X_s) - G(s, X_s(l))]dW_s \right\|_H^p$$

$$+ 3^p \left\{ E \sup_{0 \leq t \leq T} \left\| T(t)(x_0 - R(l)x_0) + \int_0^t T(t - s)[I - R(l)]F(s, X_s)ds \right. \right.$$

$$\left. \left. + \int_0^t T(t - s)[I - R(l)]G(s, X_s)dW_s \right\|_H^p \right\}$$

$$:= 3^p[I_1 + I_2 + I_3].$$

The condition (1.3.11) and Hölder's inequality imply that

$$I_1 \leq E \sup_{0 \leq t \leq T} \left(\int_0^t \left\| T(t-s)R(l)\left[F(s,X_s) - F(s,X_s(l))\right] \right\|_H ds \right)^p$$

$$\leq C_1(T) E \sup_{0 \leq t \leq T} \left\{ \int_0^t \left\| F(s,X_s) - F(s,X_s(l)) \right\|_H^p ds \right\}$$

$$\leq C_2(T) E \int_0^T \sup_{0 \leq r \leq s} \left\| X_r - X_r(l) \right\|_H^p ds,$$

where $C_1(T)$, $C_2(T)$ are two positive numbers, dependent of $T \geq 0$. On the other hand, by virtue of Theorem 1.2.5, we have for $l > 0$ large enough, there exists a real number $C_3(T) > 0$ such that

$$I_2 \leq E \sup_{0 \leq t \leq T} \left\| \int_0^t T(t-s)R(l)[G(s,X_s) - G(s,X_s(l))]dW_s \right\|_H^p$$

$$\leq C_3(T) E \int_0^T \sup_{0 \leq r \leq s} \left\| X_r - X_r(l) \right\|_H^p ds,$$

and

$$I_3 \leq 3^p \left\{ E \sup_{0 \leq t \leq T} \left\| T(t)(x_0 - R(l)x_0) \right\|_H^p \right.$$

$$+ E \sup_{0 \leq t \leq T} \left\| \int_0^t T(t-s)[I - R(l)]F(s,X_s)ds \right\|_H^p \qquad (1.3.17)$$

$$\left. + E \sup_{0 \leq t \leq T} \left\| \int_0^t T(t-s)[I - R(l)]G(s,X_s)dW_s \right\|_H^p \right\}.$$

We now estimate each term in (1.3.17),

$$E \sup_{0 \leq t \leq T} \left\| T(t)(x_0 - R(l)x_0) \right\|_H^p \leq C_4(T) \cdot E \| x_0 - R(l)x_0 \|_H^p \to 0, \quad l \to \infty,$$

where $C_4(T) > 0$ is some positive number. On the other hand, using the Lipschitz condition (1.3.11), we get

$$E \sup_{0 \leq t \leq T} \left\| \int_0^t T(t-s)[I - R(l)]F(s,X_s)ds \right\|_H^p$$

$$\leq C_5(T) E \int_0^T \| I - R(l) \|^p (1 + \| X_s \|_H^p) ds$$

$$\leq \| I - R(l) \|^2 C_6(T) \to 0, \quad \text{as} \quad l \to \infty,$$

for some $C_5(T) > 0$, $C_6(T) > 0$. In a similar manner, by using Theorem 1.2.5, it is easy to deduce that there exists a number $C_7(T) > 0$ such that

$$E \sup_{0 \leq t \leq T} \left\| \int_0^t T(t-s)[I - R(l)]G(s,X_s)dW_s \right\|_H^p$$

$$\leq C_7(T) \| I - R(l) \|^p \to 0, \quad \text{as} \quad l \to \infty.$$

Hence, we can get that there exist numbers $C(T) > 0$ and $\varepsilon(l) > 0$ such that

$$E \sup_{0 \le t \le T} \left\| X_t - X_t(l) \right\|_H^p \le C(T) \int_0^T E \sup_{0 \le r \le s} \left\| X_r - X_r(l) \right\|_H^p ds + \varepsilon(l),$$

where $\lim_{l \to \infty} \varepsilon(l) = 0$. By the well-known Gronwall's inequality, we deduce

$$E \sup_{0 \le t \le T} \left\| X_t - X_t(l) \right\|_H^p \le \varepsilon(l)C(T)T \to 0, \quad \text{as } l \to \infty. \qquad (1.3.18)$$

Now we are in a position to construct the desired sequence by a diagonal sequence trick. Indeed, for the positive integer $n = 1$, by virtue of (1.3.18), there exists a positive integer sequence $\{m_1(i)\}_{i=1}^\infty$ in $\rho(A)$ such that $X_t(l_{m_1(i)}) \to X_t$ almost surely as $i \to \infty$, uniformly with respect to $t \in [0,1]$. Now for the positive integer $n = 2$, consider the sequence $X_t(l_{m_1(i)})$, we can find a subsequence $X_t(l_{m_2(i)})$, $\{l_{m_2(i)}\} \subset \{l_{m_1(i)}\}$, such that $X_t(l_{m_2(i)}) \to X_t$ almost surely as $i \to \infty$, uniformly with respect to $t \in [0,2]$. Proceeding inductively, we find successive subsequences, $X_t(l_{m_n(i)})$ so that (a): $X_t(l_{m_{n+1}(i)})$ is a subsequence of $X_t(l_{m_n(i)})$, $\{l_{m_{n+1}(i)}\} \subset \{l_{m_n(i)}\}$, and (b): $X_t(l_{m_n(i)}) \to X_t$ almost surely as $i \to \infty$, uniformly with respect to $t \in [0,n]$. To get a subsequence converging for each n, one may take the diagonal sequence $l(n) = l_{m_n(n)}$. Then we can obtain the sequence $\{X_t(l(n))\}_{n=1}^\infty$, more simply denoted by $\{X_t^n\}_{n=1}^\infty$, which has the desired properties. $\quad\Box$

It is worth pointing out that, in general, we cannot conclude directly from Theorem 1.3.4 that the mild solution of (1.3.10) has continuous paths, a fact which makes it justifiable to consider asymptotic stability of its sample paths. However, Proposition 1.3.6 allows us to have a modification with continuous sample paths of the mild solution of (1.3.10). In particular, unless otherwise stated, we will always suppose the mild solutions considered have continuous sample paths in the sequel.

Suppose $p \ge 2$, $0 \le t \le T$, and let $L_t^p(\Omega, \mathcal{F}, P; H)$, simply $L_t^p(\Omega; H)$, be the subspace of $L^p(\Omega, \mathcal{F}, P; H)$ which consists of all $\mathcal{F}_t$-measurable random variables. For arbitrary $0 \le s \le T$, let $C_a([s,T]; L^p(\Omega, \mathcal{F}, P; H))$ be the subspace of $C([s,T]; L^p(\Omega, \mathcal{F}, P; H))$ which consists of $\{\mathcal{F}_t\}$-adapted processes. For arbitrary $0 \le t \le T$, consider the stochastic evolution equation (1.3.12), however, with initial datum $x_s \in L_s^p(\Omega; H)$, $s \le t \le T$,

$$X_t = T(t-s)x_s + \int_s^t T(t-u)F(u, X_u)du + \int_s^t T(t-u)G(u, X_u)dW_u,$$

$$X_s = x_s \in L_s^p(\Omega; H). \qquad (1.3.19)$$

In many situations, we find it convenient to restate Theorem 1.3.4 in the following form.

Theorem 1.3.5   *For any $0 \le s \le t \le T$, there exists a unique map* $U(t,s) : L_s^p(\Omega; H) \to L_t^p(\Omega; H)$ *with properties:*

(i). For any $s \leq t \leq T$, $x_s \in L_s^p(\Omega; H)$, $U(t,s)x_s$ is $\mathcal{B}([s,T]) \times \mathcal{F}$ measurable;

(ii). $U(t,s)x_s \in C_a([s,T]; L^p(\Omega, \mathcal{F}, P; H))$ for any $0 \leq T < \infty$ and it satisfies the equation (1.3.19);

(iii). $U(s,s)x_s = x_s$ for all $x_s \in L_s^p(\Omega; H)$ and $0 \leq s \leq T$;

(iv). $U(t,r)U(r,s)x_s = U(t,s)x_s$ for any $s \leq r \leq t$ and $x_s \in L_s^p(\Omega; H)$;

(v). For each fixed $s < t$, the map $(x,\omega) \in H \times \Omega \to U(t,s)x$ is $\mathcal{B}(H) \times \mathcal{F}_{s,t}^W$-measurable, where $\mathcal{F}_{s,t}^W$ is the σ-field generated by $W_r - W_s$, $s \leq r \leq t$.

In particular, if the equation (1.3.19) is linear, the map $U(t,s)$ is a linear operator.

Proof We only sketch the proof because it is quite similar to the finite dimensional case. Define $U(t,s)$ by $U(t,s)x = X_t(x)$, $x \in H$, $X_s(x_s) = x_s \in L_s^p(\Omega; H)$. The properties (i)–(iii) follow from those of $X_t(x_s)$ and (iv) follows from uniqueness. By arguments similar to those in Gihman and Skorohod [1], it is possible to show that $X_t(x)$ has a modification which is $\mathcal{B}(H) \times \mathcal{F}_{s,t}^W$-measurable as a function of $(x,\omega) \in H \times \Omega$. So we define $U(t,s)x$ by this particular choice to obtain the required (v). ☐

Let $M_b(H)$ be the space of all bounded measurable functions on H with its Borel σ-field. Define a map P_s^t, $0 \leq s \leq t \leq T$, on $M_b(H)$ by

$$(P_s^t f)(x) = E[f(U(t,s)x)], \quad x \in H, \quad f \in M_b(H), \quad 0 \leq s \leq t \leq T.$$

By Theorem 1.3.5(v), it is $\mathcal{B}(H)$-measurable and hence lies in $M_b(H)$. Define furthermore $P(s, x, t, \Gamma) = E[\chi_\Gamma(X_t(x))]$, $X_s = x \in H$, $\Gamma \in \mathcal{B}(H)$ and let χ_Γ be the indicator function of the set Γ. Using Theorem 1.3.5, we can show that $P(s, x, t, \Gamma)$ is the transition probability function of the mild solution of (1.3.19) and P_s^t is the semigroup associated with it. In particular, we have the following result:

Theorem 1.3.6 *Under the same conditions as in Theorem 1.3.4, the mild solution of (1.3.19) is a Markov process (in fact, a strong Markov process). Moreover, the semigroup P_s^t, $0 \leq s \leq t \leq T$, associated with it has the Feller property in the sense of Definition 1.1.4.*

1.4 Definitions and Methods of Stability

Stability of a system is the ability of the system to resist a small influence or disturbance unknown beforehand. A system or process is said to be stable if such disturbance does not essentially change it. Indeed, an individual predictable process can be physically realized only if it is stable in the corresponding natural sense. For instance, we know from classical control theory that, before we can consider the design of a regulatory or tracking control system, we need to make sure that such a system is stable from input to output. Problems like this naturally suggest that we should formulate stability concepts not only from a strictly mathematical viewpoint but also with practical applications in mind.

To motivate our stability ideas, let us investigate the following situation. Consider solutions $Y_t(y_0)$, $t \geq 0$, of a deterministic differential equation on the Hilbert space H,

$$\begin{cases} dY_t = f(t, Y_t)dt, & t \geq 0, \\ Y_0 = y_0 \in H, \end{cases} \tag{1.4.1}$$

where $f(\cdot, \cdot)$ is some given function. Let $\tilde{Y}_t$, $t \geq 0$, be a particular solution of (1.4.1); the corresponding system is thought of as describing a process without perturbations. The systems associated with other solutions $Y_t(y_0)$ are regarded as perturbed ones. When one talks about stability, or stability in the sense of Lyapunov, of the solution $\tilde{Y}_t$, $t \geq 0$, it means that the norm $\|Y_t - \tilde{Y}_t\|_H$ could be made small enough if some reasonable conditions are imposed, for instance, that the initial perturbation scale $\|Y_0 - \tilde{Y}_0\|_H$ is very small or time t is large enough. In practice, it is enough to investigate the stability problem for the null solution of some relevant system. Indeed, let $X_t = Y_t - \tilde{Y}_t$, then the equation (1.4.1) could be changed into

$$\begin{aligned} dX_t = dY_t - d\tilde{Y}_t &= (f(t, Y_t) - f(t, \tilde{Y}_t))dt \\ &= (f(t, X_t + \tilde{Y}_t) - f(t, \tilde{Y}_t))dt := F(t, X_t)dt, \end{aligned} \tag{1.4.2}$$

where $F(t, 0) = 0$, $t \geq 0$. Therefore, we could content ourselves with defining and studying stability for the null solution of (1.4.2).

Definition 1.4.1   The null solution of (1.4.2) is said to be *stable* if for arbitrarily given $\varepsilon > 0$, there exists $\delta = \delta(\varepsilon) > 0$ such that if $\|x_0\|_H < \delta$, then

$$\|X_t(x_0)\|_H < \varepsilon$$

for all $t \geq 0$.

Definition 1.4.2   The null solution of (1.4.2) is said to be *asymptotically stable* if it is stable and there exists $\delta > 0$ such that $\|x_0\|_H < \delta$ guarantees

$$\lim_{t \to \infty} \|X_t(x_0)\|_H = 0.$$

Definition 1.4.3   The null solution of (1.4.2) is said to be *exponentially stable* if it is asymptotically stable and there exist numbers $\alpha > 0$ and $\beta > 0$

such that

$$\|X_t(x_0)\|_H \le \beta \|x_0\|_H e^{-\alpha t}$$

for all $t \ge 0$.

There are at least three times as many definitions for the stability of stochastic systems as there are for deterministic ones. This is certainly because in a stochastic setting, there are three basic types of convergence: convergence in probability, convergence in mean and convergence in almost sure (sample path, probability one) sense. The above deterministic stability definitions can be translated into a stochastic setting by properly interpreting the notion of convergence. During the initial development (mainly, finite dimensional cases) of the theory and methods of stochastic stability, some confusion about stability concepts, their usefulness in applications and the relationship among the different concepts of stability existed. Kozin [1], Khas'minskii [1] and Arnold [1] (among others) clarified some of the confusion and provided a good foundation for further work. In what follows, we have no intention of listing all the possible definitions, but prefer to confine ourselves to those which are in our view of the greatest practical interest. We are interested in the stability of the equations (1.3.1) and (1.3.12). To this end, we assume that $A(t,0) = B(t,0) = 0$ and $F(t,0) = G(t,0) = 0$ for any $t \ge 0$.

Definition 1.4.4 (Stability in Probability) The null solution of (1.3.1) or (1.3.12) is said to be *stable in probability* if for arbitrarily given ε, $\varepsilon' > 0$, there exists $\delta = \delta(\varepsilon, \varepsilon') > 0$ such that if $\|x_0\|_H < \delta$, then

$$P\{\|X_t(x_0)\|_H > \varepsilon'\} < \varepsilon$$

for all $t \ge 0$.

Definition 1.4.5 (Stability in p-th Moment) The null solution of (1.3.1) or (1.3.12) is said to be *stable in p-th moment*, $p > 0$, if for arbitrarily given $\varepsilon > 0$, there exists $\delta = \delta(\varepsilon) > 0$ such that $\|x_0\|_H < \delta$ guarantees

$$E\|X_t(x_0)\|_H^p < \varepsilon$$

for all $t \ge 0$.

Definition 1.4.6 (Almost Sure Stability) The null solution of (1.3.1) or (1.3.12) is said to be *almost surely stable* if for each $\varepsilon > 0$, there exists a $\delta = \delta(\varepsilon) > 0$ such that $\|x_0\|_H < \delta$ guarantees

$$P\{\|X_t(x_0)\|_H < \varepsilon \ \text{ for all } \ t \ge 0\} = 1.$$

Similar statements can be made for asymptotic stability and exponential stability.

Definition 1.4.7   (Asymptotic Stability in Probability) The null solution of (1.3.1) or (1.3.12) is said to be *asymptotically stable in probability* if it is stable in probability and for each $\varepsilon > 0$, there exists $\delta = \delta(\varepsilon) > 0$ such that $\|x_0\|_H < \delta$ guarantees

$$\lim_{t \to \infty} P\{\|X_t(x_0)\|_H > \varepsilon\} = 0.$$

Definition 1.4.8   (Asymptotic Stability in p-th Moment) The null solution of (1.3.1) or (1.3.12) is said to be *asymptotically stable in p-th moment*, $p > 0$, if it is stable in p-th moment and there exists $\delta > 0$ such that $\|x_0\|_H < \delta$ guarantees

$$\lim_{t \to \infty} E\|X_t(x_0)\|_H^p = 0.$$

Definition 1.4.9   (Asymptotic Almost Sure Stability) The null solution of (1.3.1) or (1.3.12) is said to be *asymptotic almost surely stable* if it is stable in probability and there exists $\delta > 0$ such that $\|x_0\|_H < \delta$ guarantees

$$P\Big\{ \lim_{t \to \infty} \|X_t(x_0)\|_H = 0 \Big\} = 1.$$

Definition 1.4.10   (Stability in the Large) The null solution of (1.3.1) or (1.3.12) is said to be *stable in probability in the large* if it is stable in probability and furthermore for each $x_0 \in H$, $\varepsilon > 0$ and $\delta > 0$, there exists a number $T = T(x_0, \varepsilon, \delta) > 0$ such that

$$P\{\|X_t(x_0)\|_H > \varepsilon\} < \delta$$

for all $t \geq T > 0$.

A similar definition can be derived for asymptotic stability in probability and p-th moment stability in the large.

Definition 1.4.11   (p-th Moment Exponential Stability) The null solution of (1.3.1) or (1.3.12) is said to be *exponentially stable in p-th moment*, $p > 0$, if there exist positive numbers $\alpha > 0$ and $\beta > 0$ such that

$$E\|X_t(x_0)\|_H^p \leq \beta\|x_0\|_H^p e^{-\alpha t}$$

for all $t \geq 0$.

Definition 1.4.12   (Almost Sure Exponential Stability) The null solution of (1.3.1) or (1.3.12) is said to be *exponentially stable in almost sure sense* if there exist numbers $\alpha > 0$ and $\beta > 0$ such that

$$P\{\|X_t(x_0)\|_H \leq \beta\|x_0\|_H e^{-\alpha t}\} = 1$$

for all $t \geq 0$.

Remark   It is worth pointing out that we find it useful on some occasions to remove the condition $A(t, 0) = 0$, $B(t, 0) = 0$ in (1.3.1) and $F(t, 0) = 0$, $G(t, 0) = 0$ in (1.3.12), while consider exactly the same behavior of solutions as Definitions 1.4.4 to 1.4.12 (e.g., see Section 3.2). This treatment will recapture the above stability definitions when this condition is assumed to hold. As far as this extension is concerned, we intend to find conditions under which the solutions of (1.3.1) or (1.3.12) decay to zero, and say in this case that the solutions or systems are *decayable*.

It is clear that (asymptotic) stability in p-th moment of the null solution of (1.3.1) or (1.3.12) for any value of $p > 0$ implies its (asymptotic) moment stability for every smaller value than p and stability in probability. On the other hand, one can easily show that a null solution could be (asymptotically) p-th moment stable for some $p > 0$ and not (asymptotically) q-th moment stable for $q > p$. The case most often discussed in the literature is (asymptotic) p-th moment stability with $p = 2$. Henceforth, we shall also refer to this case as (asymptotic) *stability in mean square*.

Although the stability Definitions 1.4.8 and 1.4.11 above do not appear to be as strong a restriction on systems as that given in the definitions such as Definitions 1.4.4 and 1.4.7, there are significant implications in Definitions 1.4.8 and 1.4.11 for sample stability behavior. However, it is worth pointing out that stability of the moment alone does not always provide a satisfactory intuitive basis upon which to judge the stability characteristics of the systems of interest. This can be illustrated by the following simple example.

Example 1.4.1   Consider a one-dimensional Itô equation

$$dX_t = aX_t dt + bX_t dB_t, \tag{1.4.3}$$

where B_t, $t \geq 0$, is a one dimensional real Brownian motion, $X_0 = x_0 \in \mathbf{R}^1$, and a, b are two real numbers.

A direct computation yields that the solution process X_t, $t \geq 0$, is given by

$$X_t = \exp\left\{bB_t + (a - b^2/2)t\right\}x_0, \quad t \geq 0. \tag{1.4.4}$$

Hence, by using the law of iterated logarithm (cf. Revuz and Yor [1]), it is easy to deduce that the asymptotically exponential growth rate of the solution is

$$\lambda = \limsup_{t \to \infty} \frac{\log |X_t|}{t} = a - \frac{b^2}{2} \tag{1.4.5}$$

almost surely. We then conclude that the null solution is almost surely exponentially stable in the sense that

$$P\left\{\lim_{t \to \infty} X_t = 0 \text{ at an exponential decay}\right\} = 1$$

if and only if $a < b^2/2$. On the other hand, using the standard exponential martingale properties of Brownian motions, it is also easy to see for any $n \in \mathbf{N}$, $\mathbf{N} = \{1, 2, \cdots\}$,

$$EX_t^n = x_0^n \cdot \exp\left\{(a - b^2/2)nt + \frac{b^2 n^2}{2}t\right\},$$

and we conclude that the null solution is exponentially stable in n-th moment if and only if $a < \frac{b^2}{2}(1-n)$. Therefore, unlike deterministic systems, for $a < 0$, the first moment is exponentially stable, but higher moments are probably unstable. For $a < -b^2/2$, the first and second moments are exponentially stable, and higher moments are probably unstable, etc. It seems difficult to associate a physical meaning to the behavior of a system, knowing only that the first moment is stable but the second moment is unstable, or that the first n moments are stable and all higher moments are unstable. To make matters even more interesting, it is clear from (1.4.5) that the stability of the sample trajectories are determined by the algebraic sign of $a - b^2/2$ only. That is, the condition $a < b^2/2$ is necessary and sufficient for the null solution to be almost surely asymptotically stable. It is also interesting to notice that for $a < b^2/2$, the sample solution possesses almost surely asymptotic stability but it is possible that all moments will diverge exponentially. Hence, we see in the example that unlike deterministic systems, even though stability in mean square implies almost sure stability, almost sure stability need not imply the moment stability of the system.

Remark   If a system is almost surely asymptotically stable, then it is also asymptotically stable in probability. From the analogy of deterministic stability, it seems reasonable to assume in Definition 1.4.9 almost sure stability rather than stability in probability. However, it is worth pointing out this requirement is actually too strong. In fact, let us consider Example 1.4.1 once again. By (1.4.4) and the properties of Brownian motion, it is easy to see that for no positive constant $\varepsilon > 0$ does there exist a number $\delta > 0$ such that almost all the sample trajectories of the solutions originating at $x_0 \neq 0$, $|x_0| < \delta$, remain in an ε-neighborhood of zero (i.e., not almost surely stable) even if the unperturbed term is very stable (i.e., $a < 0$) and $|b|$ is very small.

In the history of the study of stability properties, there are two main approaches: the Lyapunov function method and the Lyapunov exponent method. The Lyapunov function method is probably the most effective tool to handle stochastic stability of systems. This approach, also called Lyapunov's second (direct) method, provides a powerful tool for the study of stability properties of (random) dynamical systems because the technique does not require one to solve the system equation explicitly. We shall take finite dimensional deterministic systems as an example and interpret briefly the main ideas as follows.

Consider a nonnegative continuous function $\Lambda(x)$ on $\mathbf{R}^n$ with $\Lambda(0) = 0$ and $\Lambda(x) > 0$ for $x \neq 0$. Suppose for some $m \in \mathbf{R}^1$, the set $D_m = \{x \in \mathbf{R}^n : \Lambda(x) < m\}$ is bounded and $\Lambda(x)$ has continuous first order derivatives in D_m. Let $X_t = X_t(x_0)$ be the unique solution of the initial value problem:

$$dX_t = f(X_t)dt, \quad t \geq 0,$$
$$X_0 = x_0 \in D_m \subset \mathbf{R}^n, \quad f(0) = 0, \tag{1.4.6}$$

for a suitable function $f(\cdot) \in \mathbf{R}^n$. Since $\Lambda(x)$ is continuous, the open set D_r for $r \in (0, m]$ defined by $D_r = \{x \in \mathbf{R}^n : \Lambda(x) < r\}$ contains the origin and monotonically decreases to the singleton set $\{0\}$ as $r \to 0^+$. If the total derivative $\dot{\Lambda}(x)$ of $\Lambda(x)$ (along the solution trajectory $X_t(x_0)$), which is given by

$$\dot{\Lambda}(x) = \frac{d\Lambda(x)}{dt} = f^T(x) \cdot \frac{\partial \Lambda}{\partial x} := -k(x), \tag{1.4.7}$$

satisfies $-k(x) \leq 0$ for all $x \in D_m$, where $k(x)$ is continuous and $f^T(x)$ the transpose of $f(x)$, then $\Lambda(X_t)$ is a non-increasing function of t, i.e., $\Lambda(x_0) < m$ implies $\Lambda(X_t) < m$ for all $t \geq 0$. Equivalently, $x_0 \in D_m$ implies that $X_t \in D_m$ for all $t \geq 0$. This establishes the stability of the null solution of (1.4.6) in the sense of Lyapunov and $\Lambda(x)$ is called a Lyapunov function for Equation (1.4.6). If we further assume that $k(x) > 0$ for $x \in D_m \backslash \{0\}$, $k(0) = 0$. Then $\Lambda(X_t)$, as a function of t, is strictly monotone decreasing. In this case, $\Lambda(X_t) \to 0$ as $t \to \infty$ from (1.4.7), which implies that $X_t \to 0$ as $t \to \infty$. This fact can be seen through an integration of the equation (1.4.7), i.e.,

$$0 < \Lambda(x_0) - \Lambda(X_t) = \int_0^t k(X_s)ds < \infty \quad \text{for} \quad t \in [0, \infty).$$

It is evident from the above that $X_t \to \{0\} = \{x \in D_m : k(x) = 0\}$ as t leads to infinity. This establishes the asymptotic stability for the system (1.4.6).

The Lyapunov function $\Lambda(x)$ may be regarded as a generalized energy function of the system equation (1.4.6). The above argument illustrates the physical intuition that if the energy of a physical system is always decreasing near an equilibrium state, then the equilibrium state is stable.

Since Lyapunov's original work, this direct method for stability study has been extensively investigated. The main advantage of the method is that one can obtain considerable information about the stability of a given system without explicitly solving the system equation. One major drawback of this method is that for general classes of nonlinear systems a systematic way does not exist to construct or generate a suitable Lyapunov function $\Lambda(\cdot)$, and the stability criterion with this method, which usually provides only a sufficient condition for stability, depends critically on the Lyapunov function chosen.

The first attempts to generalize the Lyapunov function method to stochastic stability for finite dimensional stochastic differential equations were made at least sixty years ago. Since then, a comprehensive study has been carried out by various researchers. A systematic presentation of this theory and applications can be found in the existing literature, for example, Kushner [1], [2] and Khas'minskii [1].

The other contribution to the study of stochastic stability (mainly in finite dimensional spaces) during the past three decades is the application of the Lyapunov exponent concept to stochastic systems. This method uses sophisticated mathematical techniques to study the solution behavior of stochastic systems and often yields necessary and sufficient conditions for stability. We present a brief summary of Lyapunov's original ideas for this method. Consider the linear initial value problem defined by

$$\begin{cases} dX_t = AX_t dt, & t \geq 0, \\ X_0 = x_0 \in \mathbf{R}^n, \end{cases} \tag{1.4.8}$$

where A is some $n \times n$ matrix. Let $X_t(x_0)$ denote the unique solution of (1.4.8) initially from $x_0 \in \mathbf{R}^n$. The Lyapunov exponent $\bar{\lambda}(x_0)$ determined by x_0 is defined by the term

$$\bar{\lambda}(x_0) = \limsup_{t \to \infty} \frac{\log \|X_t(x_0)\|_{\mathbf{R}^n}}{t}, \tag{1.4.9}$$

for (1.4.8). In particular, Lyapunov [1] proved the following fundamental results of the exponent $\bar{\lambda}(x_0)$ for the system:

(i). $\bar{\lambda}(x_0)$ is finite for all $x_0 \in \mathbf{R}^n \backslash \{0\}$;

(ii). The set of real numbers which are Lyapunov exponent for some $x_0 \in \mathbf{R}^n \backslash \{0\}$ is finite with cardinality m, $1 \leq m \leq n$,

$$-\infty < \lambda_1 < \cdots < \lambda_m < \infty, \quad \lambda_i \in \mathbf{R}^1, \quad i = 1, \cdots, m;$$

(iii). $\bar{\lambda}(cx_0) = \bar{\lambda}(x_0)$ for arbitrary $x_0 \in \mathbf{R}^n \backslash \{0\}$ and constant $c \in \mathbf{R}^1 \backslash \{0\}$. $\bar{\lambda}(\alpha x_0 + \beta y_0) \leq \max\{\bar{\lambda}(x_0), \bar{\lambda}(y_0)\}$ for $x_0, y_0 \in \mathbf{R}^n \backslash \{0\}$ and $\alpha, \beta \in \mathbf{R}^1$ with equality if $\bar{\lambda}(x_0) < \bar{\lambda}(y_0)$ and $\beta \neq 0$. The sets $L_i = \{x \in \mathbf{R}^n \backslash \{0\} : \bar{\lambda}(x) = \lambda_i\}$, $i = 1, 2, \cdots, m$, are linear subspaces of $\mathbf{R}^n$, and $\{L_i\}_{i=1}^m$ is a filtration of $\mathbf{R}^n$, i.e.,

$$\{0\} := L_0 \subset L_1 \subset L_1 \subset \cdots \subset L_m = \mathbf{R}^n$$

where $n_i := \dim(L_i) - \dim(L_{i-1})$ is called the *multiplicity* of the exponent λ_i for $i = 1, \cdots, m$, and the collection $\{(\lambda_i, n_i)\}_{i=1}^m$ is referred to as the *Lyapunov spectrum* of the system (1.4.8).

For the system, the relation (1.4.9) implies that if $\bar{\lambda}(x_0) < 0$, then the solution $X_t(x_0)$ will converge to zero at the exponential rate $|\bar{\lambda}(x_0)|$, and if $\bar{\lambda}(x_0) > 0$, then the solution cannot remain in any bounded region of $\mathbf{R}^n$. From this we see that $\{\lambda_i\}$, $i = 1, 2, \cdots, m$, contains information about stability of the system. For finite dimensional stochastic systems, a similar scheme can be carried out and under some circumstances a necessary and sufficient condition for stability can be obtained. At this stage, we have no intention of going through any further details in this respect but refer the reader to the existing literature such as Arnold [5], Arnold and Wihstutz [1], Arnold, Kliemann and Oeljeklaus [1], Khas'minskii [1], Mohammed [2] and Oseledec [1] for more details.

Although Lyapunov exponent theory produces remarkable results in the developments of stochastic stability in finite dimensional spaces, it is at present far from being a mature subject area in infinite dimensional cases. In particular, this method uses sophisticated tools from various fields such as stochastic analysis, functional analysis and stochastic partial differential equations and few satisfactory results have been obtained until now. From the next chapter on, we shall mainly focus on the Lyapunov function approach for analyzing and establishing stability theory of infinite dimensional stochastic differential equations.

1.5 Notes and Comments

Much material in Section 1.1 is classical and taken mainly from Chojnowska-Michalik and Goldys [1], Da Prato and Zabczyk [1], Métivier [1] and Rozovskii [1]. For more details of Sections 1.2 and 1.3, see also Curtain and Pritchard [1], Da Prato and Zabczyk [1], [2], Gihman and Skorohod [1], Hille and Phillips [1], Ikeda and Watanabe [1], Kallianpur [1], Kallianpur and Xiong [1], Karazas and Shreve [1], Métivier [1], Pardoux [1], Rozovskii [1] and Skorohod [1].

Theorems 1.2.5, 1.2.6 are taken from Da Prato and Zabczyk [1] and Tubaro [1]. A systematic study of the variational method for infinite dimensional (stochastic) systems was carried out by Lions [1], Pardoux [1], Krylov and Rozovskii [1] and Rozovskii [1]. As for applications of semigroup approaches to infinite dimensional stochastic systems, more detailed material can be found in the existing literature such as Da Prato and Zabczyk [1], Métivier [1] and Métivier and Pellaumail [1]. Theorems 1.3.1 and 1.3.3 are adapted from Pardoux [1], [2]. Much material in Section 1.3.2 is taken from Ichikawa [3], [5]. As to stability of differential equations, some systematic statements can be found in the existing literature, for instance, in Hahn [1], LaSalle and Lefschetz [1], Hale [1] and Hale and Lunel [1] for finite dimensional deterministic systems, in Arnold [1], Khas'minskii [1], Kolmanovskii and Myshkis [1], Kolmanovskii and Nosov [1], Kushner [1], Mao [1], Mohammed [2] and Skorohod [1] for finite dimensional stochastic systems and in Luo, Guo and Morgul [1], Pazy [1], Curtain and Pritchad [1], Curtain and Zwart [1] and Wu [1] for infinite dimensional deterministic systems.

Chapter 2

Stability of Linear Stochastic Differential Equations

The purpose of this chapter is to establish the stability of systems defined by stochastic linear evolution equations. We mainly concern ourselves with exploring some characteristic results which are a natural extension of Lyapunov's classical work in finite dimensional spaces. We begin our statements with the deterministic case in which a linear unbounded operator satisfying appropriate conditions generates a stable C_0-semigroup of bounded linear operators. Under suitable circumstances, the characterization of mean square exponential stability is established and applied at the end of the chapter to various stochastic (partial or delay) differential equations. Subsequently, we shall establish almost sure pathwise stability of stochastic systems, a case which can be most closely related to their deterministic counterparts. In some sense, it is this kind of stability that one really likes to have in practical situations.

2.1 Stable Semigroups

We start our discussion by considering the following linear system on the n-dimensional Euclidean space $\mathbf{R}^n$:

$$\frac{dX_t(x_0)}{dt} = AX_t(x_0), \qquad X_0(x_0) = x_0 \in \mathbf{R}^n,$$

where A is some $n \times n$ constant matrix. Clearly, the equation has a unique solution which is given by

$$X_t(x_0) = e^{At}x_0.$$

It is well known (cf. Hahn [1]) that for finite dimensional linear systems as above, the following statements are equivalent:

(P1) the null solution is asymptotically stable,
(P2) the null solution is exponentially stable,
(P3) for some real number $1 \leq p < \infty$, we have $\int_0^\infty \|X_t(x_0)\|_{\mathbf{R}^n}^p \, dt < \infty$.

Furthermore, if we define $T(t) = e^{At}$, the transition matrix, it is easy to see that (P2) is also equivalent to

(P4) $T(t)$ is exponentially stable, i.e., there exist positive constants M and μ such that $\|T(t)\| \leq Me^{-\mu t}$, where $\|T(t)\|$ denotes the matrix norm of $T(t) : \mathbf{R}^n \to \mathbf{R}^n$.

In this case, $T(t)$ is actually a uniformly continuous semigroup of bounded linear operators on $\mathbf{R}^n$.

For infinite dimensional systems, the stability analysis becomes much more complicated. For instance, the equivalence of (P1) and (P4) generally does not hold. To illustrate this, let us consider the following deterministic linear Cauchy problem

$$\frac{dX_t(x_0)}{dt} = AX_t(x_0), \qquad X_0(x_0) = x_0 \in S, \tag{2.1.1}$$

where the unbounded operator A is the infinitesimal generator of some C_0-semigroup $T(t)$, $t \geq 0$, on a real separable Banach space S with norm $\| \cdot \|_S$. If $x_0 \in \mathcal{D}(A)$, the domain of A, then $T(t)x_0 \in \mathcal{D}(A)$, and

$$\frac{d}{dt}(T(t)x_0) = AT(t)x_0 = T(t)Ax_0, \qquad t \geq 0.$$

Hence, $X_t = T(t)x_0$ is a solution of the linear system (2.1.1).

Definition 2.1.1   Let $T(t)$, $t \geq 0$, be a strongly continuous semigroup of bounded linear operators on the Banach space S. It is said to be *asymptotically stable* if $\lim_{t\to\infty} T(t)x = 0$ for all $x \in S$.

Clearly, the semigroup $T(t)$ of (2.1.1) is asymptotically stable if and only if the null solution of (2.1.1) is asymptotically stable. The following example shows that when the null solution or $T(t)$ of (2.1.1) is asymptotically stable, it remains possible for it to be exponentially unstable in infinite dimensions.

Example 2.1.1   Let l^2 be the Hilbert space of all square summable sequences with norm $\|a\|_{l^2}^2 = \sum_{i=1}^{\infty} a_i^2 < \infty$, $a = (a_1, \cdots, a_n, \cdots) \in l^2$. On l^2 we define the semigroup of operators

$$T(t)a = (e^{-t}a_1, e^{-t/2}a_2, \cdots, e^{-t/n}a_n, \cdots)$$

for any $a = (a_1, a_2, \cdots, a_n, \cdots) \in l^2$. Then, $T(t)$ is a C_0-semigroup on l^2, and for each t in $[0, \infty)$,

$$\|T(t)\| = \sup_{\|a\|_{l^2}=1} \|T(t)a\|_{l^2} = \sup_{\|a\|_{l^2}=1} \left(\sum_{n=1}^{\infty} e^{-2t/n} a_n^2 \right)^{1/2} = \lim_{n\to\infty} e^{-t/n} = 1.$$

However, for each a in l^2 we have

$$\lim_{t\to\infty} \|T(t)a\|_{l^2}^2 = \lim_{t\to\infty} \sum_{n=1}^{\infty} e^{-2t/n} a_n^2 = 0$$

which indicates that $T(t)$ is asymptotically stable.

Remark   Although it is generally not true that asymptotic stability of semigroups implies exponential stability in infinite dimensional spaces, there is however an important category of semigroups in which asymptotic stability is equivalent to exponential stability. In fact, it may be shown (Datko [5]) that if for some $t_0 > 0$, the operator $T(t_0)$ is compact, i.e., $T(t_0)$ maps any bounded sets in S into subsets of compact sets (thus $T(t)$, $t \geq t_0$, is compact), then the asymptotic stability of $T(t)$ implies its exponential stability.

In the following subsection, we shall carry out a Lyapunov equation and function type of argument to generalize significant results from finite to infinite dimensional spaces. In particular, we shall show that the equivalence of (P2)–(P4) remains true. These results are especially useful in dealing with stochastic stability of infinite dimensional systems.

There are some useful criteria for the exponential stability of C_0-semigroups of the following form.

Proposition 2.1.1   *Let $T(t)$ be a C_0-semigroup with infinitesimal generator A on S. The following statements are equivalent:*

(i). *$T(t)$ is exponentially stable, i.e., $\|T(t)\| \leq Me^{-\mu t}$, for $M \geq 1$, $\mu > 0$;*
(ii). *$\lim_{t\to\infty} \|T(t)\| = 0$;*
(iii). *there exists some $t_0 > 0$ such that $\|T(t_0)\| < 1$.*

Proof   It is obvious that (i) implies (ii), and (ii) implies (iii). Suppose (iii) is true, then by the well-known results of semigroup theory, the upper stability indice (cf. Yosida [1]) satisfies

$$\Gamma(A) := \lim_{n\to\infty} \frac{\log \|T(nt_0)\|}{nt_0} < 0,$$

and so there exist $M \geq 1$ and $\mu > 0$ such that

$$\|T(t)\| \leq Me^{-\mu t}, \qquad t \geq 0.$$

Hence, (iii) implies (i).   □

2.1.1   Lyapunov Functions

We start in this subsection by an investigation of the exponential stability of (2.1.1) since it guarantees the convergence rate of systems. Actually, it is immediate to see, as in the finite dimensional case, that the null solution of (2.1.1) is exponentially stable if and only if the semigroup $T(t)$, $t \geq 0$, is exponentially stable.

It was shown above that, if the state space S is finite dimensional, there are a few equivalent conditions for the null solution of (2.1.1) to be exponentially stable. The same conditions can also be based either on properties of the spectrum of the matrix A or on the existence of an appropriate Lyapunov function. More precisely, the following proposition holds (cf. Hahn [1]).

Proposition 2.1.2   *Let S be finite dimensional, for instance, $\mathbf{R}^n$, $n \geq 1$. The null solution of (2.1.1) is exponentially stable if and only if one of the following conditions holds:*

(i). All eigenvalues of the matrix A have negative real parts,

$$\max\{Re\,\lambda : \det(\lambda I - A) = 0\} < 0; \qquad (2.1.2)$$

(ii). There exists a nonnegative definite matrix, denote it by $P \geq 0$, such that

$$PA + A^T P = -I, \qquad (2.1.3)$$

where A^T is the transpose of A. In the latter case, the function $\Lambda(x) = \langle Px, x \rangle_{\mathbf{R}^n}$, $x \in \mathbf{R}^n$, is the Lyapunov function for (2.1.1) in the sense that for every X_t, $t \geq 0$, of (2.1.1)

$$\frac{d\Lambda(X_t)}{dt} = \frac{d}{dt} \langle PX_t, X_t \rangle_{\mathbf{R}^n} < 0,$$

where $\langle \cdot, \cdot \rangle_{\mathbf{R}^n}$ denotes the standard Euclidean inner product in $\mathbf{R}^n$, $n \geq 1$.

If the Banach space S is infinite dimensional, the above Proposition 2.1.2 is only partially true. For instance, the condition (2.1.2) does not imply stability of the Cauchy problem (see Example 2.1.2 below) unless some extra conditions on A are imposed and in that case, (2.1.2) is, of course, replaced by

$$\sup\{\text{Re}\,\lambda : \lambda \in \sigma(A)\} < 0 \qquad (2.1.4)$$

where $\sigma(A)$ is the spectrum of the linear operator A. This behavior is certainly a consequence of the fact that linear operators in finite dimensional Banach spaces have only pure point spectra. Since this is generally not the case in infinite dimensional spaces, one does not expect this result to remain true. As a matter of fact, in this setting, (i) of Proposition 2.1.2 may be formulated as follows:

Proposition 2.1.3 *Define for the generator A of $T(t)$, $t \geq 0$, the lower and upper stability indices:*

$$\gamma(A) = \sup \left\{ Re\, \lambda : \lambda \in \sigma(A) \right\},$$

$$\Gamma(A) = \inf \left\{ \mu : \|T(t)\| \leq M \cdot e^{\mu t} \ for \ some \ M \geq 1 \ and \ all \ t \geq 0 \right\}.$$

Then

$$\gamma(A) \leq \Gamma(A), \tag{2.1.5}$$

and therefore, if the system (2.1.1) is exponentially stable,

$$\gamma(A) < 0. \tag{2.1.6}$$

Moreover, if

(i) *the semigroup $T(t)$ is differentiable on $t \in [0, \infty)$, i.e., for every $x \in S$, $t \to T(t)x$ is differentiable for $t \geq 0$,*

or

(ii) *for some $t_0 > 0$, $T(t_0)$ is a compact operator,*

then

$$\gamma(A) = \Gamma(A), \tag{2.1.7}$$

and consequently (2.1.6) implies exponential stability. In particular, if S is finite dimensional, the equality (2.1.7) always holds.

Proof The first part of Proposition 2.1.3 is an immediate consequence of the celebrated Hille-Yosida Theorem. The proof of (i) is contained in Pazy [1] and part (ii) can be found in Hille and Phillips [1]. We shall not go into further details here because these results are well-established. ▯

It is worth mentioning that unlike finite dimensional systems, the equality of (2.1.5) is generally not true in infinite dimensional spaces, a fact which is illustrated by the example below.

Example 2.1.2 For a measurable function f defined on $[0, \infty)$, set

$$\|f\|_{exp} = \int_0^\infty e^s |f(s)| ds,$$

and let E be the space of all measurable functions $f(\cdot)$ on $[0, \infty)$ for which $\|f\|_{exp} < \infty$. Let $S = E \cap L^p(0, \infty)$, $1 < p < \infty$. It is easy to see that S, endowed with the norm $\|f\|_{e,p} = \|f\|_{exp} + \|f\|_p$, is a Banach space. In S we define a semigroup $\{T(t)\}$ by:

$$T(t)f(x) = f(x + t) \qquad for \qquad t \geq 0. \tag{2.1.8}$$

It follows readily from its definition that $\{T(t)\}$ is a C_0-semigroup on S and that $\|T(t)\| \leq 1$. Choosing $f \in S$ to be the indicator function of the interval $[t, t + \varepsilon^p]$, $\varepsilon > 0$, and letting $\varepsilon \downarrow 0$ shows that $\|T(t)\| \geq 1$ and thus $\|T(t)\| = 1$ for $t \geq 0$, i.e., $\Gamma(A) = 0$ in Proposition 2.1.3.

The infinitesimal generator A of $\{T(t)\}$ is given by

$$\mathcal{D}(A) = \left\{ u : u(t) \text{ is absolutely continuous, } \frac{du(t)}{dt} \in S \text{ almost surely} \right\},$$
(2.1.9)

and

$$Au(t) = \frac{du(t)}{dt} \quad \text{for} \quad u(\cdot) \in \mathcal{D}(A).$$

Let $f \in S$ and consider the equation

$$\lambda u(t) - Au(t) = \lambda u(t) - \frac{du(t)}{dt} = f(t), \quad t \geq 0.$$
(2.1.10)

A simple computation shows that

$$u(t) = \int_0^\infty e^{-\lambda s} f(t + s) ds = e^{\lambda t} \int_t^\infty e^{-\lambda s} f(s) ds$$
(2.1.11)

is a solution of (2.1.10). We will show that if λ satisfies $\mathrm{Re}\,\lambda > -1$, then u, given by (2.1.11), is in $\mathcal{D}(A)$ and thus $\{\lambda : \mathrm{Re}\,\lambda > -1\} \subset \rho(A)$, the resolvent set of A. To show that $u \in \mathcal{D}(A)$, it suffices by (2.1.10) to show that $u \in S$ and this follows from

$$|u(t)| \leq e^{(\mathrm{Re}\,\lambda)t} \int_t^\infty e^{-(\mathrm{Re}\,\lambda+1)s} e^s |f(s)| ds \leq e^{-t} \int_t^\infty e^s |f(s)| ds \leq e^{-t} \|f\|_{exp}$$

which implies that $u \in L^p(0, \infty)$, and

$$\|u\|_{exp} \leq \int_0^\infty \int_t^\infty e^{(\mathrm{Re}\,\lambda+1)(t-s)} e^s |f(s)| ds dt$$

$$= \int_0^\infty \left(\int_0^s e^{(\mathrm{Re}\,\lambda+1)(t-s)} dt \right) e^s |f(s)| ds$$

$$= (\mathrm{Re}\,\lambda + 1)^{-1} \int_0^\infty \left(1 - e^{-(\mathrm{Re}\,\lambda+1)s} \right) e^s |f(s)| ds$$

$$\leq (\mathrm{Re}\,\lambda + 1)^{-1} \|f\|_{exp}.$$

The set $\{\lambda : \mathrm{Re}\,\lambda > -1\}$ is therefore a subset of $\rho(A)$, $\gamma(A) = \sup\{\mathrm{Re}\,\lambda : \lambda \in \sigma(A)\} \leq -1$ while $\Gamma(A) = 0$.

In the case where S is infinite dimensional, although it is not generally guaranteed that (2.1.4) yields stable semigroups, the Lyapunov type of the stability condition (2.1.3) has its complete infinite dimensional counterpart.

Indeed, we may show the following characteristic theorem about stable semi-groups which also justifies the equivalence of (P2)–(P4) at the beginning of this section.

Theorem 2.1.1   *Consider the system (2.1.1) defined on a separable real Hilbert space H. Then the following relations are equivalent:*

(i). There exist constants $M \geq 1$, $\mu > 0$ such that

$$\|T(t)\| \leq M \cdot e^{-\mu t}, \quad t \geq 0. \tag{2.1.12}$$

(ii). There exists a nonnegative self-adjoint operator $P \geq 0$ such that

$$2\langle PAx, x \rangle_H = -\langle x, x \rangle_H \quad \text{for each} \quad x \in \mathcal{D}(A). \tag{2.1.13}$$

(iii). For every $x \in H$,

$$\int_0^\infty \|T(t)x\|_H^2 dt < \infty. \tag{2.1.14}$$

Proof   If $\|T(t)\| \leq Me^{-\mu t}$, $t \geq 0$, for some $M \geq 1$, $\mu > 0$, then for every x in H the integral $\int_0^\infty \|T(s)x\|_H^2 ds$ is convergent. Moreover, if we define a mapping P from H into itself which is given by $Px = \int_0^\infty T^*(t)T(t)x dt$, we easily see that P is a self-adjoint operator on H and $P \geq 0$. The equation (2.1.13) follows from the fact that if x and y both are in $\mathcal{D}(A)$,

$$\langle AT(t)x, PT(t)y \rangle_H + \langle PT(t)x, AT(t)y \rangle_H = \frac{d}{dt} \int_0^\infty \langle T(t+s)x, T(t+s)y \rangle_H ds$$

$$= \int_0^\infty \frac{d}{ds} \langle T(t+s)x, T(t+s)y \rangle_H ds$$

$$= \langle T(t+s)x, T(t+s)y \rangle_H \Big|_{s=0}^{s=\infty}.$$

Expanding the right hand side of the above equality and noting $\lim_{t \to \infty} T(t)x = 0$ for all $x \in H$, we obtain the relation (2.1.13). Hence (i) $\Longrightarrow$ (ii).

In order to show (ii) $\Longrightarrow$ (iii), suppose that there exists a self-adjoint operator P on H with $P \geq 0$ such that for all x in $\mathcal{D}(A)$, $2\langle PAx, x \rangle_H = -\|x\|_H^2$. For each x in H, $t \geq 0$, define the function $\Lambda(x, t) = \langle PT(t)x, T(t)x \rangle_H$. Since P is nonnegative, this implies that $\Lambda(x, t) \geq 0$ for all $x \in H$, $t \geq 0$. Let $x \in \mathcal{D}(A)$, then $T(t)x \in \mathcal{D}(A)$, $t \geq 0$, and $\Lambda(x, t)$ is differentiable with derivative

$$\frac{d\Lambda(x, t)}{dt} = 2\langle PAT(t)x, T(t)x \rangle_H = -\|T(t)x\|_H^2.$$

Hence,

$$0 \leq \Lambda(x, t) = \Lambda(x, 0) - \int_0^t \|T(s)x\|_H^2 ds,$$

which means that

$$\Lambda(x,0) \geq \int_0^t \|T(s)x\|_H^2 ds \quad \text{for all} \quad t \geq 0, \quad x \in \mathcal{D}(A). \tag{2.1.15}$$

On the other hand, by the properties of C_0-semigroups, there exist $C \geq 1$, $\lambda \in \mathbf{R}^1$ such that $\|T(t)(x_n - x)\|_H \leq Ce^{\lambda t}\|x_n - x\|_H$ for arbitrary $x_n, x \in H$ which shows that if $x_n \to x$, then $T(t)x_n \to T(t)x$ uniformly on compact intervals as $n \to \infty$. Hence, the inequality (2.1.15) is valid for all x in H since $\mathcal{D}(A)$ is dense in H. This means

$$\int_0^\infty \|T(s)x\|_H^2 ds \leq \Lambda(x,0) = \langle Px, x \rangle_H < \infty \tag{2.1.16}$$

for all x in H, which implies the desired (iii).

Finally, let us show (iii) $\Longrightarrow$ (i). We first prove that there exists a positive constant $C > 0$ such that for all $x \in H$,

$$\int_0^\infty \|T(t)x\|_H^2 dt \leq C\|x\|_H^2. \tag{2.1.17}$$

To see this, let us define a mapping $\tilde{Q} : H \to L^2(\mathbf{R}_+; H)$ by $\tilde{Q}x = T(t)x$, $x \in H$. From (2.1.14), it follows that $\tilde{Q}$ is defined on all elements of H. It is not difficult to see by the standard principle of uniform boundedness that $\tilde{Q}$ is bounded, i.e.,

$$\int_0^\infty \|T(t)x\|_H^2 dt \leq C\|x\|_H^2 \tag{2.1.18}$$

where C is some positive constant.

Let $L \geq 1$, $\lambda > 0$ be such numbers that

$$\|T(t)\| \leq Le^{\lambda t} \quad \text{for all} \quad t \geq 0.$$

Then, since for all $x \in H$,

$$\frac{1 - e^{-2\lambda t}}{2\lambda}\|T(t)x\|_H^2 = \int_0^t e^{-2\lambda s}\|T(t)x\|_H^2 ds$$

$$\leq \int_0^t e^{-2\lambda s}\|T(s)\|^2\|T(t-s)x\|_H^2 ds$$

$$\leq L^2 \int_0^t \|T(t-s)x\|_H^2 ds \leq L^2 C\|x\|_H^2,$$

we have for some $\theta > 0$ and all $t \geq 0$, $\|T(t)\| \leq \theta$. Thus

$$t\|T(t)x\|_H^2 = \int_0^t \|T(t)x\|_H^2 ds \leq \int_0^t \|T(s)\|^2\|T(t-s)x\|_H^2 ds \leq \theta^2 C\|x\|_H^2$$

for all $x \in H$. Therefore,

$$\|T(t)\| \leq \frac{\theta\sqrt{C}}{t},$$

which immediately implies $\|T(t_0)\| < 1$ for some $t_0 > 0$ sufficiently large. Then, by Proposition 2.1.1, we have (iii) $\Longrightarrow$ (i). The proof is complete. □

A C_0-semigroup $T(t)$, $t \geq 0$, satisfying (2.1.14) is called L^2-*stable*. By analogy with the finite dimensional case, the condition (2.1.14) in Theorem 2.1.1 which implies stability of semigroups may be appropriately relaxed in the following manner.

Theorem 2.1.2   *Let $T(t)$, $t \geq 0$, be a strongly continuous semigroup of bounded linear operators. If for some positive number p, $1 \leq p < \infty$,*

$$\int_0^\infty \|T(t)x\|_H^p dt < \infty \qquad \text{for every} \quad x \in H, \qquad (2.1.19)$$

then there exist constants $M \geq 1$ and $\mu > 0$ such that $\|T(t)\| \leq Me^{-\mu t}$, $t \geq 0$.

Proof   We begin by showing that (2.1.19) implies the boundedness of $t \to \|T(t)\|$. Let $\|T(t)\| \leq Le^{\lambda t}$, $t \geq 0$, where $L \geq 1$ and $\lambda > 0$. From (2.1.19), it follows that $T(t)x \to 0$ as $t \to \infty$ for every $x \in H$. Indeed, if this were false, we could find $x \in H$, $\delta > 0$ and $t_j \to \infty$ such that $\|T(t_j)x\|_H \geq \delta$. Without loss of generality, we can assume that $t_{j+1} - t_j > \lambda^{-1}$. Set $\Delta_j = [t_j - \lambda^{-1}, t_j]$, then $dis(\Delta_j) = \lambda^{-1} > 0$ and the intervals Δ_j do not overlap. For $t \in \Delta_j$, we have $\|T(t)x\|_H \geq \delta(Le)^{-1}$ and therefore

$$\int_0^\infty \|T(t)x\|_H^p dt \geq \sum_{j=0}^\infty \int_{\Delta_j} \|T(t)x\|_H^p dt \geq \left(\frac{\delta}{Le}\right)^p \sum_{j=1}^\infty dis(\Delta_j) = \infty$$

contradicting (2.1.19). Thus $T(t)x \to 0$ as $t \to \infty$ for every $x \in H$ and the principle of uniform boundedness implies $\|T(t)\| \leq R$ for some $R \geq 1$ and all $t \geq 0$.

Consider the mapping $\tilde{Q} : H \to L^p(\mathbf{R}_+; H)$ defined by $\tilde{Q}x = T(t)x$. From (2.1.19), it follows that $\tilde{Q}$ is defined on all elements of H. It is not difficult to see that $\tilde{Q}$ is closed and therefore, by the closed graph theorem, $\tilde{Q}$ is bounded, i.e.,

$$\int_0^\infty \|T(t)x\|_H^p dt \leq C^p \|x\|_H^p \qquad (2.1.20)$$

for some constant $C > 0$. Let $0 < \rho < R^{-1}$ and define $t_x(\rho)$ by

$$t_x(\rho) = \sup\left\{t : \|T(s)x\|_H \geq \rho\|x\|_H \text{ for } 0 \leq s \leq t\right\}.$$

Since $\|T(t)x\|_H \to 0$, as $t \to \infty$, $t_x(\rho)$ is finite and positive for every $x \in H$. Moreover,

$$t_x(\rho)\rho^p\|x\|_H^p \leq \int_0^{t_x(\rho)} \|T(t)x\|_H^p dt \leq \int_0^\infty \|T(t)x\|_H^p dt \leq C^p\|x\|_H^p,$$

and therefore $t_x(\rho) \leq (C/\rho)^p = t_0$. For $t > t_0$, we have

$$\|T(t)x\|_H \leq \|T(t - t_x(\rho))\| \cdot \|T(t_x(\rho))x\|_H \leq R\rho\|x\|_H = \beta\|x\|_H,$$

where $0 < \beta = R\rho < 1$. Finally, let $t_1 > t_0$ be fixed and $t = nt_1 + s, 0 \leq s < t_1$. Then

$$\|T(t)\| \leq \|T(s)\| \cdot \|T(nt_1)\| \leq R\|T(t_1)\|^n \leq R\beta^n \leq Me^{-\mu t},$$

where $M = R\beta^{-1}$ and $\mu = -(1/t_1)\log \beta > 0$. $\quad\square$

Remark   The arguments in the proof cannot go through for $0 < p < 1$. This is because p-integrable functions in general do not form a Banach space if $p < 1$. However, by a different approach, we shall show in Section 3.1 that the result in Theorem 2.1.2 remains valid for $0 < p < 1$.

If S is finite dimensional, it is known from Proposition 2.1.2 that there exist necessary and sufficient conditions for exponential stability which are stated in terms of Lyapunov functions. The following theorem extends these results to infinite dimensional Banach spaces.

Theorem 2.1.3   *Let $T(t)$, $t \geq 0$, defined in (2.1.1) be exponentially stable. Then there exists a unique continuous mapping $\Lambda(\cdot) : S \to [0, \infty)$ such that for each $x \in S$:*

   (i) the mapping

$$t \to \Lambda(T(t)x) = \bar{\Lambda}(t, x) \tag{2.1.21}$$

   from $[0, \infty) \to [0, \infty)$ has the property $\lim_{t\to\infty} \bar{\Lambda}(t, x) = 0$;
   (ii)

$$d\bar{\Lambda}(t, x)/dt = -\|T(t)x\|_S^2; \tag{2.1.22}$$

   (iii) there exists a positive constant C such that the inequality

$$\Lambda(x) \leq C\|x\|_S^2 \tag{2.1.23}$$

   holds.

Conversely, if such a mapping $\Lambda(\cdot) : S \to [0, \infty)$ satisfying (2.1.21), (2.1.22) and (2.1.23) exists, then the semigroup $T(t)$, $t \geq 0$, is exponentially stable.

Proof Suppose $T(t)$, $t \geq 0$, is exponentially stable and hence the mapping Λ defined by

$$\Lambda(x) = \int_0^\infty \|T(t)x\|_S^2 dt \qquad (2.1.24)$$

is well-defined for any $x \in S$, and by Theorem 2.1.1 it actually satisfies an inequality of the form (2.1.23). Moreover, since $T(t)$ is exponentially stable and satisfies (2.1.23), it follows that

$$\Lambda(T(t)x) \leq C\|T(t)x\|_S^2 \leq CM^2 e^{-2\mu t}\|x\|_S^2$$

for some $C > 0$, $M \geq 1$, $\mu > 0$. Hence, $\bar{\Lambda}(t, x) \to 0$ as $t \to \infty$ for any $x \in S$.

On the other hand, from (2.1.24) we see that if $t \geq 0$,

$$\Lambda(T(t)x) = \int_0^\infty \|T(s)T(t)x\|_S^2 ds = \int_t^\infty \|T(s)x\|_S^2 ds,$$

and hence

$$d\bar{\Lambda}(t, x)/dt = -\|T(t)x\|_S^2,$$

which establishes (2.1.22). Uniqueness is a consequence of the equality

$$\bar{\Lambda}(t, x) = \Lambda(x) + \int_0^t \frac{d\bar{\Lambda}(s, x)}{ds} ds = \bar{\Lambda}(0, x) - \int_0^t \|T(s)x\|_S^2 ds \qquad (2.1.25)$$

and the fact that $\bar{\Lambda}(t, x) \to 0$ as $t \to \infty$.

The converse statement is due to (2.1.23) and (2.1.25). Since from (2.1.25) and (2.1.23)

$$\int_0^\infty \|T(t)x\|_S^2 dt = \bar{\Lambda}(0, x) = \Lambda(x) \leq C\|x\|_S^2,$$

and by Theorem 2.1.1, this is sufficient for exponential stability. □

Corollary 2.1.1 *A necessary and sufficient condition that a C_0-semigroup $T(t)$, $t \geq 0$, is exponentially stable is the existence of a mapping $\Lambda(\cdot) : S \to [0, \infty)$ such that $\Lambda(x) \leq C\|x\|_S^2$ for some positive constant C and all $x \in S$, and that*

$$\frac{d\Lambda(T(t)x)}{dt} \leq -\|T(t)x\|_S^2 \qquad (2.1.26)$$

for all x in S and $t \geq 0$.

Proof If $T(t)$ is exponentially stable, then by Theorem 2.1.3, such a $\Lambda(\cdot)$ satisfying the hypotheses of the corollary exists and is given by (2.1.24).

If, on the other hand, some $\Lambda(\cdot)$ satisfying the hypotheses of the corollary exists, then for each $x \in S$

$$0 \leq \Lambda(T(t)x) = \Lambda(x) + \int_0^t \frac{d\Lambda(T(s)x)}{ds} ds$$

$$\leq \Lambda(x) - \int_0^t \|T(s)x\|_S^2 ds \leq C\|x\|_S^2 - \int_0^t \|T(s)x\|_S^2 ds.$$

Hence,

$$\int_0^\infty \|T(s)x\|_S^2 ds \leq C\|x\|_S^2.$$

By Theorem 2.1.1, this implies exponential stability of $T(t)$.   □

Corollary 2.1.2   *Let $S = H$ be a separable real Hilbert space. Then $T(t)$ defined on H will be exponentially stable if and only if there exists an operator $P \in \mathcal{L}(H)$ such that $P \geq 0$ and for all $x \in H$, $t \geq 0$,*

$$\frac{d\langle PT(t)x, T(t)x \rangle_H}{dt} = -\|T(t)x\|_H^2. \tag{2.1.27}$$

Proof   If $T(t)$, $t \geq 0$, is exponentially stable, then $\|T(t)\| \leq Me^{-\mu t}$ for some $M \geq 1$, $\mu > 0$ and all $t \geq 0$. Let $T^*(t)$ denote the adjoint of $T(t)$. Clearly, $\|T^*(t)\| \leq Me^{-\mu t}$ for all $t \geq 0$. Thus the mapping

$$P = \int_0^\infty T^*(s)T(s)ds$$

in $\mathcal{L}(H)$ is well defined. Moreover, for each $x \in H$ and $t \geq 0$,

$$\langle PT(t)x, T(t)x \rangle_H = \int_t^\infty \|T(s)x\|_H^2 ds.$$

Consequently,

$$\frac{d\langle PT(t)x, T(t)x \rangle_H}{dt} = -\|T(t)x\|_H^2.$$

On the other hand, assume P satisfies the hypotheses of the corollary and let $x \in H$, then

$$0 \leq \langle PT(t)x, T(t)x \rangle_H = \langle Px, x \rangle_H + \int_0^t \frac{d\langle PT(s)x, T(s)x \rangle_H}{ds} ds$$

$$= \langle Px, x \rangle_H - \int_0^t \|T(s)x\|_H^2 ds, \quad t \geq 0.$$

Thus,

$$\int_0^\infty \|T(t)x\|_H^2 dt \leq \langle Px, x \rangle_H \leq \|P\| \|x\|_H^2$$

which, by Theorem 2.1.1, implies that $T(t)$, $t \geq 0$, is exponentially stable.   □

2.1.2 A Useful Stability Criterion

Proposition 2.1.3 supplements conditions on the spectral set of A to yield an exponentially stable semigroup $T(t)$. In most situations, it is found convenient to give conditions on the infinitesimal generators themselves. In the case where S is infinite dimensional, the following theorem turns out to be quite useful to yield exponentially stable semigroups.

Proposition 2.1.4 *Let A be a closed, densely defined linear operator on the real separable Hilbert space H with norm $\|\cdot\|_H$ and inner product $\langle\cdot,\cdot\rangle_H$, respectively. Then there exists a real number $\alpha \in \mathbf{R}^1$ such that*

$$\langle x, Ax\rangle_H \leq \alpha\|x\|_H^2 \qquad \text{for all} \qquad x \in \mathcal{D}(A), \qquad (2.1.28)$$

and

$$\langle x, A^*x\rangle_H \leq \alpha\|x\|_H^2 \qquad \text{for all} \qquad x \in \mathcal{D}(A^*), \qquad (2.1.29)$$

where A^* is the adjoint of A if and only if A generates a strongly continuous semigroup $T(t)$ such that

$$\|T(t)\| \leq e^{\alpha t}, \qquad t \geq 0, \qquad (2.1.30)$$

for some number $\alpha \in \mathbf{R}^1$. In particular, if $\alpha < 0$, i.e., both A and its adjoint A^* are strictly dissipative (cf. Pazy [1]), the semigroup $T(t)$, $t \geq 0$, is then exponentially stable.

Proof First of all, suppose (2.1.28) and (2.1.29) both are valid. Letting $A_\alpha = A - \alpha I$, it is obvious that A_α is a densely defined, closed linear operator satisfying $\langle x, A_\alpha x\rangle_H \leq 0$ for $x \in \mathcal{D}(A_\alpha) = \mathcal{D}(A)$ and

$$\langle x, A_\alpha^* x\rangle_H = \langle x, A^*x\rangle_H - \alpha\langle x, x\rangle_H \leq 0$$

for all $x \in \mathcal{D}(A_\alpha^*) = \mathcal{D}(A^*)$. Therefore, it follows from Corollary 1.4.4 in Pazy [1] that A_α generates a C_0-semigroup of contractions $S(t)$, $t \geq 0$, and hence, A generates a semigroup $T(t) = S(t)e^{\alpha t}$, $t \geq 0$, in H with $\|T(t)\| \leq e^{\alpha t}$.

Conversely, assume A generates a semigroup $T(t)$, $t \geq 0$, with $\|T(t)\| \leq e^{\alpha t}$. Now, A is a densely defined, closed linear operator (cf. Pazy [1]) and the semigroup $S(t) = T(t)e^{-\alpha t}$, $t \geq 0$, corresponding to the generator $A_\alpha = A - \alpha I$, is contractive. Hence, $\langle x, A_\alpha x\rangle_H \leq 0$ for $x \in \mathcal{D}(A_\alpha)$. On the other hand, since H is a Hilbert space, its adjoint $S^*(t)$ is generated by A_α^* and it is also contractive as $\|S^*(t)\| = \|S(t)\| \leq 1$. Thus $\langle x, A_\alpha^* x\rangle_H \leq 0$ for $x \in \mathcal{D}(A_\alpha^*)$. That is, $\langle x, Ax\rangle_H \leq \alpha\|x\|_H^2$ for $x \in \mathcal{D}(A)$ and $\langle x, A^*x\rangle_H \leq \alpha\|x\|_H^2$ for $x \in \mathcal{D}(A^*)$. This completes the proof. ▢

As an illustration of the technique described above, we point out that many parabolic partial differential equations can be formulated as $\frac{dz(t)}{dt} = \mathcal{A}z(t)$, where $\mathcal{A} = \mathcal{A}^* \leq 0$ is some properly interpreted self-adjoint operator. The next example is typical for wave equations. Let $\alpha \in \mathbf{R}^1$ and consider

$$\frac{d^2v(t)}{dt^2} + \alpha \cdot \frac{dv(t)}{dt} + Av(t) = 0, \quad v(0) = v_0 \in H, \quad \frac{dv}{dt}(0) = v_1 \in H, \quad (2.1.31)$$

where A is a positive, self-adjoint operator on a real separable Hilbert space H with domain $\mathcal{D}(A)$, so that the so-called coercive condition

$$\langle Ax, x \rangle_H \geq C \|x\|_H^2, \quad \forall x \in \mathcal{D}(A), \quad C > 0,$$

holds. From this it is clear that A is injective, and so we obtain that its algebraic inverse exists. For each $y \in \operatorname{Ran} A$, the range of A, we have

$$\|A^{-1}y\|_H^2 \leq \frac{1}{C}\langle AA^{-1}y, A^{-1}y \rangle_H \leq \frac{1}{C}\|y\|_H\|A^{-1}y\|_H.$$

This implies that $\|A^{-1}y\|_H \leq \frac{1}{C}\|y\|_H$, and A^{-1} is bounded on its range. If $\operatorname{Ran} A$ is dense in H, then 0 is in the resolvent set of A and $A^{-1} \in \mathcal{L}(H)$. Indeed, let x be in the orthogonal complement to the range of A, i.e., for all $y \in \mathcal{D}(A)$ the following holds: $\langle y, x \rangle_H = 0$. By the definition of adjoint operators this implies that $x \in \mathcal{D}(A^*)$ and $A^*x = 0$. Since A is self-adjoint, we conclude that $Ax = A^*x = 0$. The positivity of A shows that this can only happen if $x = 0$, and so $\operatorname{Ran} A$ is dense in H.

We intend to show that the space $\mathcal{H} = \mathcal{D}(A^{1/2}) \oplus H$, equipped with the mapping $\langle \cdot, \cdot \rangle_\mathcal{H} : \mathcal{H} \times \mathcal{H} \to \mathbf{R}^1$,

$$\langle z, \tilde{z} \rangle_\mathcal{H} := \langle A^{1/2}z_1, A^{1/2}\tilde{z}_1 \rangle_H + \langle z_2, \tilde{z}_2 \rangle_H,$$

where

$$z = \begin{pmatrix} z_1 \\ z_2 \end{pmatrix}, \qquad \tilde{z} = \begin{pmatrix} \tilde{z}_1 \\ \tilde{z}_2 \end{pmatrix}$$

is actually a Hilbert space. Indeed, since $A^{1/2}$ is positive and $\langle \cdot, \cdot \rangle_H$ is the inner product on H, it is easy to see that $\langle \cdot, \cdot \rangle_\mathcal{H}$ defines an inner product on $\mathcal{H}$. Thus it remains to show that $\mathcal{H}$ with the norm $\|z\|_\mathcal{H} = \sqrt{\langle z, z \rangle_\mathcal{H}}$ is complete. To this end, let $z_n = \begin{pmatrix} z_{1,n} \\ z_{2,n} \end{pmatrix}$ be a Cauchy sequence in $\mathcal{H}$. This implies that

$$\|A^{1/2}[z_{1,n} - z_{1,m}]\|_H^2 + \|z_{2,n} - z_{2,m}\|_H^2 = \|z_n - z_m\|_\mathcal{H}^2 \to 0 \quad \text{as} \quad n, m \to \infty.$$

Hence, $z_{2,n}$ is a Cauchy sequence in H, and since H is a Hilbert space we know that $z_{2,n}$ converges to some $z_2 \in H$. Similarly, we have that $A^{1/2}z_{1,n}$ converges to some $x \in H$. Since A is boundedly invertible, so is $A^{1/2}$ and $(A^{1/2})^{-1} = (A^{-1})^{1/2} = A^{-1/2}$. So $z_{1,n} = A^{-1/2}[A^{1/2}z_{1,n}] \to A^{1/2}x$ as $n \to \infty$, and $z_1 = A^{-1/2}x \in \mathcal{D}(A^{1/2})$. Thus $\|z_n - z\|_\mathcal{H}^2 \to 0$, where $z = \begin{pmatrix} z_1 \\ z_2 \end{pmatrix}$ and $\mathcal{H}$ is complete.

Define linear operators on $\mathcal{H}$

$$\mathcal{A} = \begin{pmatrix} 0 & I \\ -A & -\alpha I \end{pmatrix} \quad \text{with domain} \quad \mathcal{D}(\mathcal{A}) = \mathcal{D}(A) \oplus \mathcal{D}(A^{1/2}),$$

and

$$Q = \begin{pmatrix} -\alpha A^{-1} & -A^{-1} \\ I & 0 \end{pmatrix},$$

then Q is a bounded linear operator on $\mathcal{H}$ with $\mathrm{Ran}\, Q = \mathcal{D}(A)$ and $AQ = I$. Thus we see that A is closed. It is also easy to show that the domain of A is dense in $\mathcal{H}$. Hence, (2.1.31) may be rewritten as a first order differential equation on $\mathcal{H}$,

$$\frac{dz(t)}{dt} = Az(t), \qquad z(0) = z_0 \in \mathcal{H}, \qquad (2.1.32)$$

where

$$z(t) = \begin{pmatrix} v(t) \\ dv(t)/dt \end{pmatrix}, \qquad z_0 = \begin{pmatrix} v_0 \\ v_1 \end{pmatrix}.$$

Note that it is immediate to deduce

$$\langle Az, z \rangle_{\mathcal{H}} = \langle Az_1, z_2 \rangle_H + \langle z_2, -Az_1 - \alpha z_2 \rangle_H = -\alpha \|z_2\|_H^2 \leq 0$$

for any $z \in \mathcal{D}(A) \oplus \mathcal{D}(A^{1/2})$. Similarly, the adjoint operator of A with respect to the Hilbert space $\mathcal{H}$ is easily shown to be

$$A^* \begin{pmatrix} z_1 \\ z_2 \end{pmatrix} = \begin{pmatrix} 0 & -I \\ A & -\alpha \end{pmatrix} \begin{pmatrix} z_1 \\ z_2 \end{pmatrix}, \qquad \mathcal{D}(A^*) = \mathcal{D}(A),$$

which immediately implies $\langle z, A^* z \rangle_{\mathcal{H}} = -\alpha \|z_2\|_H^2 \leq 0$. Therefore, by virtue of Proposition 2.1.4, we conclude that A generates a strongly continuous semigroup $\mathcal{T}(t)$, $t \geq 0$, on $\mathcal{H}$. In particular, we may obtain the following stability results by means of the Lyapunov equation method in Theorem 2.1.1 and Proposition 2.1.3.

Example 2.1.3 *(1). For every number $\alpha \in \mathbf{R}^1$,*

$$\gamma(A) = \begin{cases} -\frac{\alpha}{2} + \sqrt{\frac{\alpha^2}{4} + \gamma(-A)} & \text{if } \frac{\alpha^2}{4} + \gamma(-A) \geq 0, \\ -\frac{\alpha}{2} & \text{otherwise,} \end{cases} \qquad (2.1.33)$$

where $\gamma(\cdot)$ is the lower stability index defined in Proposition 2.1.3. Moreover,

$$\gamma(A) \geq -\sqrt{|\gamma(-A)|}; \qquad (2.1.34)$$

(2). If $\gamma(-A) = 0$, then for any α, the system (2.1.31) is not exponentially stable.

(3). If $\gamma(-A) < 0$, then for any positive $\alpha > 0$, the operator

$$P = \begin{pmatrix} I - \frac{\alpha^2}{2}(-A)^{-1} & -\frac{\alpha}{2}(-A)^{-1} \\ \frac{\alpha}{2}I & I \end{pmatrix}$$

is the unique nonnegative solution of the following Lyapunov equation,

$$2\left\langle \mathcal{P}\mathcal{A}\begin{pmatrix} z_1 \\ z_2 \end{pmatrix}, \begin{pmatrix} z_1 \\ z_2 \end{pmatrix}\right\rangle_{\mathcal{H}} = -\alpha\left\|\begin{pmatrix} z_1 \\ z_2 \end{pmatrix}\right\|_{\mathcal{H}}^2, \quad z_1 \in \mathcal{D}(A), \quad z_2 \in \mathcal{D}(A^{-1/2}).$$

(4). If $\gamma(-A) < 0$ and $\alpha > 0$, then

$$\Gamma(\mathcal{A}) \leq -\frac{2\alpha|\gamma(-A)|}{4|\gamma(-A)| + \alpha(\alpha + \sqrt{\alpha^2 + 4|\gamma(-A)|})},$$

where $\Gamma(\cdot)$ is the upper stability index defined in Proposition 2.1.3 and consequently, the system (2.1.31) is exponentially stable.

Proof   From the very definition of the resolvent sets $\rho(-A)$ and $\rho(\mathcal{A})$, we easily obtain that $\lambda \in \rho(\mathcal{A})$ if and only if $\lambda(\lambda + \alpha) \in \rho(-A)$. Now a straightforward computation yields (2.1.33). Property (2) follows from (1) and Proposition 2.1.3. Property (3) follows from the easily checked identities:

$$\left\langle \mathcal{P}\begin{pmatrix} z_1 \\ z_2 \end{pmatrix}, \begin{pmatrix} z_1 \\ z_2 \end{pmatrix}\right\rangle_{\mathcal{H}} = \|A^{1/2}z_1\|_H^2 + \frac{\alpha^2}{2}\|z_1\|_H^2 + \alpha\langle z_1, z_2\rangle_H + \|z_2\|_H^2$$

$$= \|A^{1/2}z_1\|_H^2 + \frac{1}{2}(\|\alpha z_1 + z_2\|_H^2 + \|z_2\|_H^2),$$

$$\frac{d}{dt}\left\langle \mathcal{P}\begin{pmatrix} v(t) \\ dv(t)/dt \end{pmatrix}, \begin{pmatrix} v(t) \\ dv(t)/dt \end{pmatrix}\right\rangle_{\mathcal{H}} = -\alpha(\|dv(t)/dt\|_H^2 + \|A^{-1/2}v(t)\|_H^2),$$

$$(2.1.35)$$

provided

$$v(0) \in \mathcal{D}(A), \quad \frac{dv}{dt}(0) \in \mathcal{D}(A^{1/2}).$$

Also, a direct calculation gives the following estimate:

$$\frac{1}{2}\left\|\begin{pmatrix} z_1 \\ z_2 \end{pmatrix}\right\|_{\mathcal{H}}^2 \leq \left\langle \mathcal{P}\begin{pmatrix} z_1 \\ z_2 \end{pmatrix}, \begin{pmatrix} z_1 \\ z_2 \end{pmatrix}\right\rangle_{\mathcal{H}}$$

$$\leq \frac{4|\gamma(-A)| + \alpha(\alpha + \sqrt{\alpha^2 + 4|\gamma(-A)|})}{4|\gamma(-A)|}\left\|\begin{pmatrix} z_1 \\ z_2 \end{pmatrix}\right\|_{\mathcal{H}}^2 \quad (2.1.36)$$

Combining (2.1.35) with (2.1.36), we finally obtain Property (4).   ☐

2.2   Lyapunov Equations and Stability

Assume $T(t)$, $t \geq 0$, is a strongly continuous semigroup of bounded linear operators on the real Hilbert space H with infinitesimal generator A. In

Theorem 2.1.1, we have shown that the following statements which actually play a key role in deducing stability characteristics in terms of Lyapunov functions are equivalent:

(i). $T(t)$, $t \geq 0$, is an exponentially stable semigroup.
(ii). $\int_0^\infty \|T(t)x\|_H^2 dt < \infty$ for each $x \in H$.
(iii). There exists a nonnegative, self-adjoint operator P in $\mathcal{L}(H)$ such that

$$2\langle PAx, x \rangle_H = -\langle x, x \rangle_H \quad \text{for each} \quad x \in \mathcal{D}(A).$$

One of the main objectives in this section is to establish a stochastic version of the above result in the mean square sense. Based on this, we shall study and derive sufficient conditions which are more effective and easily checked to ensure the mean square and almost sure stability of associated linear stochastic systems.

2.2.1 Characterization of Mean Square Stability

Assume a complete probability space $(\Omega, \mathcal{F}, P)$, equipped with a normal filtration $\{\mathcal{F}_t\}_{t \geq 0}$ with respect to which $\{W_t\}_{t \geq 0}$, $t \geq 0$, is some given Q-Wiener process with $trQ < \infty$ in the Hilbert space K. Consider the following linear stochastic integral equation on the Hilbert space H

$$\begin{cases} X_t = T(t)x_0 + \int_0^t T(t-s)B(X_s)dW_s, \\ X_0 = x_0 \in H, \end{cases} \tag{2.2.1}$$

where $T(t)$, $t \geq 0$, is a strongly continuous semigroup with its infinitesimal generator A on the Hilbert space H and $B \in \mathcal{L}(H, \mathcal{L}(K, H))$. It is immediate from Theorem 1.3.4 that the equation (2.2.1) has a unique (mild) solution $X_t \in C(0, \infty; L^2(\Omega; H))$, $t \geq 0$.

Our arguments to establish stability equivalences below are basically involved with a calculation of the term

$$\int_0^T E\langle MX_t, X_t \rangle_H dt + E\langle GX_T, X_T \rangle_H, \quad T \geq 0, \tag{2.2.2}$$

where $M \geq 0$, $G \geq 0$ both are in $\mathcal{L}(H)$. To this end, consider the following backward linear operator differential equation

$$\frac{d}{dt}\langle P(t)x, x \rangle_H + 2\langle Ax, P(t)x \rangle_H + \langle [M + \Delta(P(t))]x, x \rangle_H = 0,$$
$$x \in \mathcal{D}(A), \quad 0 \leq t \leq T, \tag{2.2.3}$$
$$P(T) = G,$$

or its integral version (see Proposition 2.2.1 below),

$$P(t)x = \int_t^T T^*(r-t)[M + \Delta(P(r))]T(r-t)x\,dr$$
$$+ T^*(T-t)GT(T-t)x, \quad \forall x \in H, \quad 0 \leq t \leq T, \tag{2.2.4}$$

where $\langle \Delta(P)x, y\rangle_H := tr\{B^*(y)PB(x)Q\}$, $x, y \in H$, $P \in \mathcal{L}(H)$.

Lemma 2.2.1   *There exists a unique solution $P(t) \in \mathcal{L}(H)$, $0 \le t \le T$, satisfying (2.2.4) in the class of linear nonnegative, self-adjoint strongly continuous operators on H.*

Proof   Define a sequence of strongly continuous, self-adjoint, nonnegative operators $P_n(t)$, $n \ge 0$, $0 \le t \le T$, by the equations:

$$P_n(t)x = \int_t^T T^*(r-t)[M + \Delta(P_{n-1}(r))]T(r-t)x\,dr$$
$$+ T^*(T-t)GT(T-t)x, \quad x \in H, \quad n \ge 1,$$
$$P_0(t) = 0.$$

Then for any $n \ge 2$ and $T \ge t \ge 0$, we have the estimate

$$\|P_n(t)\| \le p_1(T) + a \int_t^T \|P_{n-1}(r)\|\,dr,$$

where $p_1(T) = \sup_{0 \le t \le T} \|P_1(t)\|$ and $a = a(T) > 0$. By induction, we have

$$\|P_n(t)\| \le p_1(T)e^{a(T-t)} \quad \text{for any} \quad n \ge 1.$$

Thus, $P_n(t)$ is uniformly bounded in $t \in [0, T]$ and $n \ge 0$. Since

$$\langle \Delta(P)x, x\rangle_H = \sum_{i=1}^\infty \lambda_i \langle PB(x)e_i, B(x)e_i\rangle_H,$$

where $\{e_i\}$ is the orthonormal sequence of H consisting of eigenvectors of Q with $Qe_i = \lambda_i e_i$, $\lambda_i \ge 0$, $trQ = \sum_{i=1}^\infty \lambda_i < \infty$. Now, $P_n(t)$ is monotonically increasing, i.e., $P_n(t) \le P_{n+1}(t)$ for each $t \ge 0$. Therefore, there exists a strong limit $P(t)$, $t \ge 0$, which satisfies (2.2.4) and has the desired properties.

The uniqueness follows from that of the null solution of the integral equation

$$P(t)x = \int_t^T T^*(r-t)\Delta(P(r))T(r-t)x\,dr,$$

$0 \le t \le T$. Indeed, we have

$$\|P(t)\| \le a \int_t^T \|P(r)\|\,dr, \quad \forall T \ge t \ge 0,$$

for some $a = a(T) > 0$, which immediately implies $\|P(t)\| = 0$, $0 \le t \le T$. $\quad\square$

Proposition 2.2.1   *The equations (2.2.3) and (2.2.4) are equivalent. Moreover, there exists a unique solution $P(t)$, $0 \le t \le T$, satisfying (2.2.3) in the class of linear, self-adjoint, nonnegative strongly continuous operators on H.*

Proof Obviously, it suffices by Lemma 2.2.1 to show the equivalence of (2.2.3) and (2.2.4). Suppose first that $P(t)$, $t \geq 0$, satisfies (2.2.4). Then differentiating $\langle P(t)x, x \rangle_H$, $x \in \mathcal{D}(A)$, yields (2.2.3).

Conversely, suppose $P(t)$, $t \geq 0$, satisfies (2.2.3). Let $x \in \mathcal{D}(A)$ and $0 < s \leq t$, then $\langle P(t)T(t-s)x, T(t-s)x \rangle_H$ is differentiable in t and

$$\frac{d}{dt} \langle P(t)T(t-s)x, T(t-s)x \rangle_H = -2\langle AT(t-s)x, P(t)T(t-s)x \rangle_H$$
$$-\langle [M + \Delta(P(t))]T(t-s)x, T(t-s)x \rangle_H$$
$$+2\langle P(t)T(t-s)x, AT(t-s)x \rangle_H$$
$$= -\langle [M + \Delta(P(t))]T(t-s)x, T(t-s)x \rangle_H.$$

Integrating this from s to T, we obtain

$$\langle P(s)x, x \rangle_H = \langle GT(T-s)x, T(T-s)x \rangle_H$$
$$+ \int_s^T \langle [M + \Delta(P(t))]T(t-s)x, T(t-s)x \rangle_H dt.$$

Since $\mathcal{D}(A)$ is dense in H, (2.2.4) follows easily. $\square$

Let K_i, $i = 1$, 2, be two real separable Hilbert spaces. To calculate (2.2.2), we intend to consider the following stochastic differential equation in H,

$$\begin{cases} dX_t = AX_t dt + B(X_t)dW_t^1 + FdW_t^2, \\ X_0 = x_0 \in H, \end{cases} \tag{2.2.5}$$

where A is the infinitesimal generator of a C_0-semigroup $T(t)$, $t \geq 0$, on H, W_t^i are two K_i-valued Q-Wiener processes independent mutually with covariance operators Q_i, $trQ_i < \infty$, $i = 1$, 2, respectively, and $B \in \mathcal{L}(H, \mathcal{L}(K_1, H))$, $F \in \mathcal{L}(K_2, H)$. Using the same notations as in (2.2.2) and (2.2.3), we have:

Proposition 2.2.2 *Let X_t, $t \geq 0$, be the mild solution of (2.2.5), then the following relation holds:*

$$\int_t^T E\langle MX_s, X_s \rangle_H ds + E\langle GX_T, X_T \rangle_H = E\langle P(t)X_t, X_t \rangle_H$$
$$+ \int_t^T tr\{F^* P(s)FQ_2\}ds \tag{2.2.6}$$

for any $0 \leq t \leq T$.

Proof To prove (2.2.6), let us introduce the approximating system of (2.2.5):

$$dX_t = AX_t dt + nR(n, A)[B(X_t)dW_t^1 + FdW_t^2], $$
$$X_0 = nR(n, A)x_0, \tag{2.2.7}$$

where $0 < n_0 \leq n \in \rho(A)$ for some $n_0 \in \mathbf{N}$, and $R(n, A)$ is the resolvent of A. Take $x_0 \in \mathcal{D}(A)$ and since $nAR(n, A) = n - n^2R(n, A)$ is bounded, the conditions in Proposition 1.3.5 may be proved to hold so that there exists a unique strong solution of (2.2.7), denote it by X_t^n, $t \geq 0$. Then applying Itô's formula to $\langle P(t)X_t^n, X_t^n\rangle_H$, we obtain

$$\int_t^T E\langle MX_s^n, X_s^n\rangle_H ds + E\langle GX_T^n, X_T^n\rangle_H$$

$$= E\langle P(t)X_t^n, X_t^n\rangle_H + \int_t^T tr\{nR(n, A)F^*P(s)nR(n, A)FQ_2\}ds$$

which, passing to the limit $n \to \infty$ and using Proposition 1.3.6, immediately yields (2.2.6). But since $\mathcal{D}(A)$ is dense in H and $X_t(x_0)$ depends continuously on x_0, (2.2.6) holds for any $x_0 \in H$. The proof is complete. □

We are now in a position to establish the main results in this subsection.

Theorem 2.2.1   *Suppose $X_t(x_0)$, $t \geq 0$, is the unique solution of (2.2.1) with initial datum $x_0 \in H$. Then the following statements are equivalent:*

(i). The solution $X_t(x_0)$, $t \geq 0$, satisfies

$$\int_0^\infty E\|X_t(x_0)\|_H^2 dt < \infty \quad \text{for each} \quad x_0 \in H. \tag{2.2.8}$$

(ii). There exists a nonnegative, self-adjoint operator $P \in \mathcal{L}(H)$ such that

$$2\langle Ax, Px\rangle_H + \langle\Delta(P)x, x\rangle_H = -\langle x, x\rangle_H \quad \text{for any} \quad x \in \mathcal{D}(A), \tag{2.2.9}$$

where $\langle\Delta(P)x, x\rangle_H = tr\{B^*(x)PB(x)Q\}$.
(iii). There exist positive numbers $M \geq 1$, $\mu > 0$ such that for all $t \geq 0$,

$$E\|X_t(x_0)\|_H^2 \leq M \cdot e^{-\mu t}\|x_0\|_H^2.$$

Proof   Suppose (i) holds and let $P_T(\cdot)$ be the solution of

$$\frac{d}{dt}\langle P_T(t)x, x\rangle_H + 2\langle Ax, P_T(t)x\rangle_H + \langle[I + \Delta(P_T(t))]x, x\rangle_H = 0,$$

$$0 \leq t \leq T, \quad x \in \mathcal{D}(A), \tag{2.2.10}$$

$$P_T(T) = 0.$$

Then, by using Proposition 2.2.1, (2.2.4) and (2.2.6), we obtain

$$\int_0^T \|T(t)x_0\|_H^2 dt \leq \langle P_T(0)x_0, x_0\rangle_H = \int_0^T E\|X_t\|_H^2 dt \leq \int_0^\infty E\|X_t\|_H^2 dt < \infty.$$

Hence, by virtue of Theorem 2.1.1, $T(t)$, $t \geq 0$, is exponentially stable which implies $P_T(0)$ is monotonically increasing in T and uniformly bounded (Banach-Steinhaus theorem). Thus, there exists a limit $P \geq 0$, and in view of (2.2.10), we can conclude that P satisfies (2.2.9). This proves that (i) implies (ii).

Suppose (ii) holds. With the aid of Itô's formula, for $\Lambda(x) = \langle Px, x\rangle_H$, $x \in H$, we can establish by using (2.2.6) that

$$\langle Px_0, x_0\rangle_H = E\langle PX_t, X_t\rangle_H + \int_0^t E\|X_s(x_0)\|_H^2 ds, \qquad t \geq 0. \qquad (2.2.11)$$

Hence,

$$\int_0^\infty \|T(t)x_0\|_H^2 dt \leq \int_0^\infty E\|X_t(x_0)\|_H^2 dt \leq \langle Px_0, x_0\rangle_H < \infty,$$

which implies (i) as well as the exponential stability of $T(t)$, $t \geq 0$.

On the other hand, from (2.2.11) we obtain

$$\frac{d}{dt} E\langle PX_t, X_t\rangle_H = -E\|X_t\|_H^2 \leq -\|P\|^{-1} E\langle PX_t, X_t\rangle_H,$$

which immediately yields

$$E\langle PX_t, X_t\rangle_H \leq e^{-pt}\langle Px_0, x_0\rangle_H, \qquad p = \|P\|^{-1}.$$

Since $T(t)$, $t \geq 0$, is exponentially stable, $\|T(t)\| \leq C \cdot e^{-\gamma t}$ for some $C \geq 1$, $\gamma > 0$. Now carrying out an approximating argument as in Proposition 1.3.6 and applying Itô's formula yield that for some number $\bar{C} > 0$,

$$E\|X_t(x_0)\|_H^2 = \|T(t)x_0\|_H^2 + E\int_0^t tr\{T(t-r)B(X_r)QB^*(X_r)T^*(t-r)\}dr$$

$$\leq C^2 e^{-2\gamma t}\|x_0\|_H^2 + \bar{C}\int_0^t e^{-2\gamma(t-r)} E\|X_r\|_H^2 dr$$

$$\leq C^2 e^{-2\gamma t}\|x_0\|_H^2 - \bar{C}\int_0^t e^{-2\gamma(t-r)}\frac{d}{dr} E\langle PX_r, X_r\rangle_H dr$$

$$\leq C^2 e^{-2\gamma t}\|x_0\|_H^2 - \bar{C}E\langle PX_t, X_t\rangle_H + \bar{C}e^{-2\gamma t}\langle Px_0, x_0\rangle_H$$

$$+\frac{\bar{C}}{2\gamma}\int_0^t e^{-2\gamma(t-r)} E\langle PX_r, X_r\rangle_H dr$$

$$\leq C^2 e^{-2\gamma t}\|x_0\|_H^2 + \bar{C}e^{-2\gamma t}\langle Px_0, x_0\rangle_H$$

$$+\frac{\bar{C}}{2\gamma}\int_0^t e^{-2\gamma(t-r)}e^{-pr}\langle Px_0, x_0\rangle_H dr$$

$$\leq C^2 e^{-2\gamma t}\|x_0\|_H^2 + \bar{C}e^{-2\gamma t}\langle Px_0, x_0\rangle_H$$

$$+\frac{\bar{C}}{2\gamma}\frac{1}{2\gamma - p}(e^{-pt} - e^{-2\gamma t})\langle Px_0, x_0\rangle_H \qquad (p \neq 2\gamma)$$

$$\leq C^2 e^{-2\gamma t} \|x_0\|_H^2 + \bar{C} e^{-2\gamma t} \langle Px_0, x_0 \rangle_H$$

$$+ \frac{\bar{C}}{2\gamma(2\gamma - p)} e^{-pt} \langle Px_0, x_0 \rangle_H, \qquad (p \neq 2\gamma).$$

If $p = 2\gamma$, the last term above is replaced by $(\bar{C}/(2\gamma))te^{-2\gamma t}\langle Px_0, x_0 \rangle_H$. This implies (iii), i.e.,

$$E\|X_t(x_0)\|_H^2 \leq M \cdot e^{-\mu t} \|x_0\|_H^2 \quad \text{for some} \quad M \geq 1, \;\; \mu > 0.$$

Note that it is obvious (iii) implies (i), so the proof is complete. □

Remark   When the relation (2.2.8) holds for solutions of stochastic differential equations such as (2.2.1), it is said that the systems are L^2-*stable in mean*, a concept which seems weaker than exponential stability in mean square. Theorem 2.2.1 shows however that these two concepts are actually equivalent for the equation (2.2.1).

Corollary 2.2.1   *The equation (2.2.9) has at most one solution.*

Proof   If there exists a solution P of (2.2.9), then the null solution of (2.2.1) is exponentially stable in mean square. Therefore

$$E\langle PX_t(x_0), X_t(x_0) \rangle_H \to 0, \quad \text{as} \quad t \to \infty.$$

Hence, by virtue of (2.2.11), it is easy to deduce

$$\langle Px_0, x_0 \rangle_H = \int_0^\infty E\|X_t(x_0)\|_H^2 dt, \quad x_0 \in H,$$

which immediately implies the uniqueness. □

Corollary 2.2.2   *Assume P is a self-adjoint, nonnegative operator solution of (2.2.9), then we have*

$$\langle Px_0, x_0 \rangle_H = \int_0^\infty E\|X_s(x_0)\|_H^2 ds, \quad x_0 \in H,$$

and

$$P = \int_0^\infty T^*(t)[I + \Delta(P)]T(t)dt.$$

Proof   This is immediate by the proof of Corollary 2.2.1, the construction of P and (2.2.4). □

Remark A similar characterization in terms of Lyapunov functions to those in Theorem 2.1.3 and its corollaries may be established based on the results obtained above. However, we prefer to present hesitantly at the moment and leave its study to Sections 3.4–3.7. Over there, a detailed investigation of this topic with its application to nonlinear stochastic systems will be carried out.

If the null solution of (2.2.1) is stable, the average of second moment for the mild solution of (2.2.5) is finite. Precisely, assume $X_t(x_0)$, $t \geq 0$, is the unique mild solution of Equation (2.2.5) with initial datum $x_0 \in H$. The system (2.2.5) is said to be *mean square stable in average* if

$$\lim_{T \to \infty} \frac{1}{T} \int_0^T E\|X_t(x_0)\|_H^2 dt < \infty.$$

Corollary 2.2.3 *Let $X_t(x_0)$, $t \geq 0$, be the mild solution of (2.2.5). Then the null solution of (2.2.1) is exponentially stable in mean square if and only if the system (2.2.5) is mean square stable in average. Moreover, in that case, the solution X_t, $t \geq 0$, of (2.2.5) satisfies*

$$\lim_{T \to \infty} \frac{1}{T} \int_0^T E\|X_t(x_0)\|_H^2 dt = tr\{F^*PFQ_2\} < \infty$$

where P is the solution of (2.2.9).

Proof Let $P_T(\cdot)$ be the solution of (2.2.10), then in view of Proposition 2.2.2, we have

$$\int_0^T E\|X_t(x_0)\|_H^2 dt = \langle P_T(0)x_0, x_0 \rangle_H + \int_0^T tr\{F^*P_T(t)FQ_2\}dt.$$

However, we know by Theorem 2.2.1 that $P_T(T)$ converges monotonically to P as $T \to \infty$. Hence the assertion easily follows. □

Under some suitable conditions on the operators A and B, the results in this subsection allow an extension to the case when the operator B is unbounded. Also, it is worth pointing out that in many practical situations, it is not always easy to check the hypothesis (i) or (ii) of Theorem 2.2.1 straightforwardly. At the moment we do not try to go into further details about this and refer the reader to Da Prato and Ichikawa [1] or Da Prato and Zabczyk [1]. In the following subsection, by employing Theorem 2.2.1 we shall establish some useful and easily checked sufficient conditions to obtain stability of Equation (2.2.1). Lastly, to close this subsection we present an interesting example due to Da Prato and Zabczyk [1] (also see Zabczyk [2]) for a system of the so-called Lurie type whose stability can be completely characterized by Theorem 2.2.1.

Example 2.2.1   Consider the stochastic system

$$dX_t = AX_t dt + \sum_{i=1}^{n} b_i \langle h_i, X_t \rangle_H dB_t^i, \qquad X_0 = x_0, \qquad (2.2.12)$$

where A is the generator of an exponentially stable C_0-semigroup $T(\cdot)$. $b_i, h_i, i = 1, \cdots, n$ are some given non-zero elements in H, and B_t^i, $i = 1, \cdots, n$, are independent real-valued Wiener processes. We have here $K = \mathbf{R}^n$, $Q = I$ and $B(\cdot)$ in (2.2.1) is given by

$$B(x)k = \sum_{i=1}^{n} b_i \langle h_i, x \rangle_H k_i, \qquad k = (k_1, \cdots, k_n) \in K. \qquad (2.2.13)$$

Moreover, note that $\langle \Delta(P)x, y \rangle_H := tr\{B^*(y)PB(x)Q\}$, $x, y \in H$, and we have in this case

$$\Delta(P) = \sum_{i=1}^{n} \langle Ph_i, h_i \rangle_H b_i \otimes b_i.$$

We remark that since $T(t)$, $t \geq 0$, is exponentially stable, Equation (2.2.9) is equivalent to the following one:

$$P = \int_0^\infty T(t)\Delta(P)T^*(t)dt + \int_0^\infty T(t)T^*(t)dt$$

$$= \sum_{i=1}^{n} \langle Ph_i, h_i \rangle_H \int_0^\infty (T(t)h_i) \otimes (T(t)h_i)dt + \int_0^\infty T(t)T^*(t)dt,$$

and in order to solve it, it suffices to find a vector $\xi \in \mathbf{R}^n$, $\xi = \{\langle Ph_i, h_i \rangle_H : i = 1, \cdots, n\}$, with nonnegative components such that

$$\xi = M\xi + \eta, \qquad (2.2.14)$$

where $M = (M_{ij})$ is the $n \times n$ matrix defined as

$$M_{ij} = \int_0^\infty \langle T(t)b_i, h_j \rangle_H^2 dt, \qquad i, j = 1, \cdots, n,$$

and η is the vector

$$\eta_j = \int_0^\infty \|T^*(t)h_j\|_H^2 dt, \qquad j = 1, \cdots, n.$$

Since the entries of matrix M are all nonnegative and the components of η are strictly positive, Equation (2.2.14) has a nonnegative solution, and so the system (2.2.12) is stable if and only if the eigenvalues of M are all of modulus less than 1.

2.2.2 Almost Sure Pathwise Stability

As mentioned in the last subsection, it is not always easy to use Theorem 2.2.1 in a straightforward way. Based on Theorem 2.2.1, in this subsection we shall establish and analyse sufficient conditions for sample path stability of processes in the almost sure sense and apply them to some examples in the section 2.4.

Consider the linear stochastic evolution equation (2.2.1) again. We wish to show that under some conditions the null solution of (2.2.1) is exponentially stable in mean square, and from this fact we further deduce that the sample paths of solutions decay to zero almost surely as $t \to \infty$. To this end, we impose the following conditions which are easy to verify in most situations:

(H1). The semigroup $T(t)$, $t \geq 0$, is exponentially stable, i.e., there exist constants $C \geq 1$, $\gamma > 0$ such that $\|T(t)\| \leq C \cdot e^{-\gamma t}$, $t \geq 0$;

(H2). $\| \int_0^\infty T^*(t)\Delta(I)T(t)dt \| < 1$, where $\langle \Delta(P)x, y \rangle_H := tr[B(x)^*PB(y)Q]$ for any $P \in \mathcal{L}(H)$ and $x, y \in H$.

Theorem 2.2.2 *Assume (H1) and (H2) hold. Then there exist positive constants M, μ such that for any solution $X_t(x_0)$ of Equation (2.2.1),*

$$E\|X_t(x_0)\|_H^2 \leq M \cdot \|x_0\|_H^2 e^{-\mu t}, \qquad t \geq 0. \tag{2.2.15}$$

Proof By virtue of Theorem 2.2.1, it suffices to prove that there exists a nonnegative, self-adjoint operator P in $\mathcal{L}(H)$ which is the solution of (2.2.9). Indeed, P can be constructed by $P = \lim_{n \to \infty} P_n$, where

$$0 < P_1 := \int_0^\infty T^*(t)T(t)dt \in \mathcal{L}(H),$$

and

$$P_{n+1} := P_1 + S(P_n),$$

where

$$S(G) = \int_0^\infty T^*(t)\Delta(G)T(t)dt$$

for any $G \in \mathcal{L}(H)$. Note that $S(G) \geq 0$ if $G \geq 0$, so that P_n is nondecreasing as n tends to infinity. On the other hand,

$$\|S(G)\| = \sup_{\|x\|_H=1} \int_0^\infty \langle T^*(t)\Delta(G)T(t)x, x \rangle_H dt$$

$$= \sup_{\|x\|_H=1} \int_0^\infty tr[B(T^*(t)x)GB(T(t)x)Q]dt \leq \|G\|\|S(I)\|.$$

According to (H2), $\|S(I)\| < 1$ which immediately implies $P = \lim_{n \to} P_n$ exists, and

$$P = \int_0^\infty T^*(t)[I + \Delta(P)]T(t)dt, \tag{2.2.16}$$

or

$$2\langle PAx, x\rangle_H + \langle \Delta(P)x, x\rangle_H + \|x\|_H^2 = 0 \quad \text{for all} \quad x \in \mathcal{D}(A),$$

which, together with Theorem 2.2.1, immediately yields the required result. ☐

Remark   There exists a condition which is usually stronger but easier to use than (H1) and (H2). It implies (2.2.15) by an approximation type of argument as in Proposition 1.3.6. This condition may be described as follows: there exists a constant $\nu > 0$ such that for arbitrary $x \in \mathcal{D}(A)$,

$$2\langle Ax, x\rangle_H + \langle \Delta(I)x, x\rangle_H \leq -\nu\|x\|_H^2. \tag{2.2.17}$$

We shall not go into further details at present because some more general conditions which contain (2.2.17) as a special case will be derived to deal with nonlinear stochastic systems in Chapter 3.

In the early stages (1940s to 1960s) of the study of stability of finite dimensional stochastic systems, investigators were primarily concerned with moment stability and stability in probability. The mathematical theory for the study of almost sure (sample path) stability was not yet fully developed then. Subsequently, work along these lines gradually appeared in the literature. Over the past twenty years, pathwise stability with probability one studies of infinite dimensional stochastic systems have attracted increasing attention from researchers. This is not surprising because sample paths rather than moments or probabilities associated with trajectories are observed in real systems and the stability properties of sample paths can be most closely related to their deterministic counterpart. In what follows, we shall carry out an investigation of sample path stability in the almost sure sense. First of all, we mention the following conditions imposed on (2.2.1):

(H3). $\{T(t)\}$, $t \geq 0$, is an analytic semigroup;
(H4). There exists a real function $f(\cdot) > 0$ such that for all $t < \infty$,

$$\int_0^t f(s)^2 ds < \infty$$

and for all $x \in H$,

$$\|AT(t)B(x)\| \leq f(t)\|x\|_H.$$

In particular, the following stability result was derived in Haussmann [1]:

Theorem 2.2.3   *Assume Conditions (H3) and (H4) above hold. Suppose X_t, $t \geq 0$, is the solution of (2.2.1) satisfying (2.2.15), then there exist constants $M \geq 1$, $\mu > 0$, and random variable $0 \leq T(\omega) < \infty$ such that for all*

$t \geq T(\omega)$, $x_0 \in H$,

$$\|X_t(x_0)\|_H \leq M\|x_0\|_H e^{-\mu t} \qquad a.s. \tag{2.2.18}$$

We prefer to omit the proof here and refer the reader to Haussmann [1]. The reason is that the conditions (H3) and (H4) are somewhat restrictive so that it is not quite satisfactory for them to be applied to practical situations. In fact, a more powerful criterion is (2.2.17) which will be generalized to deal with sample path almost sure stability of nonlinear stochastic evolution systems in Chapter 3.

Another effective method to treat almost sure stability of Equation (2.2.1) is to consider its strong solution. Let V be a densely embedded subspace of the real Hilbert space H and a separable Banach space under some norm $\|\cdot\|_V$. Then $V \hookrightarrow H \hookrightarrow V^*$ with $\|\cdot\|_H \leq \beta\|\cdot\|_V$ for some $\beta > 0$. Consider the following linear stochastic system

$$\begin{cases} dX_t = AX_t + B(X_s)dW_s, \\ X_0 = x_0 \in H, \end{cases} \tag{2.2.19}$$

where $A : V \to V^*$ is a bounded mapping which is coercive in the following sense: there exist constants $\alpha > 0$, $\lambda \in \mathbf{R}^1$ such that for any $x \in V$,

$$\langle x, Ax \rangle_{V,V^*} \leq -\alpha\|x\|_V^2 + \lambda\|x\|_H^2, \tag{2.2.20}$$

and B is supposed to be an element of $\mathcal{L}(H, \mathcal{L}(K, H))$. Then, Theorem 1.3.1 immediately tells us that for any $T < \infty$, there is a unique strong solution

$$X \in L^2(\Omega \times (0,T); V) \cap L^2(\Omega; C(0,T; H)).$$

Note that under the above condition (2.2.20), A generates a strongly continuous semigroup $T(t)$, $t \geq 0$, and the strong solution is also a mild solution since (2.2.19) is linear and $\mathcal{D}(A)$ is dense in H. Hence, if we assume (H1) and (H2) hold, then, according to Theorem 2.2.2, there exist positive constants M, μ such that for the strong solution $X_t(x_0)$, $t \geq 0$, of the equation (2.2.19),

$$E\|X_t(x_0)\|_H^2 \leq M \cdot \|x_0\|_H^2 e^{-\mu t}, \qquad t \geq 0. \tag{2.2.21}$$

Lemma 2.2.2 *Suppose (2.2.21) holds. Then there exists positive constant $C > 0$ such that*

$$E\left(\sup_{0 \leq t \leq T} \|X_t(x_0)\|_H^2 \right) \leq C \cdot \|x_0\|_H^2 \tag{2.2.22}$$

for any $T \geq 0$ and $x_0 \in H$.

Proof Applying Itô's formula to $X_t(x_0)$, $t \geq 0$, yields

$$\|X_t\|_H^2 = \|x_0\|_H^2 + 2\int_0^t \langle X_s, AX_s \rangle_{V,V^*} ds + 2\int_0^t \langle X_s, B(X_s)dW_s \rangle_H$$

$$+ \int_0^t \langle \Delta(I) X_s, X_s \rangle_H ds$$

$$\leq \|x_0\|_H^2 + (2|\lambda| + \|\Delta(I)\|) \int_0^t \|X_s\|_H^2 ds + 2 \int_0^t \langle X_s, B(X_s) dW_s \rangle_H.$$

$$(2.2.23)$$

Hence,

$$E \sup_{0 \leq t \leq T} \|X_t\|_H^2 \leq \|x_0\|_H^2 + (2|\lambda| + \|\Delta(I)\|) \int_0^T E\|X_s\|_H^2 ds$$

$$(2.2.24)$$

$$+ 2E \sup_{0 \leq t \leq T} \left| \int_0^t \langle X_s, B(X_s) dW_s \rangle_H \right|.$$

However, by using Burkholder-Davis-Gundy inequality, we have

$$2E \sup_{0 \leq t \leq T} \left| \int_0^t \langle X_s, B(X_s) dW_s \rangle_H \right|$$

$$\leq 6E \left\{ \int_0^T \|X_t\|_H^2 \langle \Delta(I) X_s, X_s \rangle_H ds \right\}^{1/2}$$

$$(2.2.25)$$

$$\leq 3E \left\{ 2 \sup_{0 \leq t \leq T} \|X_t\|_H \left[\int_0^T \langle \Delta(I) X_s, X_s \rangle_H ds \right]^{1/2} \right\}$$

$$\leq 3lE \left\{ \sup_{0 \leq t \leq T} \|X_t\|_H^2 \right\} + 3l^{-1} \int_0^T E\langle \Delta(I) X_s, X_s \rangle_H ds$$

for any $l > 0$. If we take $l = 1/6$ in (2.2.25) and substitute it into (2.2.24), we obtain after using (2.2.21)

$$E \sup_{0 \leq t \leq T} \|X_t\|_H^2 \leq 2\|x_0\|_H^2 + (4|\lambda| + 38\|\Delta(I)\|) M\mu^{-1} \|x_0\|_H^2 (1 - e^{-\mu T})$$

$$\leq \left\{ 2 + M\mu^{-1} (4|\lambda| + 38\|\Delta(I)\|) \right\} \|x_0\|_H^2.$$

The proof is complete.   ☐

We are now in a position to obtain the desired pathwise exponential decay of strong solutions.

Theorem 2.2.4 *Assume the coercive condition (2.2.20) holds. If X_t, $t \geq 0$, is a strong solution of (2.2.19) satisfying (2.2.21), then there exist positive constants M, μ, and a random variable $0 \leq T(\omega) < \infty$ such that for all $t \geq T(\omega)$,*

$$\|X_t(x_0)\|_H \leq M \cdot \|x_0\|_H e^{-\mu t} \qquad a.s.$$

Proof From (2.2.20) and (2.2.23), it follows that for arbitrary $t \geq N > 0$,

$$\|X_t\|_H^2 = \|X_N\|_H^2 + 2\int_N^t \langle X_s, AX_s \rangle_{V,V^*} ds + 2\int_N^t \langle X_s, B(X_s)dW_s \rangle_H$$
$$+ \int_N^t \langle \Delta(I)X_s, X_s \rangle_H ds$$
$$\leq \|X_N\|_H^2 + (2|\lambda| + \|\Delta(I)\|)\int_N^t \|X_s\|_H^2 ds + 2\int_N^t \langle X_s, B(X_s)dW_s \rangle_H.$$

Hence, for any positive constant $\varepsilon_N > 0$ to be determined later, we have

$$P\left\{ \sup_{N \leq t \leq N+1} \|X_s\|_H \geq \varepsilon_N \right\}$$
$$\leq P\{\|X_N\|_H^2 \geq \varepsilon_N^2/3\} + P\left\{ \int_N^{N+1} \|X_t\|_H^2 dt \geq \varepsilon_N^2/3(2|\lambda| + \|\Delta(I)\|) \right\}$$
$$+ P\left\{ \sup_{N \leq t \leq N+1} \left| \int_N^t \langle X_s, B(X_s)dW_s \rangle_H \right| \geq \varepsilon_N^2/6 \right\}. \tag{2.2.26}$$

Now, from (2.2.21), (2.2.22) and (2.2.25) it follows

$$P\left\{ \sup_{N \leq t \leq N+1} \left| \int_N^t \langle X_s, B(X_s)dW_s \rangle_H \right| \geq \varepsilon_N^2/6 \right\}$$
$$\leq 6\varepsilon_N^{-2}\left\{ E \sup_{N \leq t \leq N+1} \left| \int_N^t \langle X_s, B(X_s)dW_s \rangle_H \right| \right\}$$
$$\leq 18\varepsilon_N^{-2} E\left\{ \sup_{N \leq t \leq N+1} \|X_t\|_H \right\}\left\{ \int_N^{N+1} E\langle \Delta(I)X_s, X_s \rangle_H ds \right\}^{1/2}$$
$$\leq 18\varepsilon_N^{-2} K^{1/2}\|x_0\|_H^2 \|\Delta(I)\|^{1/2} M^{1/2} e^{-\mu N/2}$$
$$\leq k_1 \|x_0\|_H^2 e^{-\mu N/2}/\varepsilon_N^2,$$

where k_1 is some positive constant. Therefore, from (2.2.26) it follows after using (2.2.21) that there exist some positive constants k_2, k_3, $k_4 > 0$ such that

$$P\left\{ \sup_{N \leq t \leq N+1} \|X_s\|_H \geq \varepsilon_N \right\}$$
$$\leq k_2\|x_0\|_H^2 e^{-\mu N}/\varepsilon_N^2 + k_3\left\{ \|x_0\|_H^2 \cdot e^{-\mu N/2} + \|x_0\|_H^2 \cdot e^{-\mu N} \right\}/\varepsilon_N^2$$
$$\leq k_4 e^{-\mu N/4}$$

if we take $\varepsilon_N^2 = 2\|x_0\|_H^2 e^{-\mu N/4}$. The well-known Borel-Cantelli lemma now implies that there exists $N_1(\omega) > 0$ such that if $N \geq N_1(\omega)$, then

$$\sup_{N \leq t \leq N+1} \|X_t\|_H^2 \leq 2\|x_0\|_H^2 e^{-\mu N/4} \qquad a.s.$$

and the required result follows.   ⧠

It is possible to relax our conditions on B in the way that B will only be required to lie in $\mathcal{L}(V, \mathcal{L}(K, H))$, a case which may be used to treat unbounded operators $B(\cdot)$. However, for our stability purpose the coercive condition (2.2.20) will be strengthened as

$$2\langle x, Ax\rangle_{V,V^*} + \langle x, \Delta(I)x\rangle_{V,V^*} \le -\alpha\|x\|_V^2 + \lambda\|x\|_H^2, \qquad \forall x \in V. \quad (2.2.27)$$

On this occasion, for arbitrary $P \in \mathcal{L}(H, H)$, $\Delta(P) \in \mathcal{L}(V, V^*)$ is defined by

$$\langle y, \Delta(P)x\rangle_{V,V^*} = tr[B(x)^* PB(y)Q], \qquad x, \ y \in V.$$

It immediately follows that in view of Theorem 1.3.1, a unique strong solution exists in $L^2(\Omega \times (0, T); V) \cap L^2(\Omega; C(0, T; H))$. In particular, we also have the following stability result similar to Theorems 2.2.2 and 2.2.4.

Theorem 2.2.5   *Assume $B \in \mathcal{L}(V, \mathcal{L}(K, H))$ satisfying (2.2.27). If the conditions (H1) and (H2) hold, then there exist constants $M \ge 1$, $\mu > 0$ such that for any strong solution X_t, $t \ge 0$, of (2.2.19)*

$$E\|X_t(x_0)\|_H^2 \le M \cdot \|x_0\|_H^2 e^{-\mu t}, \qquad t \ge 0.$$

Moreover, under the same conditions the null solution is also pathwise exponentially stable with probability one, i.e., there exist constants $L \ge 1$, $\theta > 0$, and random variable $0 \le T(\omega) < \infty$ such that for all $t \ge T(\omega)$,

$$\|X_t(x_0)\|_H \le L \cdot \|x_0\|_H e^{-\theta t} \qquad a.s.$$

Proof   Following a similar argument as in Theorems 2.2.1 and 2.2.2, it is easy to show that there exists a symmetric nonnegative operator $P \in \mathcal{L}(H)$ such that for all $x_0 \in \mathcal{D}(A)$,

$$P = \int_0^\infty T^*(t)[I + \Delta(P)]T(t)dt,$$

and

$$E\langle PX_t(x_0), X_t(x_0)\rangle_H = E\langle PX_s, X_s\rangle_H - \int_s^t E\|X_r\|_H^2 dr, \qquad \forall 0 \le s \le t. \quad (2.2.28)$$

On the other hand, Itô's formula, (2.2.23) and (2.2.27) yield

$$E\|X_t\|_H^2 \le E\|X_s\|_H^2 + \lambda \int_s^t E\|X_r\|_H^2 dr - \alpha \int_s^t E\|X_r\|_V^2 dr. \quad (2.2.29)$$

Let $g_t = \lambda E\langle PX_t, X_t\rangle_H + E\|X_t\|_H^2$ and from (2.2.28) and (2.2.29), we obtain

$$g_t \leq g_s - \alpha \int_s^t E\|X_r\|_V^2 dr \leq g_s - \alpha C_1 \int_s^t g_s ds, \quad 0 \leq s \leq t,$$

for some constant $C_1 > 0$ since

$$E\|X_r\|_V^2 \geq \beta^{-2} E\|X_r\|_H^2 \geq \beta^{-2} g_r (1 + \lambda\|P\|)^{-1}.$$

It follows that

$$\frac{dg_t}{dt} \leq -\alpha C_1 g_t := -\nu g_t,$$

which immediately implies

$$g_t \leq g_0 \cdot e^{-\nu t}, \quad t \geq 0, \tag{2.2.30}$$

and

$$E\|X_t\|_V^2 \leq -\alpha^{-1} \frac{dg_t}{dt}, \quad t \geq 0. \tag{2.2.31}$$

Note that X_t, $t \geq 0$, also satisfies (2.2.1). By using (H1) and (2.2.30), (2.2.31), we deduce

$$E\|X_t\|_H^2 \leq 2C^2 e^{-2\gamma t}\|x_0\|_H^2 + 2E\left\|\int_0^t T(t-s)B(X_s)dW_s\right\|_H^2$$

and

$$E\left\|\int_0^t T(t-s)B(X_s)dW_s\right\|_H^2$$
$$\leq trQ \int_0^t E\|T(t-s)B(X_s)\|_H^2 ds$$
$$\leq trQC^2\|B\|_{\mathcal{L}(V,\mathcal{L}(K,H))}^2 \int_0^t e^{-2\gamma(t-s)} E\|X_s\|_V^2 ds \leq C_2 \cdot e^{-C_3 t},$$

where $\|T(t)\| \leq C \cdot e^{-\gamma t}$, $t \geq 0$, C_2, C_3 are two positive constants. Hence, there exist positive constants M, μ such that

$$E\|X_t\|_H^2 \leq M \cdot \|x_0\|_H^2 e^{-\mu t}, \quad t \geq 0,$$

if $x_0 \in \mathcal{D}(A)$. But since $\mathcal{D}(A)$ is dense in H the result holds for all $x_0 \in H$.

Lastly, in order to complete our proof, note that the proofs of Lemma 2.2.2 and Theorem 2.2.4 go through with (2.2.23) and (2.2.25) changed into

$$\|X_t\|_H^2 \leq \|x_0\|_H^2 + |\lambda| \int_0^t \|X_s\|_H^2 ds + 2\int_0^t \langle X_s, B(X_s)dW_s\rangle_H,$$

and for some $C_4 > 0$,

$$2E \sup_{0 \le t \le T} \left| \int_0^t \langle X_s, B(X_s)dW_s \rangle_H \right|$$

$$\le \frac{1}{2}E\left\{ \sup_{0 \le t \le T} \|X_s\|_H^2 \right\} + 18 \int_0^t E\langle X_s, \Delta(I)X_s \rangle_{V,V^*} ds$$

$$\le \frac{1}{2}E\left\{ \sup_{0 \le t \le T} \|X_s\|_H^2 \right\} + C_4\|x_0\|_H^2,$$

where we use (2.2.30) together with

$$E\langle X_s, \Delta(I)X_s \rangle_{V,V^*} = E\left\{ tr[B(X_s)^* B(X_s)Q] \right\}$$

$$\le trQ\|B\|_{\mathcal{L}(V,\mathcal{L}(K,H))}^2 E\|X_s\|_V^2$$

$$\le -\alpha^{-1} trQ\|B\|_{\mathcal{L}(V,\mathcal{L}(K,H))}^2 \frac{dg_s}{ds}$$

by (2.2.31). $\square$

2.3   Uniformly Asymptotic Stability

In the section, we intend to study a stability of the equation (2.2.1), uniformly asymptotic stability, which is usually stronger than asymptotic stability but equivalent to exponential stability. To this end, we prefer to take the viewpoint of Theorem 1.3.5 to consider a stochastic version of (2.2.1):

$$\begin{cases} U(t,s)\xi = T(t-s)\xi + \int_s^t T(t-r)B(U(r,s)\xi)dW_r, & 0 \le s \le t, \\ U(s,s)\xi = \xi, & \xi \in L_s^2(\Omega; H), \end{cases} \quad (2.3.1)$$

where $L_s^2(\Omega; H)$ is the subspace of $L^2(\Omega; H)$ which consists of all $\mathcal{F}_s$-measurable random variables. This is certainly an operator integral equation version of (2.2.1), and by Theorem 1.3.5 the following is immediate:

Proposition 2.3.1   *There exists a unique linear operator family $U(t,s)$:*
$L_s^2(\Omega; H) \to L_t^2(\Omega; H)$, $0 \le s \le t$, *such that*

(i). $U(t,s)\xi$ *is adapted to $\mathcal{F}_{s,t}^W = \sigma\{W_r - W_s, s \le r \le t\}$ for each $\xi \in L_s^2(\Omega; H)$;*

(ii). $U(s,s) = I$, $s \ge 0$;

(iii). $U(t,r)U(r,s) = U(t,s)$, $0 \le s \le r \le t$;

(iv). $E\{U(t,s)\xi \mid \mathcal{F}_{s,r}^W\} = T(t-r)U(r,s)\xi$, $0 \le s \le r \le t$, $\xi \in L_s^2(\Omega; H)$;

(v). $U(t, s)\xi$ is mean square continuous in t, $0 \le s \le t$, $\xi \in L_s^2(\Omega; H)$;
(vi). $U(t, s)\xi$ satisfies the equation (2.3.1).

Example 2.3.1   Consider the following stochastic heat equation

$$\begin{cases} dX_t(x) = \frac{\partial^2}{\partial x^2} X_t(x)dt + \mu \cdot X_t(x)dB_t, & \mu > 0, \\ X_t(0) = X_t(1) = 0, \quad t \ge 0; \quad X_0(x) = x_0(x) \in \mathbf{R}^1, & x \in [0, 1], \end{cases} \quad (2.3.2)$$

where B_t, $t \ge 0$, is a one dimensional standard Brownian motion.
    In this case $H = L^2(0, 1)$, the C_0-semigroup $T(t)$, $t \ge 0$, is generated by

$$A = \frac{d^2}{dx^2}, \quad (2.3.3)$$

$$\mathcal{D}(A) = \left\{ u(\cdot) \in L^2(0, 1) : \frac{du(x)}{dx}, \frac{d^2 u(x)}{dx^2} \in L^2(0, 1), u(0) = u(1) = 0 \right\},$$

and $B(X) = \mu \cdot X$ in (2.2.1). It is not difficult to see that the unique mild solution is given by

$$X_t = e^{-(\mu^2/2)t + \mu B_t} T(t)x_0, \quad \forall t \ge 0,$$

which has continuous sample paths, and for arbitrary $\xi \in L_s^2(\Omega; H)$, $0 \le s \le t$,

$$U(t, s)\xi = e^{-(\mu^2/2)(t-s) + \mu(B_t - B_s)} T(t - s)\xi.$$

It is easy to see from the very definitions of stability in Section 1.4 that for the null solution of (2.2.1), the concept of exponential stability in mean square implies asymptotic stability in mean square which further implies stability in mean square. It is also known from Section 2.1 that for deterministic finite dimensional linear systems, the concept of exponential stability is equivalent to asymptotic stability although Example 2.1.1 shows that this is generally not true for linear systems in infinite dimensions. In spite of this, it is worth pointing out however that for the evolution process $U(t, s)\xi$ in (2.3.1), there exists a stability concept shown below which is somewhat stronger than asymptotic stability but equivalent to exponential stability in the mean square sense.

Definition 2.3.1   The null solution of (2.3.1) is said to be *uniformly asymptotically stable in mean square*, if for arbitrary $\xi \in L_s^2(\Omega; H)$ and number $\varepsilon > 0$, there exists $T(\varepsilon) \ge s \ge 0$ such that

$$E\|U(t, s)\xi\|_H^2 < \varepsilon \cdot E\|\xi\|_H^2$$

whenever $t \ge T(\varepsilon) \ge s$.

Before moving to our main conclusion, let us first derive a result which is also important for its own sake.

Lemma 2.3.1   *Suppose (i)–(vi) of Proposition 2.3.1 about the equation (2.3.1) hold, then there exist constants $M \geq 1$, $\tau > 0$ such that for any $\xi \in L_s^2(\Omega; H)$, $t \geq s \geq 0$,*

$$E\|U(t,s)\xi\|_H^2 \leq M \cdot E\|\xi\|_H^2 \cdot e^{\tau(t-s)}.$$

Proof   Since $T(t)$, $t \geq 0$, is a strongly continuous semigroup, $\|T(t)\| \leq C \cdot e^{\lambda t}$, $t \geq 0$, for some $C \geq 1$, $\lambda \in \mathbf{R}_+$. Now it follows from (2.3.1) that there exists some constant $\bar{C} > 0$ such that for any $t \geq s \geq 0$,

$$E\|U(t,s)\xi\|_H^2 \leq 2E\|T(t-s)\xi\|_H^2$$

$$+2\int_s^t E\Big\{tr\big[T(t-r)B(U(r,s)\xi)QB^*(U(r,s)\xi)T^*(t-r)\big]\Big\}dr$$

$$\leq 2C^2 e^{2\lambda(t-s)}E\|\xi\|_H^2 + \bar{C}\int_s^t e^{2\lambda(t-r)}E\|U(r,s)\xi\|_H^2 dr$$

which immediately implies

$$e^{-2\lambda t}E\|U(t,s)\xi\|_H^2 \leq 2C^2 e^{-2\lambda s}E\|\xi\|_H^2 + \bar{C}\int_s^t e^{-2\lambda r}E\|U(r,s)\xi\|_H^2 dr,$$

for all $0 \leq s \leq t$. Hence, by virtue of the well-known Gronwall's inequality, it follows

$$e^{-2\lambda t}E\|U(t,s)\xi\|_H^2 \leq 2C^2 e^{-2\lambda s}E\|\xi\|_H^2 \cdot e^{\bar{C}(t-s)},$$

i.e.,

$$E\|U(t,s)\xi\|_H^2 \leq 2C^2 E\|\xi\|_H^2 \cdot e^{(2\lambda+\bar{C})(t-s)} = M \cdot E\|\xi\|_H^2 \cdot e^{\tau(t-s)}$$

where $M = 2C^2 \geq 2$ and $\tau = 2\lambda + \bar{C} > 0$. $\square$

We are now in a position to show the main result in the section.

Theorem 2.3.1   *Let $U(t,s)$ be the unique evolution operator family of (2.3.1) satisfying all the properties in Proposition 2.3.1. Then, the null solution of (2.3.1) is uniformly asymptotically stable in mean square if and only if it is exponentially stable in mean square.*

Proof   Suppose the null solution of (2.3.1) is uniformly asymptotically stable in mean square and choose $\varepsilon_0 = 1/e$. Then there exists $T(\varepsilon_0) \geq s$ such that for arbitrary $\xi \in L_s^2(\Omega; H)$, $E\|U(t,s)\xi\|_H^2 < \varepsilon_0 E\|\xi\|_H^2$ for all $t \geq T(\varepsilon_0)$. Hence, for any fixed $t \in [s, \infty)$,

$$t - s = nT(\varepsilon_0) + \theta,$$

where n is some positive integer and $0 \leq \theta \leq T(\varepsilon_0)$. By virtue of (iii) in Proposition 2.3.1 and using Lemma 2.3.1, it follows that for any $\xi \in L^2_s(\Omega; H)$,

$$E\|U(t,s)\xi\|^2_H = E\left\|U(nT(\varepsilon_0) + s + \theta, nT(\varepsilon_0) + s)\right.$$

$$\cdot \prod_{k=0}^{n-1} U\left((n-k)T(\varepsilon_0) + s, (n-k-1)T(\varepsilon_0) + s\right)\xi\left.\right\|^2_H$$

$$\leq Me^{\tau\theta}(1/e)^n E\|\xi\|^2_H = M\exp\left[\tau\theta - \frac{nT(\varepsilon_0)}{T(\varepsilon_0)}\right]E\|\xi\|^2_H$$

$$= M\exp\left[\tau\theta - \frac{nT(\varepsilon_0) + \theta}{T(\varepsilon_0)} + \frac{\theta}{T(\varepsilon_0)}\right]E\|\xi\|^2_H$$

$$= M\exp[\tau\theta + \theta/T(\varepsilon_0)]\exp[-(t-s)/T(\varepsilon_0)]E\|\xi\|^2_H$$

$$\leq M\exp[\tau T(\varepsilon_0) + 1]\exp[-(t-s)/T(\varepsilon_0)]E\|\xi\|^2_H.$$

Hence, let

$$C = M\exp[\tau T(\varepsilon_0) + 1] \quad \text{and} \quad -\mu = -1/T(\varepsilon_0),$$

then

$$E\|U(t,s)\xi\|^2_H \leq Ce^{-\mu(t-s)}E\|\xi\|^2_H. \tag{2.3.4}$$

The converse statement is obvious, and the proof is now complete. ☐

2.4 Some Examples

In this section, we shall investigate several examples to illustrate the theory established in the preceding sections.

Example 2.4.1 Consider a one-dimensional rod of length π whose ends are maintained at $0°$ and whose sides are insulated. Assume that there is an exothermic reaction taking place inside the rod with heat being produced proportionally to the temperature. The temperature, denoted by $X_t(x)$, in the rod may be modelled in the following way

$$\begin{cases} \frac{\partial X_t(x)}{dt} = \frac{\partial^2 X_t(x)}{\partial x^2} + rX_t(x), & t > 0, \quad 0 < x < \pi, \\ X_t(0) = X_t(\pi) = 0, & t \geq 0, \\ X_0(x) = x_0(x), & 0 \leq x \leq \pi, \end{cases} \tag{2.4.1}$$

where $r \in \mathbf{R}^1$ depends on the rate of reaction. If we assume $r = r_0$, a constant, then we can solve the equation in an explicit way

$$X_t(x) = \sum_{n=1}^{\infty} a_n e^{-(n^2 - r_0)t} \sin nx,$$

where $x_0(x) = \sum_{n=1}^{\infty} a_n \sin nx$. Hence, we obtain exponential stability if $n^2 > r_0$ for all $n \in \mathbf{N}$, or equivalently, $r_0 < 1$. This is exactly the condition (H1) in Subsection 2.2.2. Observe that, in general, for $r_0 \geq 1$ the null solution is not stable.

Suppose now that r is random, and assume it is modelled as $r = r_0 + r_1 \dot{B}_t$, so that (2.4.1) becomes

$$dX_t(x) = \left(\frac{\partial^2}{\partial x^2} + r_0 \right) X_t(x) dt + r_1 X_t(x) dB_t, \qquad (2.4.2)$$

where B_t is a standard one-dimensional Brownian motion. We substitute this into our formulation (2.2.1) by letting $K = \mathbf{R}^1$, $H = L^2(0, \pi)$, $A = \frac{\partial^2}{\partial x^2} + r_0$, $B(X_t) = r_1 X_t$. Now, by a simple computation $\Delta(P)$ defined in (2.2.3) can be shown to be $r_1^2 P$ so that (H2) in Subsection 2.2.2 becomes $r_1^2 < 2(1 - r_0)$. This is exactly (2.2.17). Hence, if the unperturbed system (2.4.1) is very stable, i.e., r_0 is sufficiently less than one, then the perturbations (i.e., r_1) can be fairly large and according to Theorem 2.2.2, we still have $E\|X_t(x_0)\|_H^2 \leq M \cdot \|x_0\|_H^2 e^{-\mu t}$, $t \geq 0$, for some constants $M \geq 1$, $\mu > 0$.

In order to deduce pathwise almost sure stability, note that the conditions in Theorem 2.2.3 are not satisfied in this situation. However, we can apply the theory of strong solutions. Precisely, we set for any u, $v \in V$,

$$V = W_0^{1,2} = H_0^1, \quad \langle u, Av \rangle_{V,V^*} = \int_0^\pi \left(-\frac{\partial u(x)}{\partial x} \frac{\partial v(x)}{\partial x} + r_0 u(x) v(x) \right) dx.$$

Then $\langle u, Au \rangle_{V,V^*} = -\|u\|_V^2 + r_0 \|u\|_H^2$, so $\alpha = 1$ and $\lambda = r_0$ in (2.2.20). Then by Theorem 2.2.4, it follows that there exist constants $M > 0$, $\mu > 0$ and a random variable $0 \leq T(\omega) < \infty$ such that for all $t \geq T(\omega)$,

$$\|X_t\|_H \leq M \cdot \|x_0\|_H e^{-\mu t} \qquad a.s.$$

Example 2.4.2   For the next example we suppose (2.4.2) is replaced by

$$dX_t(x) = \left(\frac{\partial^2}{\partial x^2} + r_0 \frac{\partial}{\partial x} \right) X_t(x) dt + \gamma(x) \frac{\partial X_t(x)}{\partial x} dB_t, \qquad (2.4.3)$$

where $\gamma(x) \in L^\infty(0, \pi; \mathbf{R}^1)$, i.e., we are observing heat diffusion in a rod relative to an origin moving with velocity $r_0 + \gamma(\cdot) \dot{B}_t$. Let K, V and H be formulated as in Example 2.4.1, then

$$\langle u, Au \rangle_{V,V^*} = \int_0^\pi \left(-\frac{\partial u}{\partial x} \frac{\partial u}{\partial x} + r_0 \frac{\partial u}{\partial x} u \right) dx, \qquad \forall u \in V,$$

$$\langle u, \Delta(I)u \rangle_{V,V^*} = \int_0^\pi \gamma(x)^2 \frac{\partial u}{\partial x} \frac{\partial u}{\partial x} dx, \qquad \forall u \in V.$$

Hence, (2.2.27) becomes

$$2\langle u, Au \rangle_{V,V^*} + \langle u, \Delta(I)u \rangle_{V,V^*} \leq -2\|u\|_V^2 + \|\gamma(\cdot)\|_\infty^2 \|u\|_V^2,$$

which immediately yields that for arbitrary r_0 and $\|\gamma(\cdot)\|_\infty^2 < 2$, the null solution is exponentially stable both in the mean square and almost sure sense.

Example 2.4.3 Also as an application of Theorem 2.2.2, let us investigate the stability of a class of stochastic second order equations generalizing (2.1.31). To this end, let us recall the formulation of the deterministic second order system (2.1.31) with $\alpha > 0$,

$$\frac{d^2v}{dt^2} + \alpha \cdot \frac{dv}{dt} + Av = 0, \quad v(0) = v_0, \quad \frac{dv}{dt}(0) = v_1, \qquad (2.4.4)$$

where A is a strictly positive, self-adjoint operator on some real separable Hilbert space H with domain $\mathcal{D}(A)$, so that

$$\langle Ax, x\rangle_H \geq C\|x\|_H^2, \quad \forall x \in \mathcal{D}(A), \quad C > 0.$$

Then $\mathcal{H} = \mathcal{D}(A^{1/2}) \oplus H$ is a real separable Hilbert space under the inner product

$$\langle z, \tilde{z}\rangle_\mathcal{H} = \langle A^{1/2}z_1, A^{1/2}\tilde{z}_1\rangle_H + \langle z_2, \tilde{z}_2\rangle_H,$$

where

$$z = \begin{pmatrix} z_1 \\ z_2 \end{pmatrix}, \quad \tilde{z} = \begin{pmatrix} \tilde{z}_1 \\ \tilde{z}_2 \end{pmatrix}.$$

Define the closed, linear operator on $\mathcal{H}$

$$\mathcal{A} = \begin{pmatrix} 0 & I \\ -A & -\alpha I \end{pmatrix} \quad \text{with domain} \quad \mathcal{D}(\mathcal{A}) = \mathcal{D}(A) \oplus \mathcal{D}(A^{1/2}).$$

Thus (2.4.4) may be rewritten as a first order differential equation on $\mathcal{H}$,

$$\frac{dz(t)}{dt} = \mathcal{A}z(t), \quad z(0) = z_0, \qquad (2.4.5)$$

where

$$z(t) = \begin{pmatrix} v(t) \\ dv(t)/dt \end{pmatrix}, \quad z_0 = \begin{pmatrix} v_0 \\ v_1 \end{pmatrix}.$$

We have already known from Section 2.1 that $\mathcal{A}$ generates a strongly continuous, exponentially stable semigroup $\mathcal{T}(t)$, $t \geq 0$, on $\mathcal{H}$.

We would like to consider the second order stochastic partial differential equation

$$d\left[\left(\frac{\partial v}{\partial t}\right) + \alpha v\right] - \frac{\partial^2 v}{\partial x^2}dt + c\frac{\partial v}{\partial x}dB_t = 0, \quad t > 0, \quad x \in [0, 1],$$
$$v(0, t) = v(1, t) = 0, \quad t \geq 0,$$

where B_t, $t \geq 0$, is a standard one-dimensional Brownian motion and α, c are real constants with $\alpha > 0$. Let

$$H = L^2(0,1), \qquad A = -\frac{\partial^2}{\partial x^2},$$

$$\mathcal{D}(A) = \left\{ u \in H : u_x', u_{xx}'' \in H \text{ and } u(0) = u(1) = 0 \right\}.$$

Suppose

$$\mathcal{A} = \begin{pmatrix} 0 & I \\ -A & -\alpha I \end{pmatrix} \quad \text{with domain} \quad \mathcal{D}(\mathcal{A}) = \mathcal{D}(A) \oplus \mathcal{D}(A^{1/2}).$$

Then, by virtue of Example 2.1.3, $|\gamma(-A)| = \pi^2$ and $\|T(t)\| \leq e^{-\mu t}$, where

$$\mu \geq \frac{2\alpha\pi^2}{4\pi^2 + \alpha(\alpha + \sqrt{\alpha^2 + 4\pi^2})}.$$

Now

$$\left\| \int_0^\infty T^*(t)\Delta(I)T(t)dt \right\| \leq \int_0^\infty \|T(t)\|^2 \|\Delta(I)\| dt \leq \int_0^\infty e^{-2\mu t} c^2 dt$$

$$= \frac{c^2}{2\mu}.$$

Hence, by Theorem 2.2.2 we have the mean square stability of the null solution if

$$c^2 < \frac{4\alpha\pi^2}{4\pi^2 + \alpha(\alpha + \sqrt{\alpha^2 + 4\pi^2})}.$$

Example 2.4.4   Consider the following stochastic second order model

$$d\left[\left(\frac{\partial v}{\partial t} \right) + \alpha v \right] + \frac{\partial^4 v}{\partial x^4} dt + \theta \frac{\partial^2 v}{\partial x^2} dt + c \frac{\partial^2 v}{\partial x^2} dB_t = 0, \quad t > 0, \quad x \in [0,1],$$

$$v(0,t) = v(1,t) = 0, \quad t \geq 0,$$

where B_t, $t \geq 0$, is a standard one-dimensional Brownian motion and α, θ, c are real constants with $\alpha > 0$. Let

$$H = L^2(0,1), \qquad A = \frac{\partial^4}{\partial x^4} + \theta \frac{\partial^2}{\partial x^2},$$

$$\mathcal{D}(A) = \left\{ \begin{array}{l} h \in H : h_x', h_{xx}'', h_{xxx}''', h_{xxxx}'''' \in H \\ \text{and } h(0) = h(1) = 0, \, h_{xx}''(0) = h_{xx}''(1) = 0 \end{array} \right\}.$$

For A to be self-adjoint and positive, we require $\theta^2 < \pi^2$ and obtain by Example 2.1.3 that $|\gamma(-A)| = \pi^2(\pi^2 - \theta^2)$, and $\mathcal{A} = \begin{pmatrix} 0 & I \\ -A & -\alpha I \end{pmatrix}$ is stable if $\theta^2 < \pi^2$ and $\alpha > 0$, with $\|T(t)\| \leq e^{-\mu t}$ where

$$\mu \geq \frac{2\alpha\pi^2(\pi^2 - \theta^2)}{4\pi^2(\pi^2 - \theta^2) + \alpha\sqrt{\alpha^2 + 4\pi^2(\pi^2 - \theta^2)}}.$$

As in the previous example,

$$\left\| \int_0^\infty T^*(t)\Delta(I)T(t)dt \right\| \le c^2/2\mu$$

and the sufficient condition for the mean square stability is $c^2 < 2\mu$.

Example 2.4.5   Consider an n-dimensional stochastic delay differential equation of the form

$$\begin{cases} dX_t + \int_{-r}^0 X_{t+s}dN(s)dt = G(X_t)dB_t, & t \ge 0, \\ X_t = 0, & -r \le t \le 0, \quad r > 0, \end{cases} \quad (2.4.6)$$

where $X_t \in \mathbf{R}^n$, $B_t = (B_t^1, \cdots, B_t^n)$, $t \ge 0$, is a given n-dimensional standard Brownian motion, and $N(\cdot)$ is a left continuous function of bounded variation defined on $[-r, 0]$ into the space of $n \times n$ matrices. Let $H = \mathbf{R}^n \times L^2(-r, 0; \mathbf{R}^n)$ and $\mathcal{D}(A) = W^{1,2}(-r, 0; \mathbf{R}^n)$. Then $\mathcal{D}(A)$ can be embedded in H as $\mathcal{D}(A) = \{(f(0), f(\cdot)) \in H : f \in W^{1,2}(-r, 0; \mathbf{R}^n)\}$ and moreover $\mathcal{D}(A)$ is dense in H. Note that $f \in \mathcal{D}(A)$ is continuous and we write $\dot{f}$ for its generalized derivative. Define

$$Af = \left(\int_{-r}^0 f(s)dN(s), -\dot{f}(\cdot) \right), \quad f \in \mathcal{D}(A).$$

We shall assume that $-A$ generates a strongly continuous semigroup $T(t)$, $t \ge 0$, in H. In Delfour, McCalla and Mitter [1], it was shown that this assumption is natural and satisfied for many delay systems. Now if we set $X^t = (X_t, X_{t+\cdot}) \in H$, $G(X^t) = (G(X_t), 0)$, then (2.4.6) implies

$$dX^t + AX^t dt = G(X^t)dB_t.$$

Now, (H1) in Subsection 2.2.2 becomes

$$\sup\left\{ \text{Re}\,\lambda : \det\left(\int_{-r}^0 e^{\lambda s}dN(s) + \lambda I \right) = 0 \right\} < 0. \quad (2.4.7)$$

In order to justify (H2), let us be more specific and assume $K = \mathbf{R}^p$, $G(x)k = \sum_{j=1}^p G_j x k^j$, $k = \{k^j\} \in K$, where G_j is an $n \times n$ matrix. Then

$$\langle \Delta(I)x, x \rangle_H = \left\langle \sum_{j=1}^p G_j^* G_j x, x \right\rangle_H$$

and (H2) becomes

$$\sup_{\|f\|_H=1} \int_0^\infty \sum_{j=1}^p \|G_j y_t(f)\|_H^2 dt < 1, \quad (2.4.8)$$

where $y_t(f)$ is the solution at time t of equation

$$\frac{dy_t}{dt} + \int_{-r}^{0} y_{t+s} dN(s) = 0, \qquad t \geq 0,$$

with $y_t(f) = f(t)$ for $t \leq 0$, $f \in \mathcal{D}(A)$. Note that A is not usually coercive. However, Theorem 2.2.2 gives the exponential decay of the second moment, and (2.4.6) is actually a finite dimensional equation, i.e., $X_t \in \mathbf{R}^n$. Hence, we can proceed to estimate

$$P\Big\{ \sup_{N \leq t \leq N+1} \int_N^t \Big\| \int_{-r}^0 X_{h+s} dN(s) \Big\|_H dh > \varepsilon \Big\}$$

$$\leq P\Big\{ \int_N^{N+1} \Big\| \int_{-r}^0 X_{t+s} dN(s) \Big\|_H dt > \varepsilon \Big\}$$

$$\leq \int_N^{N+1} E\Big\| \int_{-r}^0 X_{t+s} dN(s) \Big\|_H^2 dt \Big/ \varepsilon^2.$$

Now we assume that $H = \mathbf{R}^n \times L^2(-r, 0; \mathbf{R}^n)$ where either L^2 is taken under the measure $d|N(\cdot)|$, or L^2 is taken under the usual Lebesgue measure and

$$dN(s) = \Big[\sum_{i=1}^m c_i \delta_{s_i}(s) + n(s) \Big] ds,$$

where $\delta_{s_i}(s)$ is the delta function at s_i, and $\int_{-r}^0 |n(s)|^2 ds < \infty$. In either case,

$$\int_N^{N+1} E\Big\| \int_{-r}^0 X_{t+s} dN(s) \Big\|_H^2 dt \leq \bar{C} \sup_{N-r \leq t \leq N+1} E\|X^t\|_H^2 \leq C\alpha E\|X^0\|_H^2 e^{-\mu N}$$

for some positive constants C and μ, so that the null solution is exponentially stable in the almost sure sense.

2.5   Notes and Comments

There exist various concepts of stability for infinite-dimensional (stochastic) systems, and the one in which we are especially interested in this chapter is that of exponential stability. Its relation for the deterministic case to the existence of a nonnegative self-adjoint operator solution to the Lyapunov equation (Theorem 2.1.1) has been shown in Datko [1]. The generalization from L^2-stability to L^p-stability, i.e., Theorem 2.1.2, was derived in Pazy [3] but the arguments here are taken from Pazy [1]. The idea of Example 2.1.2 is mainly based on Greiner, Voigt and Wolff [1]. The contents of Lyapunov functions in infinite dimensional spaces (Theorem 2.1.3 and subsequent corollaries) are mainly taken from Datko [2]. The material in Section 2.3 seems new, but the corresponding idea for the deterministic case goes back at least to Datko [2]. See also Benchimol [1], Daletskii and Krein [1], Ichikawa and Pritchard [1], Pazy [1] and Yosida [1] for related topics in the deterministic situation.

There is an extensive literature in an attempt to establish stability theory for stochastic differential equations in infinite dimensions now. Theorem 2.2.1 is well known in finite dimensional spaces (cf. Wonham [1]). If the semigroup $T(t)$, $t \geq 0$, is generated by a bounded infinitesimal operator on Banach spaces, this result in the deterministic case could be found in Daletskii and Krein [1]. An infinite dimensional stochastic version of Theorem 2.1.1 was first derived by Zabczyk [2] in a slightly different form from Theorem 2.2.1. The main proofs of Theorem 2.2.1 employed here are taken from Ichikawa [1], [2] and Haussmann [1]. The material in Section 2.2.2 is principally due to Haussmann [1]. But it is worth pointing out that the corresponding ideas of almost sure stability study in finite dimensional spaces should go back at least to, for instance, Kozin [1], [2]. See also Da Prato and Ichikawa [1], Ichikawa [1], [7], and Skorohod [1] for some related topics on stability of linear stochastic evolution equations.

Examples 2.4.3 and 2.4.4 are taken from Curtain [1]. In connection with this, the reader is also referred to Curtain and Pritchard [1], Curtain and Zwart [1] and Pritchard and Zabczyk [1] to find some more detailed research for deterministic systems.

Chapter 3

Stability of Nonlinear Stochastic Differential Equations

The purpose of this chapter is to investigate stability of nonlinear stochastic differential equations. We first formulate this problem as a comparison of the quadratic functionals of two stochastic differential equations. Then, coercivity types of criteria, actually, Lyapunov's function methods are presented to deal with stability properties for strong and mild solutions. A Lyapunov function programme is also carried out to handle, apart from stability, ultimate boundedness and associated existence and uniqueness of invariant measures. We investigate the decay rate of mild solutions for a class of nonlinear stochastic evolution equations. Lastly, based on the viewpoint of perturbations of infinite dimensional deterministic systems, the so-called stabilization by white noise sources of systems is studied.

3.1 Equivalence of L^p-Stability and Exponential Stability

Let $T(t)$, $t \geq 0$, be a strongly continuous semigroup of bounded linear operators on a real separable Hilbert space H with norm $\| \cdot \|_H$ and inner product $\langle \cdot, \cdot \rangle_H$. Let the operator A be the infinitesimal generator of $T(t)$, $t \geq 0$. Then recall that $T(t)x_0$, $x_0 \in \mathcal{D}(A)$, is the unique solution of the differential equation in H

$$\frac{dX_t}{dt} = AX_t, \qquad X_0 = x_0 \in \mathcal{D}(A). \tag{3.1.1}$$

It was shown in Theorems 2.1.1 and 2.1.2 that the two statements below are equivalent:

(a). $\|T(t)\| \leq M \cdot e^{-\mu t}$, $t \geq 0$, for some $M \geq 1$ and $\mu > 0$;
(b). $\int_0^\infty \|T(t)x\|_H^p dt \leq C\|x\|_H^p$, $x \in H$, for some $p \geq 1$ and $C > 0$.

It is natural therefore to ask whether the equivalence of (a) and (b) above still holds for some nonlinear semigroups, or more generally, nonlinear stochastic systems. To illustrate this more precisely, consider the following autonomous semilinear stochastic differential equation:

$$dX_t = [AX_t + F(X_t)]dt + G(X_t)dW_t, \quad X_0 = x_0 \in H,$$

where both $F : H \to H$, $G : H \to \mathcal{L}(K, H)$ are nonlinear, and they satisfy $F(0) = 0$, $G(0) = 0$ and the usual Lipschitz and linear growth conditions. By Theorem 1.3.4, it immediately follows that the equation has a unique mild solution $X_t = X_t(x_0)$, $t \geq 0$, for arbitrary $x_0 \in H$. In other words, X_t, $t \geq 0$, satisfies

$$X_t = T(t)x_0 + \int_0^t T(t-s)F(X_s)ds + \int_0^t T(t-s)G(X_s)dW_s. \quad (3.1.2)$$

The question is whether or not the two statements below are equivalent for each $p \geq 1$:

(a'). $E\|X_t(x_0)\|_H^p \leq Me^{-\mu t}\|x_0\|_H^p$, $x_0 \in H$, for some $M \geq 1$ and $\mu > 0$;
(b'). $\int_0^\infty E\|X_t(x_0)\|_H^p dt \leq C\|x_0\|_H^p$, $x_0 \in H$, for some $C > 0$.

It is worth pointing out that the question above is also important even in the deterministic case, i.e., $G(\cdot) = 0$ in (3.1.2), as one may expect from the viewpoint of Lyapunov's function methods. The point is that there exist some situations where one may easily find Lyapunov functions $\Lambda(\cdot)$ which are not strict positive definite (i.e., $\Lambda(\cdot) \not\geq c\| \cdot \|_H^p$ for any $c > 0$ but ensure p-th moment integrality, i.e., (b) or (b') holds. To be specific, let $P \geq 0$ satisfy (2.1.13). One may find conditions on the nonlinear perturbation $F(\cdot)$ such that

$$2\langle PAx, Ax + F(x)\rangle_H \leq -\lambda\|x\|_H^2, \quad x \in \mathcal{D}(A) \text{ for some } \lambda > 0. \quad (3.1.3)$$

Then, similarly to finite dimensional case, applying the usual Lyapunov function type of arguments to (3.1.2) with $G(\cdot) = 0$ and $\Lambda(x) = \langle Px, x\rangle_H$, it is possible to obtain

$$\Lambda(X_t(x_0)) = \langle PX_t(x_0), X_t(x_0)\rangle_H$$
$$\leq Me^{-\mu t}\|x_0\|_H^2 \text{ for some } M > 0 \text{ and } \mu > 0.$$

However, it does not generally follow immediately that

$$\|X_t(x_0)\|_H \leq C \cdot e^{-\nu t}\|x_0\|_H \text{ for some } C \geq 1 \text{ and } \nu > 0,$$

although it is indeed true if $\Lambda(\cdot)$ is strictly positive definite in the sense $\Lambda(\cdot) \geq c\| \cdot \|_H^2$ for some constant $c > 0$ (also see Example 3.1.1 below). The reason is that for some A, for instance, if A is the generator of an analytic, exponentially stable semigroup, $\langle Px, x\rangle_H$ cannot be equivalent to $\|x\|_H^2$. Indeed, suppose the contrary is true. It can be deduced that there exist $t_0 > 0$ and a constant $C > 0$ such that $\|T(t_0)x\|_H \geq C\|x\|_H$, $x \in H$ (see Pazy [2]). Therefore, if $x \in \mathcal{D}(A)$, we have $\|T(t_0)Ax\|_H \geq C\|Ax\|_H$. But since $T(t)$, $t \geq 0$, is analytic, $AT(t_0)$ is a bounded operator, and therefore,

$$\|Ax\|_H \leq \frac{1}{C}\|AT(t_0)\|\|x\|_H \quad \text{for every} \quad x \in \mathcal{D}(A).$$

Since $\mathcal{D}(A)$ is dense, this holds for every $x \in H$, and A is thus bounded, a conclusion which is generally untrue.

In spite of the difficulty mentioned above, fortunately, we can deduce in most situations from (3.1.3) the following L^2-stability

$$\int_0^\infty \|X_t(x_0)\|_H^2 dt \leq C\|x_0\|_H^2 < \infty, \quad x_0 \in H \text{ for some } C > 0. \qquad (3.1.4)$$

In what follows, we shall show that under certain circumstances, the condition (3.1.4) or (b') above actually implies exponential stability of some nonlinear differential equations in the square or mean square sense, which is a direct nonlinear generalization of Theorem 2.2.1. At the moment, we will not go into further details of Lyapunov's function methods, especially those related to the above arguments. In Sections 3.4 and 3.5, based on the so-called first order linear approximation approaches, a more detailed investigation of the construction of proper Lyapunov's functions will be presented to deal with stochastic stability of nonlinear differential equations.

3.1.1 An Extension of Linear Stability Criteria

We intend to begin our discussions by considering a general two-parameter evolution process similar to Theorem 1.3.5 in Section 1.3. Let Z be the space of random variables in the Hilbert space H defined in $(\Omega, \mathcal{F}, P)$ and $S(t, s)$, $t \geq s \geq 0$, be a family of nonlinear operators with domain $\mathcal{D}_s \subset Z$ satisfying the properties:

(1). $S(t, s)\mathcal{D}_s \subset \mathcal{D}_t$, $t \geq s$;
(2). $S(s, s)\xi = \xi$, $s \geq 0$, $\xi \in \mathcal{D}_s$;
(3). $S(t, u)S(u, s) = S(t, s)$ on $\mathcal{D}_s$, $s \leq u \leq t$;
(4). $S(\cdot, s)\xi$, for each $s \geq 0$ and $\xi \in \mathcal{D}_s$, is measurable on $[s, \infty) \times \Omega$.

We need to use the following fundamental lemma.

Lemma 3.1.1   Let $0 < r < 1$, $L > 0$ and n be a nonnegative integer. Then $nL \leq t \leq (n+1)L$ implies $e^{-at} \leq r^n \leq (1/r)e^{-at}$, $a = -(\log r)/L > 0$.

Proof   Note that $r^n = e^{n\log r}$ and $\log r < 0$, then the result is easily deduced.
$\square$

The theorem below could be regarded as a nonlinear version of Theorem 2.2.1.

Theorem 3.1.1   *Let $g(\cdot)$ be a positive continuous function on $[0, \infty)$ and $p > 0$. Suppose $S(t, s)$, $t \geq s \geq 0$, defined above satisfies the following:*

$$E\|S(t, s)z\|_H^p \leq g(t - s)E\|z\|_H^p, \qquad z \in \mathcal{D}_s(p) := \{z \in \mathcal{D}_s : E\|z\|_H^p < \infty\}.$$
$$(3.1.5)$$

Then two conditions below are equivalent:

(i). $\int_s^\infty E\|S(t, s)z\|_H^p dt \leq C \cdot E\|z\|_H^p$, $z \in \mathcal{D}_s(p)$, *for some $C > 0$;*
(ii). $E\|S(t, s)z\|_H^p \leq M \cdot e^{-\mu(t-s)} E\|z\|_H^p$, $z \in \mathcal{D}_s(p)$, *for some $M \geq 1$ and $\mu > 0$.*

Proof   It is sufficient to show that (i) implies (ii). This can be obtained by generalizing the proof of Theorem 2.2.1. For any $0 \leq s < t$ and $z \in \mathcal{D}_s(p)$, we have by Condition (3) above and the relation (3.1.5)

$$E\|S(t, s)z\|_H^p \int_s^t g^{-p}(t - u)du = \int_s^t g^{-p}(t - u)E\|S(t, s)z\|_H^p du$$

$$= \int_s^t g^{-p}(t - u)E\|S(t, u)S(u, s)z\|_H^p du$$

$$\leq \int_s^t g^{-p}(t - u)g^p(t - u)E\|S(u, s)z\|_H^p du$$

$$= \int_s^t E\|S(u, s)z\|_H^p du,$$

which, together with the condition (i), immediately implies

$$E\|S(t, s)z\|_H^p \int_s^t g^{-p}(t - u)du \leq C \cdot E\|z\|_H^p. \qquad (3.1.6)$$

Let $\tilde{L} > 0$ be an arbitrary but fixed number. Define $J > 0$ by

$$J = \int_0^{\tilde{L}} g^{-p}(u)du.$$

Then $J > 0$ and for any $t - s \geq \tilde{L}$, we have from (3.1.6) that $E\|S(t, s)z\|_H^p \leq (C/J)E\|z\|_H^p$. This, together with (3.1.5), implies the existence of some constant $\bar{C} > 0$ such that

$$E\|S(t, s)z\|_H^p \leq \bar{C} \cdot E\|z\|_H^p \quad \text{for any} \quad t \geq s \geq 0 \quad \text{and} \quad z \in \mathcal{D}_s(p). \quad (3.1.7)$$

Now let $t > s$, $z \in \mathcal{D}_s(p)$. By virtue of the condition (3) at the beginning of this subsection and (3.1.7), it follows that

$$(t - s)E\|S(t, s)z\|_H^p = \int_s^t E\|S(t, s)z\|_H^p du \leq \bar{C} \int_s^t E\|S(u, s)z\|_H^p du$$

$$\leq (\bar{C}C)E\|z\|_H^p.$$

Hence, it follows that $E\|S(t,s)z\|_H^p \leq (\bar{C}C)E\|z\|_H^p/(t-s)$, $t > s$, $z \in \mathcal{D}_s(p)$. So for each $0 < r < 1$, we can choose a number $L = L(r) > 0$ such that

$$E\|S(t,s)z\|_H^p \leq r \cdot E\|z\|_H^p, \quad z \in \mathcal{D}_s(p) \quad \text{whenever} \quad t - s \geq L. \quad (3.1.8)$$

Let $t - s \geq L$, then there is an integer $n \geq 1$ such that $nL \leq t - s \leq (n+1)L$. Using the semigroup property (3) of $S(\cdot, \cdot)$ and (3.1.8) n times and then (3.1.7), we obtain $E\|S(t,s)z\|_H^p \leq r^n RE\|z\|_H^p$, $z \in \mathcal{D}_s(p)$ for some constant $R > 0$. Now Lemma 3.1.1 yields

$$E\|S(t,s)z\|_H^p \leq \tilde{M} \cdot e^{-\mu(t-s)} E\|z\|_H^p, \quad z \in \mathcal{D}_s(p) \quad \text{for any} \quad t - s \geq L,$$

where $\tilde{M} = R/r$ and $\mu = -(\log r)/L > 0$. Combining this with (3.1.7), we conclude that

$$E\|S(t,s)z\|_H^p \leq M \cdot e^{-\mu(t-s)} E\|z\|_H^p, \quad z \in \mathcal{D}_s(p), \quad \forall t \geq s,$$

where $M = \max\{\tilde{M}, Re^{\mu L}\}$. $\qquad\qquad\qquad\qquad\qquad\qquad\qquad\qquad$ □

If we apply Theorem 3.1.1 to specific (deterministic or stochastic) situations, we shall obtain some important stability results. For instance, we may derive:

Corollary 3.1.1 *Suppose that $\mathcal{D}_s(p) = L_s^p(\Omega, \mathcal{F}, P; H)$ for any $s \geq 0$, $p > 0$, as in the notations of Section 1.3. It is known that $S(t,s) = U(t-s, 0)$, $t \geq s \geq 0$, defined as there satisfies the conditions at the beginning of this subsection. Then two statements below are equivalent:*

(i). $\int_0^\infty E\|U(t,0)x_0\|_H^p dt \leq C \cdot \|x_0\|_H^p$, $x_0 \in H$ for some $C > 0$;
(ii). $E\|U(t,0)x_0\|_H^p \leq M \cdot e^{-\mu t}\|x_0\|_H^p$, $x_0 \in H$ for some $M \geq 1$ and $\mu > 0$.

Remark This result improves some of those in Theorem 2.1.2 in which the positive constant p was supposed to be greater than or equal to one. Moreover, the continuity assumption on $S(t,s)$ and g can be relaxed. We may only assume measurability and local boundedness. A typical example of $g(t)$, $t \geq 0$, is $Ce^{\nu t}$, $C > 0$, $\nu > 0$, as in the linear case.

Corollary 3.1.2 *Suppose $S(t,s) : \mathcal{D}_s(q) \to \mathcal{D}_t(q)$ for some $q \geq 2$ and let $0 < p \leq q$. Then the conclusion of Theorem 3.1.1 holds provided that $\mathcal{D}_s(p)$ is replaced by $\mathcal{D}_s(q)$. In particular, (i) of Theorem 3.1.1 implies $E\|S(t,s)z\|_H^p \leq M \cdot e^{-\mu(t-s)}[E\|z\|_H^q]^{p/q}$, $z \in \mathcal{D}_s(q)$ for some $M \geq 1$ and $\mu > 0$.*

As an application of Theorem 3.1.1, we intend to consider Equation (3.1.2). First of all, it is clear that $U(t,0)x_0 = X_t(x_0)$ is the unique solution of (3.1.2).

Proposition 3.1.1 *Let $F : H \to H$ and $G : H \to \mathcal{L}(K, H)$ in (3.1.2) both be Lipschitz continuous with $F(0) = 0$ and $G(0) = 0$. Suppose that there* *exists a nonnegative twice Fréchet differentiable (except possibly at the zero) function* $\Lambda(x)$ *on* H *such that for some* $c > 0$ *and* $p \geq 2$,

$$\Lambda(x) + \|x\|_H \|\Lambda'(x)\|_H + \|x\|_H^2 \|\Lambda''(x)\| \leq c\|x\|_H^p, \quad \text{for arbitrary } x \in H,$$
(3.1.9)

and for some $d > 0$,

$$(\mathbf{L}\Lambda)(x) := \langle \Lambda'(x), Ax + F(x) \rangle_H + \frac{1}{2} tr[G(x)QG^*(x)\Lambda''(x)] \leq -d\|x\|_H^p,$$
(3.1.10)

for arbitrary $x \in \mathcal{D}(A)$. *Then for arbitrary* $x_0 \in H$,

$$E\|X_t(x_0)\|_H^p \leq M \cdot \|x_0\|_H^p e^{-\mu t} \quad \text{for some} \quad M \geq 1, \quad \mu > 0.$$

For the proof, we need to consider the following approximation systems of strong solutions:

$$dX_t^n = AX_t^n + R(n)F(X_t^n)dt + R(n)G(X_t^n)dW_t,$$
$$X_0^n = R(n)x_0 \in \mathcal{D}(A),$$
(3.1.11)

where $0 < n_0 \leq n \in \rho(A)$ for some $n_0 \in \rho(A)$, the resolvent set of A, and $R(n) = nR(n, A)$, $R(n, A)$ is the resolvent of A.

Proof   Applying Itô's formula to the function $\Lambda(x)$, $x \in H$, and the strong solution X_t^n of (3.1.11) yields

$$\Lambda(X_t^n) - \Lambda(X_0^n)$$
$$= \int_0^t \langle \Lambda'(X_s^n), AX_s^n + R(n)F(X_s^n) \rangle_H ds + \int_0^t \langle \Lambda'(X_s^n), R(n)G(X_s^n)dW_s \rangle_H$$
$$+ \frac{1}{2} \int_0^t tr\Big(R(n)G(X_s^n)Q[R(n)G(X_s^n)]^*\Lambda''(X_s^n)\Big)ds.$$
(3.1.12)

Therefore, by virtue of (3.1.10) we can deduce that

$$E\Lambda(X_t^n) \leq \Lambda(X_0^n) - d\int_0^t E\|X_s^n\|_H^p ds + \int_0^t E\Big\{\langle \Lambda'(X_s^n), (R(n) - I)F(X_s^n) \rangle_H$$
$$+ \frac{1}{2}tr\Big[R(n)G(X_s^n)Q\big(R(n)G(X_s^n)\big)^*\Lambda''(X_s^n)$$
$$- G(X_s^n)QG(X_s^n)^*\Lambda''(X_s^n)\Big]\Big\}ds.$$
(3.1.13)

By virtue of Proposition 1.3.6, there exists a subsequence of $\{n\} \in \rho(A)$, still denote it by $\{n\}$, such that $X_t^n \to X_t$ in $C(0, T; H)$ almost surely for any $T \geq 0$, as $n \to \infty$. Consequently, letting $n \to \infty$ in (3.1.13), together with (3.1.9), immediately yields that for arbitrary $t > 0$

$$E\Lambda(X_t) \leq \Lambda(x_0) - d\int_0^t E\|X_s\|_H^p ds.$$

This implies

$$\int_0^\infty E\|X_s(x_0)\|_H^p ds \leq \frac{c}{d} \cdot \|x_0\|_H^p < \infty.$$

Hence, by Corollary 3.1.1 it follows immediately that

$$E\|X_t(x_0)\|_H^p \leq M \cdot \|x_0\|_H^p e^{-\mu t}, \quad x_0 \in H, \ t \geq 0,$$

for some $M \geq 1$, $\mu > 0$. ☐

For the purpose of establishing stability, Proposition 3.1.1 is quite useful in constructing less restrictive Lyapunov functions, for instance, those without being strictly positive definite as mentioned at the beginning of this section. In some situations, this may produce very good results. To see this, let us apply it to the following example.

Example 3.1.1   Consider the following stochastic heat equation

$$\begin{cases} dy(x,t) = \partial^2/\partial x^2 y(x,t)dt + b(x)f(y(\cdot,t))dB_t, & t \geq 0, \ 0 < x < 1, \\ y(0,t) = y(1,t) = 0, \ t \geq 0; \quad y(x,0) = y_0(x), \ 0 \leq x \leq 1, \end{cases}$$

$$(3.1.14)$$

where B_t, $t \geq 0$, is a real standard Brownian motion, $b(\cdot) \in L^2(0,1)$ and f is a real Lipschitz continuous function on $L^2(0,1)$ satisfying $|f(u)| \leq c\|u\|_H$, $u \in L^2(0,1)$, $c > 0$. In this example, we take $H = L^2(0,1)$ and $A = d^2/dx^2$ with $\mathcal{D}(A) = H_0^1(0,1) \cap H^2(0,1)$. Then

$$\langle u, Au \rangle_H \leq -\pi^2 \|u\|_H^2 \quad \text{for any} \quad u \in \mathcal{D}(A).$$

For the sake of simplicity, we assume $\|b(\cdot)\|_H = 1$. Let $\Lambda(u)$, $u \in H$, be a twice Fréchet differentiable function on H and define **L** by

$$(\mathbf{L}\Lambda)(u) = \langle \Lambda'(u), Au \rangle_H + (1/2)\langle \Lambda''(u)b, b \rangle_H |f(u)|^2, \quad \forall u \in \mathcal{D}(A).$$

Then

$$\mathbf{L}\|u\|_H^2 = 2\langle u, Au \rangle_H + \|b\|_H^2 |f(u)|^2 \leq -(2\pi^2 - c^2)\|u\|_H^2, \quad \forall u \in \mathcal{D}(A).$$

Hence, if $c^2 < 2\pi^2$, the null solution of (3.1.14) is exponentially stable in mean square by Proposition 3.1.1.

On the other hand, consider $P \in \mathcal{L}(H)$ given by

$$P = \sum_{n=1}^\infty (1/2n^2\pi^2)e_n \otimes e_n$$

where $e_n = \sqrt{2}\sin n\pi x$ and for each g, $h \in H$, $g \otimes h \in \mathcal{L}(H)$ is defined by $(g \otimes h)u = g\langle h, u \rangle_H \in H$, $u \in H$. Then P is a self-adjoint positive nuclear operator and in fact the solution of (2.1.13). Obviously, P is not strictly positive. But we have

$$2\langle Pu, Au\rangle_H \leq -\|u\|_H^2, \qquad \forall u \in \mathcal{D}(A).$$

Now, let $\Lambda(u) = \langle Pu, u\rangle_H$ and then

$$\begin{aligned}(\mathbf{L}\Lambda)(u) &= 2\langle Pu, Au\rangle_H + \langle Pb, b\rangle_H |f(u)|^2 \\ &\leq -\left[1 - \langle Pb, b\rangle_H c^2\right]\|u\|_H^2, \quad \forall u \in \mathcal{D}(A).\end{aligned}$$

Therefore, in view of Proposition 3.1.1 we obtain the region of exponential stability in mean square: $c^2 < 1/\langle Pb, b\rangle_H$. This is larger than $\{c^2 < 2\pi^2\}$ if $\langle Pb, b\rangle_H < 1/2\pi^2$. If $b = e_m$, $m > 1$, then $\langle Pb, b\rangle_H = 1/2m^2\pi^2$. The system (3.1.14) is exponentially stable in the mean square sense if $c^2 < 2m^2\pi^2$. In other words, the function $\Lambda(u) = \langle Pu, u\rangle_H$ now plays the role of a Lyapunov function even though $\langle Pu, u\rangle_H$ is not equivalent to $\|u\|_H^2$.

3.1.2   Comparison Approaches

In this subsection, we shall employ Theorem 3.1.1 or Corollary 3.1.1 to deal with stability of nonlinear stochastic evolution equations. In particular, it will be shown that mean square stability of a class of nonlinear stochastic evolution equations is equivalent to the same stability of some linear stochastic evolution equations provided noise terms in the former are dominated by those of the latter. To this end, we first establish the following results on the trace of nuclear operators.

Lemma 3.1.2   *Let $Q \in \mathcal{L}(K)$ and $P \in \mathcal{L}(H)$ be two self-adjoint nonnegative operators and assume $G \in \mathcal{L}(K, H)$. If Q is a nuclear (or trace class) operator, then*

$$tr[GQG^*P] = tr[G^*PGQ] \geq 0.$$

Proof   Note that Q has a representation

$$Q = \sum_{i=1}^{\infty} \lambda_i e_i \otimes e_i$$

for some $\lambda_i \geq 0$ with $\sum_{i=1}^{\infty} \lambda_i < \infty$ and an orthonormal sequence $\{e_i\}$ in the real separable Hilbert space K, where $(g \otimes h)k = g\langle h, k\rangle_K$ for any g, h, $k \in K$. Let $\{f_i\}$ be any orthonormal basis in H. Then

$$tr[GQG^*P] = \sum_{j=1}^{\infty}\langle f_j, GQG^*Pf_j\rangle_K = \sum_{j=1}^{\infty}\sum_{i=1}^{\infty}\lambda_i\langle f_j, G(e_i \otimes e_i)G^*Pf_j\rangle_K$$

$$= \sum_{j=1}^{\infty} \sum_{i=1}^{\infty} \lambda_i \langle f_j, Ge_i \rangle_K \langle PGe_i, f_j \rangle_K = \sum_{i=1}^{\infty} \lambda_i \langle Ge_i, PGe_i \rangle_K$$

$$= \sum_{i=1}^{\infty} \langle e_i, G^* PGQe_i \rangle_K = tr[G^* PGQ].$$

On the other hand, since $P \geq 0$, we have

$$tr[GQG^* P] = tr[G^* PGQ] = \sum_{i=1}^{\infty} \lambda_i \langle Ge_i, PGe_i \rangle_K \geq 0$$

as required. The proof is now complete. ▯

Lemma 3.1.3   *Let $T \in \mathcal{L}(H)$ be a self-adjoint nuclear operator. Then $T \geq 0$ if and only if $tr(TP) \geq 0$ for any self-adjoint nonnegative operator $P \in \mathcal{L}(H)$.*

Proof   Since the necessity is shown in Lemma 3.1.2 (take $K = H$ and $G = I_H$, the identity operator on H), we need only prove the converse part. Let $h \in H$ be an arbitrary vector in H and $P = h \otimes h$. Then

$$0 \leq tr(TP) = tr[T(h \otimes h)] = \langle Th, h \rangle_H.$$

Hence, T is nonnegative. ▯

We intend to consider the following stochastic differential equations

$$dX_t = AX_t dt + G_i(X_t) dW_t^i, \quad X_0 = x_0 \in H, \quad i = 1, 2, \quad (3.1.15)$$

where A generates a strongly continuous semigroup $T(t)$, $t \geq 0$, on H, $G_i : H \to \mathcal{L}(K_i, H)$ is Lipschitz continuous with $G_i(0) = 0$ and W_t^i is a Wiener process defined on some real separable Hilbert space K_i with incremental covariance operator Q_i, $trQ_i < \infty$, $i = 1, 2$. Then Theorem 1.3.4 is still valid for (3.1.15). For our stability analysis, we hope to compare quadratic functionals of mild solutions under the following condition:

$$G_1(x) Q_1 G_1^*(x) \geq G_2(x) Q_2 G_2^*(x) \quad \text{for any} \quad x \in H. \quad (3.1.16)$$

By Lemmas 3.1.2 and 3.1.3, this is equivalent to

$$tr[G_1^*(x) PG_1(x) Q_1] \geq tr[G_2^*(x) PG_2(x) Q_2] \quad (3.1.17)$$

for any self-adjoint nonnegative operator $P \geq 0$ in $\mathcal{L}(H)$ and $x \in H$.

Let $X_t^i(x_0)$, $t \geq 0$, $i = 1, 2$, be the mild solutions of (3.1.15). Assume F and M are two self-adjoint nonnegative operators in $\mathcal{L}(H)$.

Theorem 3.1.2   *Suppose one of G_i is linear, i.e., $G_i \in \mathcal{L}(H, \mathcal{L}(K_i, H))$, $i = 1$ or 2. Then the condition (3.1.16) implies the inequality*

$$E\langle FX_T^2, X_T^2 \rangle_H + \int_0^T E\langle MX_t^2, X_t^2 \rangle_H dt$$
$$\leq E\langle FX_T^1, X_T^1 \rangle_H + \int_0^T E\langle MX_t^1, X_t^1 \rangle_H dt \tag{3.1.18}$$

for any $T \geq 0$.

Proof   (i). Let $G_1 \in \mathcal{L}(H, \mathcal{L}(K_1, H))$. Then by virtue of Proposition 2.2.1, there exists a unique strongly continuous, self-adjoint solution $P_1(t) \geq 0$ to the following:

$$\frac{d}{dt}\langle P(t)x, x \rangle_H + 2\langle P(t)x, Ax \rangle_H + \langle Mx, x \rangle_H + tr[G_1^*(x)P(t)G_1(x)Q_1] = 0,$$
$$x \in \mathcal{D}(A),$$
$$P(T) = F, \quad 0 \leq t \leq T. \tag{3.1.19}$$

Moreover, by Proposition 2.2.2 we have

$$\langle P_1(0)x_0, x_0 \rangle_H = E\langle FX_T^1, X_T^1 \rangle_H + \int_0^T E\langle MX_t^1, X_t^1 \rangle_H dt.$$

On the other hand, by (3.1.17) and (3.1.19) we have

$$\frac{d}{dt}\langle P_1(t)x, x \rangle_H + 2\langle P_1(t)x, Ax \rangle_H + \langle Mx, x \rangle_H$$
$$+ tr[G_2^*(x)P_1(t)G_2(x)Q_2] \leq 0, \quad x \in \mathcal{D}(A). \tag{3.1.20}$$

Hence, by a similar argument to the proof of Proposition 2.2.2 (using the approximation procedure), applying Itô's formula to (3.1.15) and $\langle P_1(t)x, x \rangle_H$, we get

$$\langle P_1(0)x_0, x_0 \rangle_H \geq E\langle FX_T^2, X_T^2 \rangle_H + \int_0^T E\langle MX_t^2, X_t^2 \rangle_H dt.$$

(ii). Similarly, we can show (3.1.18) when G_2 is linear. $\qquad\square$

Corollary 3.1.3   *Under the same conditions as in Theorem 3.1.2, we have*

$$\int_0^T E\|X_t^2(x_0)\|_H^2 dt \leq \int_0^T E\|X_t^1(x_0)\|_H^2 dt \quad \text{for all} \quad T \geq 0,$$

$$E\|X_t^2(x_0)\|_H^2 \leq E\|X_t^1(x_0)\|_H^2 \quad \text{for all} \quad t \geq 0.$$

Proof   These follow immediately by taking $F = 0$, $M = I$ and $F = I$, $M = 0$ in (3.1.18), respectively.   ⬜

Corollary 3.1.4   *Suppose G_1 is linear and the null solution of the corresponding (3.1.15) is exponentially stable in mean square. Then the condition (3.1.16) implies that the null solution of the other (3.1.15) is exponentially stable in mean square. Moreover, for any $x_0 \in H$,*

$$E\|X_t^2(x_0)\|_H^2 \leq M \cdot e^{-\mu t}\|x_0\|_H^2 \quad \text{for some} \quad M \geq 1 \quad \text{and} \quad \mu > 0.$$

Proof   This is immediate by using Corollary 3.1.3.   ⬜

Sometimes, it is more convenient to restate our results in a slightly different manner. Let U be a class of objects satisfying certain required conditions. We say that a stochastic system is *uniformly stable* in the class U if it is stable for any $g \in U$. For instance, in this notation we can restate Corollary 3.1.4 in the following way. Let $\mathcal{U}$ be the set of triples (K_2, W_t^2, G_2), which satisfies the condition (3.1.16).

Corollary 3.1.4*   *Let G_1 be linear and the null solution of the corresponding (3.1.15) be exponentially stable in mean square. Then the stochastic system*

$$dX_t = AX_t dt + G_2(X_t)dW_t^2, \qquad X_0 = x_0, \qquad (3.1.21)$$

is uniformly stable in the class $\mathcal{U}$ in the mean square sense. In other words, the mild solution, $X_t^2(x_0)$, $t \geq 0$, of (3.1.21) satisfies

$$E\|X_t^2(x_0)\|_H^2 \leq M \cdot e^{-\mu t}\|x_0\|_H^2 \quad \text{for some} \quad M \geq 1 \quad \text{and} \quad \mu > 0.$$

In order to illustrate our theory derived above, let us now consider a class of stochastic differential equations in which $K_i = \mathbf{R}^1$, $W_t^i = B_t$, $i = 1, 2$, $G_1(x) = kb\langle a, x\rangle_H$, $k > 0$, b, $a \in H$ and $G_2(x) = bg(\langle a, x\rangle_H)$ where $g(\cdot)$: $\mathbf{R}^1 \to \mathbf{R}^1$, $g(0) = 0$, is some real continuous function. Then (3.1.15) turns into two stochastic differential equations:

$$dX_t = AX_t dt + kb\langle a, X_t\rangle_H dB_t, \qquad X_0 = x_0, \qquad (3.1.22)$$

$$dX_t = AX_t dt + bg(\langle a, X_t\rangle_H)dB_t, \qquad X_0 = x_0. \qquad (3.1.23)$$

Here B_t, $t \geq 0$, is a one-dimensional standard Brownian motion. Now, let L_k be the set of all real Lipschitz continuous functions $g(\cdot)$ satisfying $|g(x)| \leq k|x|$, $k > 0$, for any $x \in \mathbf{R}^1$. If $g \in L_k$, then (3.1.17) turns out to be

$$tr[G_1^*(x)PG_1(x)] = k^2\langle Pb, b\rangle_H\langle a, x\rangle_H^2$$
$$\geq \langle Pb, b\rangle_H|g(\langle a, x\rangle_H)|^2 = tr[G_2^*(x)PG_2(x)], \qquad x \in H.$$

Hence, the nonlinear stochastic system (3.1.23) is uniformly stable in the class L_k if and only if the linear system (3.1.22) is stable. And if the system (3.1.22) is stable in mean square, the mild solution of (3.1.23), of course, satisfies

$$E\|X_t(x_0)\|_H^2 \leq Ce^{-\gamma t}\|x_0\|_H^2, \quad x_0 \in H \quad \text{for some} \quad C \geq 1 \text{ and } \gamma > 0.$$
$$(3.1.24)$$

Therefore, all we need to do is to find conditions to assure the stability of (3.1.22). Since (3.1.22) is a linear equation, we have already derived some stability results such as those in Chapter 2. However, owing to the special structure of (3.1.22), we can actually find some more delicate sufficient conditions for the stability of (3.1.22) and hence of (3.1.23). In particular, by virtue of Example 2.2.1 we have the following:

Proposition 3.1.2   *The null solution of (3.1.22) is exponentially stable in mean square if and only if the following statements hold:*

    (i). The semigroup $T(t)$, $t \geq 0$, is exponentially stable, i.e., there exist numbers $C \geq 1$, $\gamma > 0$ such that $\|T(t)\| \leq C \cdot e^{-\gamma t}$, $t \geq 0$;
    (ii). $k^2 \int_0^\infty \langle T(t)b, a\rangle_H^2 dt < 1$.

Proof   It is immediate by virtue of Example 2.2.1.   ⬚

Corollary 3.1.5   *Suppose that A is stable, i.e., A generates an exponentially stable, strongly continuous semigroup $T(t)$, $t \geq 0$, and*

$$\langle T(t)b, a\rangle_H^2 \leq Me^{-2\mu t}, \quad M > 0, \ \mu > 0 \quad \text{with} \quad k^2 M < 2\mu.$$

Then the system (3.1.23) is uniformly stable in the class L_k in mean square and moreover (3.1.24) holds. If, in particular, $\langle T(t)b, a\rangle_H = 0$, $t \geq 0$, then it is uniformly stable in mean square in the class of all Lipschitz continuous functions $g(\cdot)$ with $g(0) = 0$ and (3.1.24) holds.

Proof   The first part is immediate by (ii) of Proposition 3.1.2. The second part also follows since we can take an arbitrarily large $\mu > 0$.   ⬚

Proposition 3.1.3   *Suppose that A is stable and that $a \in \mathcal{D}(A^*)$, $A^*a = -pa$, $p > 0$. If $2p - k^2\langle b, a\rangle_H^2 > 0$, then the system (3.1.23) is stable in mean square for any $g \in L_k$ and (3.1.24) holds. If, in particular, $\langle b, a\rangle_H = 0$, then the same is true for any Lipschitz continuous function $g(\cdot)$ with $g(0) = 0$.*

Proof   Note that

$$2\langle a, x\rangle_H \langle a, Ax\rangle_H + k^2\langle b, a\rangle_H^2 \langle a, x\rangle_H^2 = -(2p - k^2\langle b, a\rangle_H^2)\langle a, x\rangle_H^2.$$

Thus Itô's formula applied to (3.1.22) and the function $\langle a, x \rangle_H^2$ yields

$$E\langle a, X_t \rangle_H^2 \le e^{-ct} \langle a, x_0 \rangle_H^2, \qquad c = 2p - k^2 \langle b, a \rangle_H^2.$$

$\Box$

Example 3.1.2 Consider the stochastic heat equation

$$dy(x,t) = \partial^2/\partial x^2 y(x,t)dt + b(x)g(\langle a(\cdot), y(\cdot, t) \rangle)dB_t, \quad 0 < x < 1, \quad t \ge 0,$$
$$y(0,t) = y(1,t) = 0, \quad t \ge 0; \quad y(x,0) = y_0(x), \quad 0 \le x \le 1,$$

where $\langle a(\cdot), y(\cdot, t) \rangle = \int_0^1 a(x)y(x,t)dx$, $t \ge 0$, and B_t is a one-dimensional standard Brownian motion.

In this example, we take $H = L^2(0,1)$ and $A = d^2/dx^2$ with domain $\mathcal{D}(A) = \{y \in H : y, y'$ are absolutely continuous, $y'' \in H$, $y(0) = y(1) = 0\}$. If we take $b(x) = \sin \pi x$ and $a(\cdot) = 1$, then $\int_0^\infty \langle T(t)b, a \rangle_H^2 dt = 2/\pi^4$. Thus by Proposition 3.1.2, we have exponential stability in mean square for any $g \in L_k$ with $k^2 < \pi^4/2$. If we take $b(x) = \cos \pi x$ and $a(x) = \sin \pi x$, then $\langle T(t)b, a \rangle_H = 0$. Hence, we have the same stability for any Lipschitz continuous functions $g(\cdot)$ with $g(0) = 0$.

Example 3.1.3 Consider the stochastic delay differential equation in $\mathbf{R}^1$,

$$\begin{cases} dx(t) = [-d_1 x(t) + d_2 x(t - r)]dt + bg(cx(t))dB_t, \\ x(0) = x_0, \quad x(s) = 0, \quad -r \le s < 0, \end{cases} \tag{3.1.25}$$

where $d_1 > 0$, $r > 0$ and $d_2, b, c \in \mathbf{R}^1$. B_t, $t \ge 0$, is a one-dimensional standard Brownian motion. In this case, we take $H = \mathbf{R}^1 \times L^2(-r, 0)$ with inner product $\langle \cdot, \cdot \rangle_H$ and define A with $\mathcal{D}(A) = \{(f(0), f(\cdot)) \in H : f \in W^{1,2}(-r, 0)\}$ by

$$A \begin{pmatrix} f(0) \\ f(\cdot) \end{pmatrix} = \begin{pmatrix} -d_1 f(0) + d_2 f(-r) \\ df(\cdot)/ds \end{pmatrix}$$

It can be shown (cf. Delfour, McCalla and Mitter [1]) that A generates a strongly continuous semigroup $T(t)$, $t \ge 0$ on H, and then the infinite dimensional model for (3.1.25) is

$$dy(t) = Ay(t)dt + \begin{pmatrix} b \\ 0 \end{pmatrix} g\left(\left\langle \begin{pmatrix} c \\ 0 \end{pmatrix}, y \right\rangle_H \right) dB_t, \quad y(0) = \begin{pmatrix} x_0 \\ 0 \end{pmatrix}. \tag{3.1.26}$$

Then the system (3.1.26) is exponentially stable in mean square for any $g \in L_k$ if and only if $T(t)$, $t \ge 0$, is exponentially stable and

$$k^2 \int_0^\infty \left\langle T(t) \begin{pmatrix} b \\ 0 \end{pmatrix}, \begin{pmatrix} c \\ 0 \end{pmatrix} \right\rangle_H^2 dt < 1.$$

This inequality can be rewritten as

$$k^2 \int_0^\infty c^2 z^2(t)dt < 1, \qquad (3.1.27)$$

where $z(t)$ is the solution of

$$dz(t)/dt = -d_1 z(t) + d_2 z(t-r), \quad z(0) = b, \quad z(s) = 0, \quad -r \leq s < 0.$$

On the other hand, it is well known by the theory of ordinary functional differential equations that if $d_1 > |d_2| > 0$, then $T(t)$, $t \geq 0$, is exponentially stable and further by the Lyapunov method (cf. Hale [1], p. 108) that

$$2(d_1 - |d_2|) \int_0^\infty z^2(t)dt \leq b^2.$$

Thus

$$k^2 \int_0^\infty c^2 z^2(t)dt \leq \frac{k^2 b^2 c^2}{2(d_1 - |d_2|)}.$$

If $k^2 < 2(d_1 - |d_2|)/b^2 c^2$, then (3.1.27) is satisfied and we have exponential stability in mean square of (3.1.26) and hence of (3.1.25) for any $g \in L_k$.

3.2   A Coercive Decay Condition

In the remainder of this chapter, we intend to study decay and relevant large time properties of general non-autonomous, nonlinear stochastic differential equations. In this section and Section 3.3, we are mainly interested in the exponential decay of strong and mild solutions. Particularly, we will no longer impose the restrictions $A(t,0) = 0$, $B(t,0) = 0$ in (1.3.1) and $F(t,0) = 0$, $G(t,0) = 0$ in (1.3.10). The reason for removing this restriction will become clear when we consider some large time properties such as ultimate boundedness of solutions of non-autonomous stochastic differential equations.

In the previous sections, we have shown in Theorems 2.2.1 and 3.1.1 that for linear stochastic differential equations (2.2.1) or a class of nonlinear stochastic differential equations (3.1.2), the L^2-stability in mean of systems is equivalent to their exponential stability in mean square. When ones focus on non-autonomous, nonlinear stochastic systems, the situation becomes quite different. For instance, ones should not expect that a similar generalization to Theorem 3.1.1 remains true even in finite dimensional cases. Indeed, to see this, let us consider a one-dimensional stochastic differential equation

$$dX_t = -\frac{p}{1+t} X_t dt + (1+t)^{-p} dB_t, \qquad t \geq 0,$$

with initial datum $X_0 = x_0 \in \mathbf{R}^1$, where $p > 1$ is a positive constant and B_t, $t \geq 0$, is a one-dimensional standard Brownian motion. It is easy to obtain its explicit solution

$$X_t = (1+t)^{-p}(x_0 + B_t), \qquad t \geq 0,$$

which, by standard properties of Brownian motion, immediately yields that the Lyapunov exponent

$$\lim_{t \to \infty} \frac{\log E|X_t(x_0)|^2}{t} = 0.$$

That is, the solution is not exponentially decayable in mean square. However, by a direct computation we have

$$\int_0^\infty E|X_t(x_0)|^2 dt = \frac{1+x_0^2}{2p-1} + \frac{1}{2p-2} < \infty.$$

In the following two sections, we shall investigate the exponential decay of strong and mild solutions of general non-autonomous stochastic differential equations in a different way from that in the preceding sections.

First of all, recall that V is a real, separable Banach space and H, K are two real, separable Hilbert spaces such that

$$V \hookrightarrow H \equiv H^* \hookrightarrow V^*,$$

where the injections " $\hookrightarrow$ " are continuous and dense such that $\|x\|_H \leq \beta\|x\|_V$, $x \in V$, for some constant $\beta > 0$.

Consider the following nonlinear stochastic differential equation:

$$X_t = x_0 + \int_0^t A(s, X_s)ds + \int_0^t B(s, X_s)dW_s \qquad (3.2.1)$$

where $A(\cdot, \cdot) : \mathbf{R}_+ \times V \to V^*$ is assumed to be a measurable family of nonlinear operators and $B(\cdot, \cdot) : \mathbf{R}_+ \times V \to \mathcal{L}(K, H)$, is the measurable family of all nonlinear operators from V into $\mathcal{L}(K, H)$, satisfying:

$$\|B(t, y) - B(t, x)\| \leq L\|y - x\|_V, \quad \forall x, y \in V, \quad \forall t \geq 0, \qquad (3.2.2)$$

for some constant $L > 0$.

Being interested in stability analysis, we always assume that for each $x_0 \in H$, there exists a global strong solution to (3.2.1). In particular, we assume the conditions (a)–(e) in Subsection 1.3.1 to ensure the existence and uniqueness of the strong solution of (3.2.1).

The following coercive condition will play the role of a stability criterion.

(H5). There exist constants $\alpha > 0$, $\lambda \in \mathbf{R}^1$, and a nonnegative continuous function $\gamma(t)$, $t \in \mathbf{R}_+$, such that

$$2\langle v, A(t, v)\rangle_{V,V^*} + \|B(t, v)\|_{\mathcal{L}_2^0}^2 \leq -\alpha\|v\|_V^p + \lambda\|v\|_H^2 + \gamma(t), \quad v \in V,$$

where $\|B(t,v)\|_{\mathcal{L}_2^0}^2 = tr(B(t,v)QB(t,v)^*)$, $p \geq 2$, and $\gamma(t)$, $t \geq 0$, satisfies that there exists $\mu > 0$ such that $\gamma(t)e^{\mu t}$ is integrable on $[0, \infty)$.

Note that this assumption is compatible with the existence of the strong solution to (3.2.1) formulated in Section 1.3.

Theorem 3.2.1   *Assume the condition (H5) holds. Let $X_t(x_0)$, $t \geq 0$, be a global strong solution of the equation (3.2.1), then there exist constants $\tau > 0$, $C = C(x_0) > 0$ such that*

$$E\|X_t(x_0)\|_H^2 \leq C(x_0) \cdot e^{-\tau t}, \qquad \forall t \geq 0, \tag{3.2.3}$$

if either one of the following hypotheses holds

(*i*). $\lambda < 0$, ($\forall p \geq 2$). *On this occasion, τ can be taken as $(-\lambda) \wedge \mu$;*
(*ii*). $\lambda\beta^2 - \alpha < 0$, *(particularly, for $p = 2$). On this occasion, τ can be taken as $\nu \wedge \mu$, where $\nu = (\alpha - \lambda\beta^2)/\beta^2$.*

Proof   We omit this proof and refer the reader to Theorem 4.3.1 since a similar formulation which contains the above theorem as a special case is presented there.   ◻

The following result shows that the condition (H5) not only justifies exponential decay of the equation (3.2.1) in mean square, but it also ensures pathwise exponential decay with probability one.

Theorem 3.2.2   *Assume the hypotheses in Theorem 3.2.1 hold. Then there exist a subset $\Omega_0 \subset \Omega$ with $P(\Omega_0) = 0$ and random variable $T(\omega) \geq 0$ such that for each $\omega \in \Omega\backslash\Omega_0$,*

$$\|X_t(x_0)\|_H \leq M(x_0) \cdot e^{-\gamma t}, \qquad \forall t \geq T(\omega), \tag{3.2.4}$$

for some positive constants $M = M(x_0) > 0$ and $\gamma > 0$.

Proof   We only prove the case (ii). Case (i) can be shown in a similar manner. Firstly, applying Itô's formula to the strong solution $X_t(x_0)$, $t \geq 0$, of (3.2.1) immediately yields for any $t \geq N > 0$,

$$\|X_t(x_0)\|_H^2 - \|X_N(x_0)\|_H^2 = 2\int_N^t \langle X_s, A(s, X_s)\rangle_{V,V^*} ds + \int_N^t \|B(s, X_s)\|_{\mathcal{L}_2^0}^2 ds$$

$$+ 2\int_N^t \langle X_s, B(s, X_s)dW_s\rangle_H, \tag{3.2.5}$$

where N is a positive integer, which, together with the coercive condition (H5), implies

$$\|X_t\|_H^2 \leq \|X_N\|_H^2 + |\lambda| \int_N^t \|X_s\|_H^2 ds + \int_N^t \gamma(s)ds + \left| 2 \int_N^t \langle X_s, B(s, X_s)dW_s \rangle_H \right|. \tag{3.2.6}$$

Let I_N denote the interval $[N, N+1]$, $N > 0$, then from (3.2.6) we can get

$$E\left[\sup_{t \in I_N} \|X_t\|_H^2 \right] \leq E\|X_N\|_H^2 + |\lambda| \int_N^{N+1} E\|X_s\|_H^2 ds + \int_N^{N+1} \gamma(s)ds \tag{3.2.7}$$
$$+ E\left[\sup_{t \in I_N} \left| 2 \int_N^t \langle X_s, B(s, X_s)dW_s \rangle_H \right| \right].$$

Now, according to the Burkholder-Davis-Gundy lemma, we can estimate the last term in (3.2.7) (in what follows, K_1, K_2, $\cdots$ denote some proper positive constants).

$$E\left[\sup_{t \in I_N} \left| 2 \int_N^t \langle X_s, B(s, X_s)dW_s \rangle_H \right| \right]$$
$$\leq K_1 E\left[\int_N^{N+1} \|X_s\|_H^2 \|B(s, X_s)\|_{\mathcal{L}_2^0}^2 ds \right]^{1/2}$$
$$\leq K_1 E\left[\sup_{t \in I_N} \|X_t\|_H \left(\int_N^{N+1} \|B(s, X_s)\|_{\mathcal{L}_2^0}^2 ds \right)^{1/2} \right]$$
$$\leq \frac{1}{2} E\left[\sup_{t \in I_N} \|X_t\|_H^2 \right] + K_2 \int_N^{N+1} E\|B(s, X_s)\|_{\mathcal{L}_2^0}^2 ds, \tag{3.2.8}$$

which, by virtue of (3.2.7) and Markov's inequality, immediately implies for any $\epsilon_N > 0$,

$$P\left\{ \sup_{t \in I_N} \|X_t\|_H^2 \geq \epsilon_N^2 \right\} \leq \epsilon_N^{-2} E\left[\sup_{t \in I_N} \|X_t\|_H^2 \right]$$
$$\leq K_3 \epsilon_N^{-2} \left[E\|X_N\|_H^2 + \int_N^{N+1} E\|X_s\|_H^2 ds \tag{3.2.9} \right.$$
$$\left. + \int_N^{N+1} \gamma(s)ds + \int_N^{N+1} E\|B(s, X_s)\|_{\mathcal{L}_2^0}^2 ds \right].$$

Note that there exists a positive constant $K_4 > 0$ such that for any $N > 0$,

$$\int_N^{N+1} \gamma(s)ds \leq e^{-\tau N} \int_N^{N+1} \gamma(s)e^{\mu s} ds < K_4 \cdot e^{-\tau N},$$

which immediately implies that there exists a constant $K_5 > 0$ such that

$$E\|X_N\|_H^2 + \int_N^{N+1} E\|X_s\|_H^2 ds + \int_N^{N+1} \gamma(s)ds \leq K_5 \cdot e^{-\tau N}. \tag{3.2.10}$$

For the last term on the right hand side of (3.2.9), assume that the following claim which will be proved below holds.

Claim: There exists a positive constant $K_6 > 0$ such that for $N > 0$ large enough,

$$\int_N^{N+1} E\|B(s, X_s)\|_{\mathcal{L}_2^0}^2 ds \leq K_6 \cdot e^{-\tau N/2}. \tag{3.2.11}$$

Then (3.2.9), (3.2.10) and (3.2.11), together with letting $\epsilon_N^2 = e^{-\tau N/4}$, imply that there exists $K_7 > 0$ such that

$$P\left\{ \sup_{t \in I_N} \|X_t\|_H^2 \geq \epsilon_N^2 \right\} \leq K_7 \cdot \epsilon_N^{-2} e^{-\tau N/2} = K_7 \cdot e^{-\tau N/4}, \tag{3.2.12}$$

and a Borel-Cantelli's lemma type argument completes the proof.

Let us finally prove our claim (3.2.11). Indeed, for the parameter $\tau > 0$ in Theorem 3.2.1, using (H5) we obtain from Itô's formula that there exists an increasing sequence of stopping times $\{\tau_n\}$, $n \geq 1$, exactly as in Theorem 3.2.1 such that

$$Ee^{\frac{\tau}{2}(t \wedge \tau_n)} \|X_{t \wedge \tau_n}\|_H^2$$
$$\leq \|x_0\|_H^2 + \frac{\tau}{2} E \int_0^{t \wedge \tau_n} e^{\frac{\tau}{2}s} \|X_s\|_H^2 ds - \alpha E \int_0^{t \wedge \tau_n} e^{\frac{\tau}{2}s} \|X_s\|_V^2 ds$$
$$+ \lambda E \int_0^{t \wedge \tau_n} e^{\frac{\tau}{2}s} \|X_s\|_H^2 ds + E \int_0^{t \wedge \tau_n} \gamma(s) e^{\frac{\tau}{2}s} ds,$$

which, by virtue of (3.2.3), implies that there exists $K_8 > 0$ such that

$$E \int_0^{t \wedge \tau_n} e^{\frac{\tau}{2}s} \|X_s\|_V^2 ds$$
$$\leq \frac{1}{\alpha} \left[\|x_0\|_H^2 + \left(\frac{\tau}{2} + |\lambda| \right) \int_0^t e^{\frac{\tau}{2}s} E\|X_s\|_H^2 ds + \int_0^t \gamma(s) e^{\frac{\tau}{2}s} ds \right]$$
$$\leq K_8 < \infty.$$

Consequently, letting n tend to infinity, it is easy to deduce

$$\int_s^t E\|X_u\|_V^2 du \leq \int_s^t e^{\frac{\tau}{2}(u-s)} E\|X_u\|_V^2 du$$
$$\leq K_8 \cdot e^{-\frac{\tau}{2}s} \quad \text{for any} \quad 0 \leq s \leq t.$$

Now, taking into account (3.2.2) and (H5), we can get

$$\int_N^{N+1} E\|B(u, X_u)\|_{\mathcal{L}_2^0}^2 du$$
$$\leq 2 \int_N^{N+1} E\|B(u, X_u) - B(u, 0)\|_{\mathcal{L}_2^0}^2 du + 2 \int_N^{N+1} E\|B(u, 0)\|_{\mathcal{L}_2^0}^2 du$$
$$\leq 2L^2 trQ \left(\int_N^{N+1} E\|X_u\|_V^2 du \right) + 2 \int_N^{N+1} \gamma(u) du$$
$$\leq K_6 \cdot e^{-\tau N/2}$$

for some $K_6 > 0$. Now the proof is complete.   □

Remarks   (1). In the case that $B(t, \cdot) : H \to \mathcal{L}(K; H)$, and Condition (3.2.2) turns out to be

$$\|B(t, u) - B(t, v)\| \leq L\|u - v\|_H, \quad \forall u, \ v \in V, \quad L > 0,$$

the last claim follows immediately since the upper bound on $\int_s^t E\|X_u\|_H^2 du$ follows easily from (3.2.3).

(2). The exponential decay imposed on $\gamma(t)$, $t \geq 0$, in (H5) is essential. Indeed, to interpret this, let us study the following one-dimensional linear equation (note that on this occasion $V = H = \mathbf{R}^1$):

Example 3.2.1   Assume X_t, $t \geq 0$, satisfies the stochastic differential equation

$$dX_t = -pX_t dt + (1 + t)^{-q} dB_t, \qquad t \geq 0,$$

with initial datum $X_0 = 0$, where p, $q > 0$ are two positive constants and B_t, $t \geq 0$, is a one-dimensional standard Brownian motion.

Clearly, the corresponding coercive condition (H5) now becomes

$$2\langle v, A(t, v)\rangle_{\mathbf{R}^1} + \|B(t, v)\|_{\mathcal{L}_2^0}^2 = -2pv^2 + (1 + t)^{-2q}, \quad \forall v \in \mathbf{R}^1.$$

However, under this condition the solution does not decay exponentially. Indeed, it is easy to obtain the explicit solution

$$X_t = e^{-pt} \int_0^t e^{ps} \cdot (1 + s)^{-q} dB_s =: e^{-pt} M_t, \qquad t \geq 0,$$

which immediately implies that for arbitrarily given $q > 0$, the Lyapunov exponent

$$\lim_{t \to \infty} \frac{\log E|X_t|^2}{t} = 0.$$

Similarly, by virtue of the well known iterated logarithmic law

$$\limsup_{t \to \infty} \frac{M_t}{\sqrt{2[M_t] \log \log[M_t]}} = 1 \qquad a.s.$$

where $[M_t]$ is the quadratic variation of M_t, and

$$\limsup_{t \to \infty} \frac{\log\left(\int_0^t e^{2ps}(1 + s)^{-2q} ds\right)}{t} = 2p,$$

we therefore get the almost sure Lyapunov exponent

$$\limsup_{t \to \infty} \frac{\log |X_t|}{t} = 0 \qquad a.s.$$

In other words, despite the obvious stability of the ordinary differential equation

$$dX_t = -pX_t dt,$$

the polynomial decay type of additive noises is not sufficient to guarantee the exponential decay of its stochastically perturbed system. In fact, we will show in Section 3.8 that the solution has a slower, polynomial, decay than the exponential one.

As an immediate consequence of Theorems 3.2.1 and 3.2.2, we may formulate the following fractional power type of coercive condition which is quite useful in dealing with certain nonlinear stochastic systems.

(H5*). There exist constants $\alpha > 0$, $\lambda \in \mathbf{R}^1$, $0 \le \sigma \le 1$ and nonnegative continuous functions $\gamma(t)$, $\zeta(t)$, $t \in \mathbf{R}_+$, such that

$$2\langle u, A(t,u)\rangle_{V,V^*} + \|B(t,u)\|^2_{\mathcal{L}_2^0}$$
$$\le - \alpha\|u\|^p_V + \lambda\|u\|^2_H + \zeta(t)\|u\|^{2\sigma}_H + \gamma(t), \quad u \in V,$$
$$\tag{3.2.13}$$

where $p \ge 2$, and $\gamma(t)$, $\zeta(t)$ satisfy that there exist $\theta > 0$, $\mu > 0$ such that $\gamma(t)e^{\mu t}$ and $\zeta(t)e^{\theta t}$ both are integrable on $[0, \infty)$.

Corollary 3.2.1   *Suppose that (H5*) and (3.2.2) hold. Let $X_t(x_0)$, $t \ge 0$, be a global strong solution to the equation (3.2.1). Then there exist constants $\tau > 0$, $C = C(x_0) > 0$ such that*

$$E\|X_t(x_0)\|^2_H \le C(x_0) \cdot e^{-\tau t}, \qquad \forall t \ge 0, \tag{3.2.14}$$

if either one of the following hypotheses holds:

(i). $\lambda < 0$, $(\forall p \ge 2)$. *On this occasion, τ can be taken as $(-\lambda) \wedge \mu \wedge \theta$;*
(ii). $\lambda\beta^2 - \alpha < 0$, *(particularly, for $p = 2$). On this occasion, τ can be taken as $\nu \wedge \mu \wedge \theta$, where $\nu = (\alpha - \lambda\beta^2)/\beta^2$.*

Moreover, there exist positive numbers $M = M(x_0) > 0$, $r > 0$, subset $\Omega_0 \subset \Omega$ with $P(\Omega_0) = 0$ and a positive random variable $T(\omega)$ such that for each $\omega \in \Omega \setminus \Omega_0$,

$$\|X_t(x_0)\|_H \le M(x_0) \cdot e^{-rt}, \qquad \forall t \ge T(\omega).$$

Proof   Observe that the case $\sigma = 0$ or $\sigma = 1$ is trivial by using Theorems 3.2.1 and 3.2.2. For $0 < \sigma < 1$, by virtue of Young's inequality

$$a \cdot b \le \frac{a^p}{p} + \frac{b^q}{q} \quad \text{for any} \quad a \ge 0,\ b \ge 0,\ p,\ q > 1 \quad \text{with} \quad 1/p + 1/q = 1,$$

we have for arbitrary $\varepsilon > 0$, the third term on the right hand side of (3.2.13) turns out to be

$$\zeta(t)\|u\|_H^{2\sigma} \leq \sigma\varepsilon^{1/\sigma}\|u\|_H^2 + (1-\sigma)\varepsilon^{\frac{1}{1-\sigma}}\zeta(t)^{\frac{1}{1-\sigma}},$$

which, together with (3.2.13), implies that

$$2\langle u, A(t,u)\rangle_{V,V^*} + \|B(t,u)\|_{\mathcal{L}_2^0}^2 \leq -\alpha\|u\|_V^p + \lambda\|u\|_H^2 + \sigma\varepsilon^{1/\sigma}\|u\|_H^2$$
$$+ \gamma(t) + (1-\sigma)\varepsilon^{\frac{1}{1-\sigma}}\zeta(t)^{\frac{1}{1-\sigma}}, \quad u \in V.$$

In the case (ii), by virtue of Theorems 3.2.1 and 3.2.2, it is easy to deduce that if $\nu > \sigma\varepsilon^{1/\sigma}$, the solution is exponentially decayable in the mean square and almost sure senses. Note that $\varepsilon > 0$ is an arbitrary number, then the proof of (ii) is complete letting $\varepsilon \to 0$. The result (i) can be shown similarly.

□

Lastly, let us investigate an example to close this section.

Example 3.2.2 Let $\mathcal{O}$ be an open, bounded subset in $\mathbf{R}^n$, $n \geq 1$, with regular boundary and let $2 < p < \infty$. Consider the Sobolev space $V = W_0^{1,p}(\mathcal{O})$, $H = L^2(\mathcal{O})$ with their usual norms, and the monotone operator $A : V \to V^*$ is defined as for any $u, v \in V$,

$$\langle v, Au\rangle_{V,V^*} = -\sum_{i=1}^n \int_{\mathcal{O}} \left|\frac{\partial u(x)}{\partial x_i}\right|^{p-2} \frac{\partial u(x)}{\partial x_i}\frac{\partial v(x)}{\partial x_i}dx - \int_{\mathcal{O}} a(x)u(x)v(x)dx,$$

where $a \in L^\infty(\mathcal{O}; \mathbf{R}^1)$ satisfies $a(x) \geq \tilde{a} > 0$, $x \in \mathcal{O}$. Here $\tilde{a}$ is some constant. We also consider $B(t,u) = g(u(x))$, $u \in V$, in (3.2.1) where $g : \mathbf{R}^1 \to \mathbf{R}^1$ is some Lipschitz continuous function satisfying

$$|g(x) - g(y)| \leq L|x - y|, \quad \forall x, y \in \mathbf{R}^1,$$

for some constant $L > 0$ such that $L^2 < 2\tilde{a}$ and $g(0) = 0$. Let B_t, $t \geq 0$, be a standard real Brownian motion (so, $K = \mathbf{R}^1$ and $Q = I$). In this case, we can claim that there exists a unique strong solution of the equation (3.2.1). In particular, (H5) holds with $\gamma(\cdot) = 0$, $\lambda = -(2\tilde{a} - L^2) < 0$, $p > 2$, $\alpha = 2$. Consequently, using Theorems 3.2.1 and 3.2.2, we may obtain the exponential decay both in the mean square and almost sure senses of the equation which is interpreted as:

$$dX_t(x) = \left[-\sum_{i=1}^n \frac{\partial}{\partial x_i}\left(\left|\frac{\partial X_t(x)}{\partial x_i}\right|^{p-2}\frac{\partial X_t(x)}{\partial x_i}\right) - a(x)X_t(x)\right]dt$$
$$+ g(X_t(x))dB_t, \quad t > 0, \ x \in \mathcal{O};$$
$$X_0(x) = x_0(x) \in H \text{ in } \mathcal{O};$$
$$X_t(x) = 0 \text{ almost surely in } (0, \infty) \times \partial\mathcal{O}.$$

3.3   Stability of Semilinear Stochastic Evolution Equations

In this section, we shall consider the following stochastic evolution equation: for any $0 \le t < \infty$,

$$\begin{cases} dX_t = T(t)x_0 + \int_0^t T(t-s)F(s, X_s)ds + \int_0^t T(t-s)G(s, X_s)dW_s, \\ X_0 = x_0 \in H, \end{cases}$$
(3.3.1)

where $T(t)$, $t \ge 0$, is some C_0-semigroup of bounded linear operators on H with its infinitesimal generator A. Also, W_t, $t \ge 0$, is some given K-valued, Q-Wiener process with $trQ < \infty$, and $F(t, \cdot)$ and $G(t, \cdot)$ are in general nonlinear measurable mappings from H to H and H to $\mathcal{L}(K, H)$, respectively, satisfying the following Lipschitz and linear growth conditions

$$\|F(t, y) - F(t, z)\|_H + \|G(t, y) - G(t, z)\| \le L\|y - z\|_H, \quad t \in [0, \infty),$$

$$\|F(t, y)\|_H + \|G(t, y)\| \le L(1 + \|y\|_H), \quad t \in [0, \infty), \tag{3.3.2}$$

for some constant $L > 0$ and arbitrary $y, z \in H$. Then by virtue of Theorem 1.3.4, there exists a unique solution X_t, $t \ge 0$, of (3.3.1) for arbitrary $x_0 \in H$. It is also known by Proposition 1.3.6 that one can always find a modification with continuous sample paths of the solution. We now consider the exponential decay of (3.3.1) by means of a Lyapunov function type of argument.

Theorem 3.3.1   *Let $\Lambda(x) : H \to \mathbf{R}^1$ satisfy that:*

(i). $\Lambda(x)$ is twice Fréchet differentiable and $\Lambda'(x)$, $\Lambda''(x)$ are continuous in H and $\mathcal{L}(H)$, respectively, and

$$|\Lambda(x)| + \|x\|_H \|\Lambda'(x)\|_H + \|x\|_H^2 \|\Lambda''(x)\| \le c\|x\|_H^p \atop \text{for some } p \ge 2 \text{ and } c > 0; \tag{3.3.3}$$

(ii). There exist constant $\alpha > 0$ and a nonnegative continuous function $\gamma(t)$, $t \in \mathbf{R}_+$, such that

$$(\mathbf{L}\Lambda)(t, x) \le -\alpha\Lambda(x) + \gamma(t), \quad x \in \mathcal{D}(A), \tag{3.3.4}$$

where

$$(\mathbf{L}\Lambda)(t, x) := \langle \Lambda'(x), Ax + F(t, x) \rangle_H + 1/2 \cdot tr(\Lambda''(x)G(t, x)QG^*(t, x)),$$
$$x \in \mathcal{D}(A), \quad t \ge 0,$$

and $\gamma(t)$ satisfies that there exists a constant $\mu > 0$ such that $\gamma(t)e^{\mu t}$ is integrable on $[0, \infty)$.

Then there exist constants $\tau > 0$, $C = C(x_0) > 0$ such that for the solution $X_t(x_0)$, $t \geq 0$, of (3.3.1),

$$E\Lambda(X_t(x_0)) \leq C(x_0) \cdot e^{-\tau t}, \qquad \forall t \geq 0. \tag{3.3.5}$$

In particular, τ can be taken as $\alpha \wedge \mu$.

To prove this theorem, we introduce the following approximating systems of (3.3.1) and use Proposition 1.3.6:

$$\begin{aligned} dX_t &= AX_t dt + R(n)F(t, X_t)dt + R(n)G(t, X_t)dW_t, \\ X_0 &= R(n)x_0 \in \mathcal{D}(A), \end{aligned} \tag{3.3.6}$$

where $0 < n_0 \leq n \in \rho(A)$ for some $n_0 \in \mathbf{N}$, the resolvent set of A and $R(n) = nR(n, A)$, $R(n, A)$ is the resolvent of A.

Proof Applying Itô's formula to the function $v(t, x) = e^{\mu t}\Lambda(x)$ and the strong solution X_t^n of (3.3.6) yields that for any $t \geq 0$,

$$\begin{aligned} e^{\mu t}&\Lambda(X_t^n) - \Lambda(X_0^n) \\ &= \mu \int_0^t e^{\mu s}\Lambda(X_s^n)ds + \int_0^t e^{\mu s}\langle \Lambda'(X_s^n), AX_s^n + R(n)F(s, X_s^n)\rangle_H ds \\ &\quad + \int_0^t e^{\mu s}\langle \Lambda'(X_s^n), R(n)G(s, X_s^n)dW_s\rangle_H \\ &\quad + \frac{1}{2} \cdot \int_0^t e^{\mu s}tr\Big(R(n)G(s, X_s^n)Q[R(n)G(s, X_s^n)]^*\Lambda''(X_s^n)\Big)ds. \end{aligned} \tag{3.3.7}$$

Therefore, by virtue of (3.3.3), (3.3.4) and taking expectations in (3.3.7), we can deduce

$$\begin{aligned} e^{\mu t}E\Lambda(X_t^n) \leq{}& \Lambda(X_0^n) + (\mu - \alpha)\int_0^t e^{\mu s}E\Lambda(X_s^n)ds + \int_0^t \gamma(s)e^{\mu s}ds \\ &+ \int_0^t e^{\mu s}E\Big\{\langle \Lambda'(X_s^n), (R(n) - I)F(s, X_s^n)\rangle_H \\ &+ \frac{1}{2}tr\Big[R(n)G(s, X_s^n)Q\big(R(n)G(s, X_s^n)\big)^*\Lambda''(X_s^n) \\ &- G(s, X_s^n)QG(s, X_s^n)^*\Lambda''(X_s^n)\Big]\Big\}ds. \end{aligned} \tag{3.3.8}$$

Consequently, letting $n \to \infty$ in (3.3.8) and using Proposition 1.3.6 immediately yields

$$e^{\mu t}E\Lambda(X_t) \leq \Lambda(x_0) + (\mu - \alpha)\int_0^t e^{\mu s}E\Lambda(X_s)ds + \int_0^t \gamma(s)e^{\mu s}ds. \tag{3.3.9}$$

Then, we can carry out a similar argument to that in the proof of Theorem 4.3.1 to obtain that there exist positive constants $\tau > 0$ and $C = C(x_0) > 0$

such that

$$E\Lambda(X_t) \leq C(x_0) \cdot e^{-\tau t}.$$

In particular, the constant τ can be taken as $\alpha \wedge \mu$. The proof is complete now.   □

As an immediate application of Theorem 3.3.1, we may consider stability of the following stochastic differential equation to obtain an autonomous version

$$dX_t = [AX_t + F(X_t)]dt + G(X_t)dW_t, \qquad X_0 = x_0 \in H, \qquad (3.3.10)$$

where $F(\cdot)$, $G(\cdot)$ both satisfy the corresponding Lipschitz continuous conditions as in (3.3.2) and $F(0) = 0$, $G(0) = 0$.

Corollary 3.3.1   *Let $\Lambda(x) : H \to \mathbf{R}^1$ satisfy the following:*

(i). $\Lambda(x)$ is twice Fréchet differentiable and $\Lambda'(x)$, $\Lambda''(x)$ are continuous in H and $\mathcal{L}(H)$, respectively, and

$$|\Lambda(x)| + \|x\|_H \|\Lambda'(x)\|_H + \|x\|_H^2 \|\Lambda''(x)\| \leq c\|x\|_H^p$$
$$\text{for some} \quad p \geq 2 \quad \text{and} \quad c > 0;$$

(ii). There exists a constant $\alpha > 0$ such that

$$(\mathbf{L}\Lambda)(x) \leq -\alpha\Lambda(x), \quad x \in \mathcal{D}(A),$$

where $(\mathbf{L}\Lambda)(x) := \langle \Lambda'(x), Ax + F(x) \rangle_H + 1/2 \cdot tr(\Lambda''(x)G(x)QG^*(x))$, $x \in \mathcal{D}(A)$.

Then the mild solution $X_t(x_0)$, $t \geq 0$, of (3.3.10) satisfies

$$E\Lambda(X_t(x_0)) \leq c\|x_0\|_H^p \cdot e^{-\alpha t}, \qquad \forall t \geq 0.$$

Moreover, if $k\Lambda(x) \geq \|x\|_H^p$ for some $k > 0$, then $E\|X_t(x_0)\|_H^p \leq k \cdot c\|x_0\|_H^p e^{-\alpha t}$.

Next, we will prove that under the same assumptions as in Theorem 3.3.1, the solution of (3.3.1) does decay exponentially in the almost sure sense. To this end, we first derive a lemma as follows.

Lemma 3.3.1   *Suppose that there exists a function $\Lambda(x)$ satisfying (i)–(ii) in Theorem 3.3.1 and $k\Lambda(x) \geq \|x\|_H^p$ for some $k > 0$. Let $X_t(x_0)$, $t \geq 0$, be the solution of (3.3.1) satisfying (3.3.2). Then for any $0 \leq t \leq T$,*

(a). $\Lambda(X_t) \leq C_1(x_0) + \int_0^t \langle \Lambda'(X_r), G(r, X_r)dW_r \rangle_H$;
(b). $E \sup_{0 \leq t \leq T} \Lambda(X_t) \leq C_2(x_0)$, where $C_1(x_0) > 0$, $C_2(x_0) > 0$ are independent of T.

Proof Note that $(\mathbf{L}\Lambda)(t,x) \le -\alpha\Lambda(x) + \gamma(t)$, $t \ge 0$, and $\int_0^\infty \gamma(t)dt < \infty$. Now we apply Itô's formula to the function $\Lambda(x)$ and the process X_t^n given by (3.3.6), then

$$
\begin{aligned}
\Lambda(X_t^n) \le{}& \Lambda(R(n)x_0) + \int_0^t \langle \Lambda'(X_r^n),\, R(n)F(r,X_r^n) - F(r,X_r^n)\rangle_H dr \\
&+ \int_0^\infty \gamma(r)dr + \frac{1}{2}\int_0^t \Big[tr(R(n)G(r,X_r^n)QG^*(r,X_r^n)R(n)^*\Lambda''(X_r^n)) \\
&\quad - tr(G(r,X_r^n)QG^*(r,X_r^n)\Lambda''(X_r^n))\Big] dr \\
&+ \int_0^t \langle \Lambda'(X_r^n),\, R(n)G(r,X_r^n)dW_r\rangle_H.
\end{aligned}
$$

In terms of (3.3.3), we can let n tend to infinity as in Proposition 1.3.6 and pass on the limit in the inequality above to obtain (a). To prove (b), note that

$$
\begin{aligned}
E\Big[\sup_{0\le t\le T} \Big| \int_0^t \langle \Lambda'(X_r),\, G(r,X_r)dW_r\rangle_H \Big|\Big] \\
\le E\Big[trQ \int_0^T \|\Lambda'(X_t)\|_H^2 \|G(t,X_t)\|^2 dt\Big]^{1/2} \\
\le \frac{1}{2}E \sup_{0\le t\le T} \Lambda(X_t) + c\int_0^T E\Lambda(X_t)dt
\end{aligned}
$$

for some $c > 0$. Thus from (a) and Theorem 3.3.1, it follows that

$$
E \sup_{0\le t\le T} \Lambda(X_t) \le 2C_1(x_0) + 2c\int_0^T C(x_0)e^{-\tau t}dt = C_2(x_0)
$$

for some $C_2(x_0) > 0$. $\qquad\square$

Theorem 3.3.2 *Under the same conditions as in Lemma 3.3.1, there exists a random variable $0 \le T(\omega) < \infty$ and a constant $M = M(x_0) > 0$ such that for all $t > T(\omega)$,*

$$
\|X_t(x_0)\|_H \le M(x_0) \cdot e^{-\tau t/4p} \qquad a.s.
$$

Proof We only sketch the proofs because they are quite similar to those in Theorem 3.2.2. By Lemma 3.3.1 and some obvious modifications, we can deduce for any $t \ge N$ (an arbitrary positive integer),

$$
\Lambda(X_t) \le \Lambda(X_N) + \int_N^t \gamma(s)ds + \int_N^t \langle \Lambda'(X_s),\, G(s,X_s)dW_s\rangle_H.
$$

By the Burkholder-Davis-Gundy inequality, we can deduce for any $\varepsilon_N > 0$,

$$
P\Big\{ \sup_{N\le t\le N+1} \Big| \int_N^t \langle \Lambda'(X_s),\, G(s,X_s)dW_s\rangle_H \Big| \ge \varepsilon_N/2 \Big\} \le (C_1(x_0)/\varepsilon_N)e^{-\tau N/2}
$$

for some $C_1(x_0) > 0$. Hence, a similar argument to Theorem 3.2.2 yields

$$P\left\{ \sup_{N \leq t \leq N+1} \Lambda(X_t) \geq \varepsilon_N \right\} \leq (C_2(x_0)/\varepsilon_N)e^{-\tau N/2}$$

for some $C_2(x_0) > 0$. If we take $\varepsilon_N = C_2(x_0)e^{-\tau N/4}$ and carry out a Borel-Cantelli lemma type of argument, it can be concluded that there exists a random variable $0 \leq T(\omega) < \infty$ such that if $t \geq T(\omega)$,

$$\|X_t(x_0)\|_H \leq M(x_0) \cdot e^{-\tau t/4p} \qquad a.s.$$

for some $M(x_0) > 0$. ▯

Example 3.3.1   Let us consider the following semilinear stochastic partial differential equation:

$$\begin{cases} dY_t(x) = \frac{\partial^2}{\partial x^2}Y_t(x)dt + e^{-\mu t}\alpha(Y_t(x))dB_t, & t > 0, \quad x \in (0, 1), \\ Y_0(x) = y_0(x), \quad 0 \leq x \leq 1; \quad Y_t(0) = Y_t(1) = 0, & t \geq 0, \end{cases} \quad (3.3.11)$$

where B_t, $t \geq 0$, is a real standard Wiener process (so, $K = \mathbf{R}^1$). $\alpha(\cdot) : \mathbf{R}^1 \to \mathbf{R}^1$ is some bounded, Lipschitz continuous function, $\alpha(0) = 0$, and μ a certain positive number. We can specialise our formulation to this example by taking $H = L^2[0, 1]$, and $A = \frac{\partial^2}{\partial x^2}$ with

$$\mathcal{D}(A) = \Big\{ y \in H : y, \ y'' \text{ are absolutely continuous with}$$
$$y', \ y'' \in H, \ y(0) = y(1) = 0 \Big\},$$

and $F(t, y) = 0$, $G(t, y) = e^{-\mu t}\alpha(y)$.

Clearly, the operator $G(t, \cdot)$ satisfies Condition (3.3.2). On the other hand, let $\Lambda(x) = \|x\|_H^2$, then it is easy to deduce that for arbitrary $y \in \mathcal{D}(A)$,

$$2\langle y, Ay + F(t, y) \rangle_H + \|G(t, y)\|^2 \leq -2\pi\|y\|_H^2 + Me^{-2\mu t}, \quad (3.3.12)$$

where M is some positive constant. Therefore, since the hypotheses in Theorems 3.3.1 and 3.3.2 are fulfilled, we may deduce that the null solution of the equation is exponentially stable in mean square, that is, there exist positive constants $\tau > 0$, $C > 0$ such that

$$E\|Y_t(y_0)\|_H^2 \leq C\|y_0\|_H^2 \cdot e^{-\tau t}, \qquad \forall t \geq 0,$$

and meanwhile it is also exponentially stable in the almost sure sense.

Example 3.3.2   Consider the semilinear stochastic heat equation

$$dY_t(x) = \left[\frac{\partial^2}{\partial x^2}Y_t(x) - \frac{Y_t(x)}{1 + |Y_t(x)|} \right]dt + \frac{\sigma Y_t(x)}{1 + |Y_t(x)|}dB_t, \quad t > 0, \quad x \in (0, 1),$$
$$Y_0(x) = y_0(x), \quad 0 \leq x \leq 1; \quad Y_t(0) = Y_t(1) = 0, \quad t \geq 0, \quad (3.3.13)$$

where $\sigma > 0$ is some constant. The space H and process B_t, $t \geq 0$, are defined exactly as in Example 3.3.1, $K = \mathbf{R}^1$,

$$F(y) = -\frac{G(y)}{\sigma} = -\frac{y}{1 + \|y\|_H},$$

and $A = \frac{\partial^2}{\partial x^2}$ with $\mathcal{D}(A) = \{y \in H : y, y'$ are absolutely continuous, $y', y'' \in H, y(0) = y(1) = 0\}$. Then $\langle y, Ay + F(y)\rangle_H \leq -\|y\|_H^2$, $y \in \mathcal{D}(A)$, and

$$\mathbf{L}\|y\|_H^p \leq -p\left[1 - \frac{1}{2}\sigma^2(p-1)\right]\|y\|_H^p.$$

Thus if $2 \leq p < 1 + 2/\sigma^2$, by virtue of Corollary 3.3.1 the null solution is exponentially stable in p-th moment. In this case, we also have exponential stability of sample paths with probability one.

There are some occasions where we only need a weaker notion of stability. In these situations, it is often possible to relax conditions such as those in Corollary 3.3.1 to ensure stability.

Theorem 3.3.3 *Suppose there exists a function $\Lambda(x)$, $x \in H$, with properties:*

i). *$|\Lambda(x)| + \|\Lambda'(x)\|_H + \|\Lambda''(x)\| \leq c(1 + \|x\|_H^p)$ for some $c > 0$ and $p > 0$;*
ii). *$\Lambda(0) = 0$, $\Lambda(x) > 0$ for $0 < \|x\|_H < R < \infty$, $R > 0$, and $b(r) = \inf_{\|x\|_H = r} \Lambda(x) > 0$ for any $r > 0$;*
iii). *$(\mathbf{L}\Lambda)(x) \leq u(x)$, for any $x \in \mathcal{D}(A)$, where $u(\cdot)$ is a continuous function such that $u(x) \leq 0$ for $0 \leq \|x\|_H \leq R$ and $|u(x)| \leq k(1 + \|x\|_H^q)$ for some $k > 0$ and $q > 0$.*

Then the null solution of (3.3.10) is stable in probability.

Proof Carrying out a similar argument to Lemma 3.3.1 and using (iii), we may deduce that

$$\Lambda(X_t(x_0)) \leq \Lambda(x_0) + \int_0^t u(X_s(x_0))ds + \int_0^t \langle\Lambda'(X_s(x_0)), G(X_s(x_0))dW_s\rangle_H.$$

Now let

$$0 < \|x_0\|_H < r < R \quad \text{and} \quad \tau = \tau(r) = \inf\{t > 0 : \|X_t(x_0)\|_H > r\}.$$

Then

$$\Lambda(X_{t\wedge\tau}(x_0)) \leq \Lambda(x_0) + \int_0^{t\wedge\tau} \langle\Lambda'(X_s(x_0)), G(X_s(x_0))dW_s\rangle_H.$$

Therefore,

$$EΛ(X_{t∧τ}(x_0)) ≤ Λ(x_0).$$

This yields

$$P\{τ < t\} \inf_{\|x\|_H = r} Λ(x) ≤ Λ(x_0).$$

But $b(r) > 0$ for any $r > 0$, and

$$P\{τ < t\} ≤ \frac{Λ(x_0)}{b(r)}.$$

Hence

$$P\{τ < ∞\} ≤ \frac{Λ(x_0)}{b(r)}.$$

Since $Λ(0) = 0$ and $Λ(x)$ is continuous, for each $ε > 0$ there exists a $δ = δ(r, ε) > 0$ such that $Λ(x_0)/b(r) < ε$ if $\|x_0\|_H < δ$. Thus

$$P\{\|X_t(x_0)\|_H > r \text{ for some } t ≥ 0\} ≤ P\{τ < ∞\} < ε \quad \text{if} \quad \|x_0\|_H < δ.$$

The proof is now complete. □

Corollary 3.3.2   *Suppose the conditions of Theorem 3.3.3 hold for $R = ∞$. Assume also $Λ(x) → ∞$ as $\|x\|_H → ∞$. Then $X_t(x_0)$ is bounded almost surely for any $x_0 ∈ H$, i.e.,*

$$P\left\{ \sup_{0≤t<∞} \|X_t(x_0)\|_H < ∞ \right\} = 1.$$

Proof   Note that

$$P\left\{ \sup_{0≤t<∞} \|X_t(x_0)\|_H < ∞ \right\} = P\left\{ \bigcap_{n=1}^{∞} \left(\sup_t \|X_t(x_0)\|_H > n \right) \right\}$$

$$≤ P\{τ(n) < ∞\} ≤ Λ(x_0)/b(n) → 0, \quad \text{as } n → ∞.$$

□

Under somewhat restricted conditions, we may also consider asymptotic almost sure stability of (3.3.10) described in the following manner.

Theorem 3.3.4   *Suppose there exists a function $Λ(x)$ with properties:*

i). $|Λ(x)| + \|Λ'(x)\|_H + \|Λ''(x)\| ≤ c(1 + \|x\|_H^p)$ for some $c > 0$ and $p > 0$;
ii). $Λ(0) = 0$, $Λ(x) > 0$ for $x ≠ 0$;
iii). $(\mathbf{L}Λ)(x) ≤ -αΛ(x)$ for any $x ∈ \mathcal{D}(A)$ and some constant $α > 0$;

iv). $b(r) = \inf_{\|x\|_H = r} \Lambda(x) > 0$ for any $r > 0$.

Then, the null solution of (3.3.10) is asymptotically stable in the almost sure sense.

Proof   Note that the stability in probability follows from Theorem 3.3.3. We can show as in Theorem 3.3.2 that for $0 < \tau < \alpha$,

$$\Lambda(X_t(x_0)) \le e^{-\tau t}\Lambda(x_0) \quad \text{for} \quad t > T_0(\omega)$$

almost surely for some random variable $0 \le T_0(\omega) < \infty$. Using iv) we conclude

$$X_t(x_0) \to 0 \quad \text{almost surely} \quad \text{as} \quad t \to \infty.$$

$\square$

Remark   The results derived in Theorem 3.3.3 of course are weaker than exponential stability. However, the restriction on the Lyapunov function here is more relaxed. For instance, we do not require its strict positive definiteness, a case which is useful in many practical situations such as Example 3.1.1.

Lastly, we take here the difference of two solutions of (3.3.10) with different initial conditions and consider their asymptotic behavior. The results below, whose proofs are fairly straightforward and therefore omitted here, will play an important role in the investigation of uniqueness of invariant measures later on.

Proposition 3.3.1   *Suppose that $\langle x, Ax \rangle_H \le -\alpha\|x\|_H^2$, $\alpha > 0$, for any $x \in \mathcal{D}(A)$,*

$$\|F(y) - F(z)\|_H \le L_1\|y - z\|_H, \quad \|G(y) - G(z)\| \le L_2\|y - z\|_H, \quad y, \ z \in H,$$

for two positive constants L_1, $L_2 > 0$, and let

$$\tau = 2\alpha - 2L_1 - L_2^2 trQ > 0.$$

Then for the mild solution of (3.3.10), we have that for some $C > 0$,

$$\|X_t(x_1) - X_t(x_2)\|_H^2 \le C\|x_1 - x_2\|_H^2 e^{-\tau t} \quad for \quad \forall t \ge T(\omega)$$

almost surely for some random variable $0 \le T(\omega) < \infty$.

We also define the following operator

$$(\mathbf{L}_d\Lambda)(y - z) = \langle \Lambda'(y - z), A(y - z) + F(y) - F(z) \rangle_H$$
$$+ (1/2)tr\Big([G(y) - G(z)]^*\Lambda''(y - z)[G(y) - G(z)]Q\Big), \quad y, \ z \in \mathcal{D}(A).$$

Proposition 3.3.2   *Suppose that there exists a function $\Lambda(x)$ with properties:*

    i). $|\Lambda(x)| + \|\Lambda'(x)\|_H + \|\Lambda''(x)\| \leq c(1 + \|x\|_H^p)$ *for some $c > 0$ and $p > 0$;*

    ii). $\Lambda(0) = 0$, $\Lambda(x) > 0$, $x \neq 0$;

    iii). $(\mathbf{L}_d\Lambda)(y - z) \leq -\alpha\Lambda(y - z)$ *for any y, $z \in \mathcal{D}(A)$ and some $\alpha > 0$;*

    iv). $\inf_{\|x\|_H \geq r} \Lambda(x) = b(r) > 0$ *for all $r > 0$.*

Then, for the mild solution $X_t(\cdot)$ of (3.3.10), we have $X_t(x_1) - X_t(x_2) \to 0$, as $t \to \infty$, almost surely.

3.4   Lyapunov Functions for Strong Solutions

In this section, using the same notions and notations as in Section 1.3, we intend to consider strong solutions of the following infinite-dimensional stochastic differential equation in V^*:

$$\begin{cases} X_t(\omega) \in M^2(t_0, T; V), \\ dX_t = A(t, X_t)dt + B(t, X_t)dW_t, \qquad t \in [t_0, T], \\ X_{t_0} = x_0 \in H, \end{cases} \tag{3.4.1}$$

where $T \geq t_0 \geq 0$ and $M^2(t_0, T; V)$ denotes the space of all V-valued, $\mathcal{F}_t$-adapted processes $X_t(\cdot)$ satisfying $E \int_{t_0}^{T} \|X_t(\omega)\|_V^2 dt < \infty$. The main objective is to provide a necessary and sufficient condition for exponential decay of strong solutions in the mean square sense in terms of the existence of Lyapunov functions.

It was shown in Proposition 2.1.2 that the usefulness of Lyapunov's theorem in linear ordinary differential equations is that it allows for an explicit representation of a Lyapunov function as a positive definite quadratic form. Using this representation one may then, for instance, study the effects of perturbations about stability on linear ordinary differential equations (cf. Section 3.1). It is essential, as we mentioned in Section 2.1, that for $\langle Px, x \rangle_H = \Lambda(x) = \int_0^\infty \|T(t)x\|_H^2 dt$ to be a Lyapunov function, we have an estimate of the form $\langle Px, x \rangle_H \geq c\|x\|_H^2$ for some $c > 0$. In other words, the form P has to define an equivalent norm on H. If H is finite dimensional, this is always the case. However, if H is infinite dimensional, this could be false as shown by the example at the beginning of Section 3.1. One of the methods of overcoming this difficulty is to find conditions to ensure L^2-stability and then use Theorem 3.1.1 to obtain exponential stability as shown in Section 3.1. However, it is still possible to construct Lyapunov functions with the desired norm equivalence by a different method. This method provides sufficient and necessary conditions, thus is quite powerful in dealing with nonlinear systems by means of the so-called first order linear approximation procedure.

In this and next sections, using a different method from the above we shall establish Lyapunov functions for strong and mild solutions. To this end, let us first study the following linear partial differential equation which motivates the construction of a desirable Lyapunov function,

$$\frac{\partial Y_t(x)}{\partial t} = a^2 \frac{\partial^2 Y_t(x)}{\partial x^2} + b\frac{\partial Y_t(x)}{\partial x} + cY_t(x) \qquad (3.4.2)$$

where a, b and c are real numbers, with the initial conditions

$$Y_0(x) = y_0(x), \qquad y_0(\cdot) \in L^2(-\infty, \infty) \cap L^1(-\infty, \infty).$$

Let $\tilde{y}_0(\lambda)$ be the Fourier transform of $y_0(x)$ and $\tilde{Y}_t(\lambda)$ of the solution $Y_t^{y_0}(x)$. Then (3.4.2) has the form, for each λ,

$$\begin{aligned}
\frac{d\tilde{Y}_t(\lambda)}{dt} &= -a^2\lambda^2\tilde{Y}_t(\lambda) + (ib\lambda + c)\tilde{Y}_t(\lambda) \\
&= \tilde{Y}_t(\lambda)(c + ib\lambda - a^2\lambda^2).
\end{aligned}$$

By the well-known Plancherel theorem, we get

$$\|Y_t^{y_0}(\cdot)\|_H^2 = \|\tilde{Y}_t(\cdot)\|_H^2$$

with $H = L^2(-\infty, \infty)$, which immediately yields

$$\|Y_t^{y_0}(\cdot)\|_H^2 = \int_{-\infty}^{\infty} \|\tilde{y}_0(\lambda)\|_H^2 \exp\{(-2a^2\lambda^2 + 2c)t\}d\lambda.$$

Assume $c < 0$, then $\|Y_t^{y_0}(\cdot)\|_H^2 \le \|y_0\|_H^2 \cdot e^{2ct}$. If we further define the Lyapunov function as we usually do in finite dimensional cases in the following manner,

$$\begin{aligned}
\Lambda(y_0) &:= \int_0^{\infty} \|Y_t^{y_0}(\cdot)\|_H^2 dt = \int_0^{\infty}\int_{-\infty}^{\infty} \|\tilde{y}_0(\lambda)\|_H^2 \exp\{(-2a^2\lambda^2 + 2c)t\}d\lambda dt \\
&= \int_{-\infty}^{\infty} \frac{\|\tilde{y}_0(\lambda)\|_H^2}{2a^2\lambda^2 - 2c}d\lambda,
\end{aligned}$$

then $\Lambda(\cdot)$ does not satisfy the condition $\Lambda(y) \ge d\|y\|_H^2$, $y \in H$, for some number $d > 0$. However, let $V \hookrightarrow L^2(-\infty, \infty)$ be the usual Sobolev spaces such that the injection $\hookrightarrow$ is continuously embedded, then if $a \ne 0$, we may get $t \to \|Y_t^{y_0}(\cdot)\|_V^2$ is continuous and in this case

$$\Lambda(y_0) = \int_0^{\infty} \|Y_t^{y_0}(\cdot)\|_V^2 dt \ge d'\|y_0\|_H^2 \quad \text{for some} \quad d' > 0.$$

Therefore, this example reminds us that in infinite dimensional spaces, we might need $\|\cdot\|_V^2$ in the definition of the Lyapunov function $\Lambda(\cdot)$.

In this section, in addition to those in Subsection 1.3.1 ensuring existence and uniqueness of strong solutions, we shall impose the following coercive condition, which is compatible with (1.3.2), for our stability purpose, i.e., we shall assume that there exist constants $\alpha > 0$, $\gamma \geq 0$, $\mu \geq 0$ and $\lambda \in \mathbf{R}^1$ such that:

$$2\langle u, A(t,u)\rangle_{V,V^*} + \|B(t,u)\|^2_{\mathcal{L}^0_2} \leq -\alpha\|u\|^2_V + \lambda\|u\|^2_H + \gamma \cdot e^{-\mu t}, \quad \forall u \in V, \ t \geq 0.$$
$$(3.4.3)$$

In order to obtain our main results, we need to use the following fundamental result.

Lemma 3.4.1   *Assume function $g(\cdot) \geq 0$ almost surely, and $g(\cdot) \in L^1[0,T]$ for arbitrary $T \geq 0$, then*

$$\lim_{\Delta \to 0} \int_0^T \frac{\int_t^{t+\Delta} g(s)ds}{\Delta} dt = \int_0^T g(t)dt.$$

Proof   By virtue of Fubini theorem, it is easy to deduce

$$\int_0^T \frac{\int_t^{t+\Delta} g(s)ds}{\Delta} dt = \frac{1}{\Delta} \int_0^T \left(\int_t^{t+\Delta} g(s)ds \right) dt$$

$$= \frac{1}{\Delta} \Big[\int_0^\Delta \Big(\int_0^s g(s)dt \Big) ds + \int_\Delta^T \Big(\int_{s-\Delta}^T g(s)dt \Big) ds$$

$$+ \int_T^{T+\Delta} \Big(\int_{s-\Delta}^T g(s)dt \Big) ds \Big]$$

$$= \frac{1}{\Delta} \Big[\int_0^\Delta sg(s)ds + \int_\Delta^T g(s)\Delta ds + \int_T^{T+\Delta} g(s)(T+\Delta - s)ds \Big]$$

$$\leq \frac{1}{\Delta} \Big[\Delta \int_0^\Delta g(s)ds + \Delta \int_\Delta^T g(s)ds + \Delta \int_T^{T+\Delta} g(s)ds \Big]$$

$$= \int_0^\Delta g(s)ds + \int_\Delta^T g(s)ds + \int_T^{T+\Delta} g(s)ds.$$

The first and third terms go to zero as $\Delta \to 0$, so

$$\lim_{\Delta \to 0} \int_0^T \frac{\int_t^{t+\Delta} g(s)ds}{\Delta} dt \leq \int_0^T g(t)dt.$$

The other direction of the inequality follows easily from Fatou's lemma. This proves the lemma.   □

Theorem 3.4.1   *Suppose X_t, $t \geq t_0 \geq 0$, is the strong solution of (3.4.1) and assume the coercive condition (3.4.3) holds. If there exists a function $\Lambda(\cdot, \cdot) : \mathbf{R}_+ \times H \to \mathbf{R}^1$ which satisfies the following:*

(a). $\Lambda(\cdot,\cdot)$ satisfies all the conditions in Remark at the end of Subsection 1.3.1;

(b). $c_1\|x\|_H^2 - k_1 e^{-\mu_1 t} \leq \Lambda(t,x) \leq c_2\|x\|_H^2 + k_2 e^{-\mu_2 t}, \quad \forall x \in V, \quad t \geq 0;$

(c). $(\mathbf{L}\Lambda)(t,x) \leq -c_3\Lambda(t,x) + k_3 e^{-\mu_3 t}, \quad \forall x \in V, \quad t \geq 0,$

where $c_i > 0$, $k_i \geq 0$, $\mu_i \geq 0$, $i = 1, 2, 3$, are some constants and $\mathbf{L}$ is the infinitesimal generator of the Markov process X_t, i.e., $(\mathbf{L}\Lambda)(t,x) := \Lambda'_t(t,x) + \langle \Lambda'_x(t,x), A(t,x) \rangle_{V,V^*} + 1/2 \cdot tr(\Lambda''_{xx}(t,x)B(t,x)QB(t,x)^*), x \in V, t \geq 0$, then X_t, $t \geq t_0 \geq 0$, satisfies

$$E\|X_t(x_0)\|_H^2 \leq \alpha_1\|x_0\|_H^2 \cdot e^{-\beta_1(t-t_0)} + \alpha_2 \cdot e^{-\beta_2 t}, \quad x_0 \in H, \quad t \geq t_0, \quad X_{t_0} = x_0,$$
$$(3.4.4)$$

where $\alpha_1 > 0$, $\alpha_2 \geq 0$, $\beta_1 > 0$ and $\beta_2 \geq 0$ are constants.

Conversely, suppose (3.4.4) holds and define

$$\Lambda(t_0, x_0) = \int_{t_0}^{T+t_0} \left(\int_{t_0}^u E\|X_s(x_0)\|_V^2 ds \right) du \quad \text{for all} \quad x_0 \in H, \quad t_0 \geq 0,$$
$$(3.4.5)$$

where $X_{t_0} = x_0$ and T is some proper positive constant and assume $\Lambda(\cdot,\cdot)$ satisfies all the conditions in Remark at the end of Subsection 1.3.1, then there exist constants $c_i > 0$, $k_i \geq 0$, $\mu_i \geq 0$, $i = 1, 2, 3$, such that Conditions (b) and (c) above hold.

Proof   Firstly, suppose the conditions (a), (b) and (c) hold. A direct application of Itô's formula to the strong solution $X_t(x_0)$, $t \geq t_0 \geq 0$, and the nonlinear functional $\Lambda(\cdot,\cdot)$ and then taking expectations immediately yield ($\forall t \geq s \geq t_0$)

$$E\Lambda(t, X_t(x_0)) - E\Lambda(s, X_s(x_0)) = E\int_s^t (\mathbf{L}\Lambda)(u, X_u(x_0))du$$

$$\leq \int_s^t \left[-c_3 E\Lambda(u, X_u(x_0)) + k_3 \cdot e^{-\mu_3 u} \right] du.$$

Let $\phi(t) = E\Lambda(t, X_t(x_0))$ and note that $\phi(t)$ is continuous in t, we get

$$\frac{d\phi(t)}{dt} \leq -c_3\phi(t) + k_3 \cdot e^{-\mu_3 t}$$

which immediately yields that

$$\phi(t) \leq \phi(t_0)e^{c_3 t_0} \cdot e^{-c_3 t} + k_3 e^{-c_3 t} \int_{t_0}^t e^{(c_3-\mu_3)u}du. \tag{3.4.6}$$

If $c_3 = \mu_3$, it is easy to deduce from (3.4.6) and (b) that

$$\phi(t) \leq c_2\|x_0\|_H^2 \cdot e^{-c_3(t-t_0)} + k_2 e^{-(\mu_2-c_3)t_0 - c_3 t} + k_3 \cdot (t-t_0)e^{-c_3 t}$$

$$\leq c_2\|x_0\|_H^2 \cdot e^{-c_3(t-t_0)} + (k_2 + 2k_3/c_3) \cdot e^{-(\frac{c_3}{2} \wedge \mu_2)t}.$$

On the other hand, if $c_3 \neq \mu_3$, a straightforward computation yields from (3.4.6) that

$$\phi(t) \leq \left[\phi(t_0)e^{-c_3 t_0} + \frac{k_3 \cdot e^{-(\mu_3 - c_3)t_0}}{|c_3 - \mu_3|}\right]e^{-c_3 t} + \frac{k_3 e^{-\mu_3 t}}{|c_3 - \mu_3|}$$

$$\leq c_2 \|x_0\|_H^2 \cdot e^{-c_3(t-t_0)} + \left(k_2 + \frac{2k_3}{|c_3 - \mu_3|}\right)e^{-(\mu_2 \wedge \mu_3 \wedge c_3)t}.$$

Combining the above results, we can derive that there exist constants $\tilde{c} > 0$, $\tilde{M} \geq 0$, $\tilde{\beta}_1 > 0$ and $\tilde{\beta}_2 \geq 0$ such that

$$\phi(t) \leq \tilde{c}\|x_0\|_H^2 e^{-\tilde{\beta}_1(t-t_0)} + \tilde{M} \cdot e^{-\tilde{\beta}_2(t-t_0)},$$

which, together with (b) in Theorem 3.4.1, immediately implies that

$$E\|X_t(x_0)\|_H^2 \leq \frac{1}{c_1}\phi(t) + \frac{k_1}{c_1}e^{-\mu_1 t} \leq \alpha_1\|x_0\|_H^2 e^{-\beta_1(t-t_0)} + \alpha_2 \cdot e^{-\beta_2 t}$$

where $\alpha_1 = \tilde{c} = c_2/c_1 > 0$, $\alpha_2 = \tilde{M} + k_1/c_1 \geq 0$, $\beta_1 = \tilde{\beta}_1 = c_3 > 0$ and $\beta_2 = \tilde{\beta}_2 \wedge \mu_1 \geq 0$ are four constants. The first part is proved.

Now suppose (3.4.4) holds. Define

$$\Lambda(t_0, x_0) = \int_{t_0}^{T+t_0}\left(\int_{t_0}^{u} E\|X_s(x_0)\|_V^2 ds\right)du, \quad x_0 \in H, \quad t \geq t_0, \quad (3.4.7)$$

where T is some positive constant to be determined later. Applying Itô's formula to $\|X_t(x_0)\|_H^2$, $t \geq t_0$, and using the coercive condition (3.4.3). For the sake of simplicity, suppose at present $\mu > 0$, $\beta_2 > 0$ (the results in other situations may be proved similarly), we then obtain for arbitrary $u \geq t_0$,

$$E\|X_u(x_0)\|_H^2 - \|x_0\|_H^2$$
$$= \int_{t_0}^{u} E\mathbf{L}\|X_s(x_0)\|_H^2 ds$$
$$\leq \lambda \int_{t_0}^{u} E\|X_s(x_0)\|_H^2 ds - \alpha \int_{t_0}^{u} E\|X_s(x_0)\|_V^2 ds + \gamma \int_{t_0}^{u} e^{-\mu s} ds,$$

which, in addition to (3.4.4), immediately implies that

$$\int_{t_0}^{u} E\|X_s(x_0)\|_V^2 ds$$
$$\leq \frac{1}{\alpha}\left(\|x_0\|_H^2 + \lambda \int_{t_0}^{u} E\|X_s(x_0)\|_H^2 ds + \gamma \int_{t_0}^{u} e^{-\mu s} ds\right)$$
$$\leq \frac{1}{\alpha}\left(\|x_0\|_H^2 + \alpha_1|\lambda|\|x_0\|_H^2 \int_{t_0}^{u} e^{-\beta_1(s-t_0)} ds + |\lambda|\alpha_2 \int_{t_0}^{u} e^{-\beta_2 s} ds \quad (3.4.8)\right.$$
$$\left. + \gamma \int_{t_0}^{u} e^{-\mu s} ds\right)$$
$$\leq \frac{1}{\alpha}\left(\|x_0\|_H^2 + \frac{\alpha_1|\lambda|\|x_0\|_H^2}{\beta_1} + \frac{|\lambda|\alpha_2}{\beta_2}e^{-\beta_2 t_0} + \frac{\gamma}{\mu}e^{-\mu t_0}\right).$$

Therefore, by virtue of (3.4.7) it follows that

$$\Lambda(t_0, x_0) \leq \frac{1}{\alpha}\left[\left(T + \frac{\alpha_1 T |\lambda|}{\beta_1}\right)\|x_0\|_H^2 + \left(\frac{|\lambda|\alpha_2 T}{\beta_2} + \frac{\gamma T}{\mu}\right) \cdot e^{-(\beta_2 \wedge \mu)t_0}\right]. \quad (3.4.9)$$

On the other hand, for arbitrary $v \in V$, $t \geq t_0$,

$$(\mathbf{L}\|v\|_H^2)(t, v) = 2\langle v, A(t, v)\rangle_{V,V^*} + tr(B(t, v)QB(t, v)^*)$$

which, together with (1.3.3), (1.3.5) and (3.4.3), immediately implies that

$$
\begin{aligned}
|(\mathbf{L}\|v\|_H^2)(t, v)| &\leq (4\beta^2 + 4c^2)\|v\|_V^2 + \|B(t, v)\|_{\mathcal{L}_2^0}^2 \\
&\leq (4\beta^2 + 4c^2)\|v\|_V^2 + 2\|B(t, v) - B(t, 0)\|_{\mathcal{L}_2^0}^2 + 2\|B(t, 0)\|_{\mathcal{L}_2^0}^2 \\
&\leq (4\beta^2 + 4c^2)\|v\|_V^2 + 2L^2\|v\|_V^2 + 2\gamma e^{-\mu t} \\
&= \theta\|v\|_V^2 + 2\gamma e^{-\mu t}
\end{aligned}
$$

where $\theta = 4\beta^2 + 4c^2 + 2L^2 > 0$. Hence

$$(\mathbf{L}\|v\|_H^2)(t, v) \geq -\theta\|v\|_V^2 - 2\gamma e^{-\mu t}.$$

Applying now Itô's formula to $\|X_t(x_0)\|_V^2$, $t \geq t_0$, and taking expectations, we get that for all $u \geq t_0$,

$$
\begin{aligned}
E\|X_u(x_0)\|_H^2 - \|x_0\|_H^2 &= \int_{t_0}^u E(\mathbf{L}\|X_s(x_0)\|_H^2)(s, X_s(x_0))ds \\
&\geq -\theta \int_{t_0}^u E\|X_s(x_0)\|_V^2 ds - 2\gamma \int_{t_0}^u e^{-\mu s}ds,
\end{aligned}
$$

which, together with (3.4.4), immediately implies that for arbitrary $u \geq t_0$,

$$
\begin{aligned}
\theta \int_{t_0}^u E\|X_s(x_0)\|_V^2 ds &\geq \|x_0\|_H^2 - E\|X_u(x_0)\|_H^2 - 2\gamma \int_{t_0}^u e^{-\mu s}ds \\
&\geq \left(1 - \alpha_1 \cdot e^{-\beta_1 u + \beta_1 t_0}\right)\|x_0\|_H^2 - \alpha_2 \cdot e^{-\beta_2 u} - \frac{2\gamma}{\mu}e^{-\mu t_0}.
\end{aligned}
$$

Therefore, we get

$$
\begin{aligned}
\Lambda(t_0, x_0) &= \int_{t_0}^{T+t_0}\left(\int_{t_0}^u E\|X_s(x_0)\|_V^2 ds\right)du \\
&\geq \frac{1}{\theta}\int_{t_0}^{T+t_0}\left[\left(1 - \alpha_1 \cdot e^{-\beta_1 u + \beta_1 t_0}\right)\|x_0\|_H^2 - \alpha_2 \cdot e^{-\beta_2 u} - \frac{2\gamma}{\mu}e^{-\mu t_0}\right]du \\
&\geq \frac{1}{\theta}\left(T - \frac{\alpha_1}{\beta_1}\right)\|x_0\|_H^2 - \frac{1}{\theta}\left(\frac{\alpha_2}{\beta_2} + \frac{2\gamma T}{\mu}\right)e^{-(\beta_2 \wedge \mu)t_0}. \quad (3.4.10)
\end{aligned}
$$

On the other hand, note that

$$E\Lambda(t, X_t(x_0)) = E \int_{t_0}^{T+t_0} \left[\int_{t_0}^u E\Big(\|X_s(X_t(x_0))\|_V^2 \ \Big| \ X_t(x_0)\Big) ds \right] du.$$

However, by the Markov property of the solution of (3.4.1), this equals

$$\int_{t_0}^{T+t_0} \int_{t_0}^u E\Big\{ E\Big(\|X_s(X_t(x_0))\|_V^2 \ \Big| \ \mathcal{F}_t^X\Big)\Big\} ds du.$$

where $\mathcal{F}_t^X = \sigma\{X_u(x_0), t_0 \le u \le t\}$. The uniqueness of the solution implies

$$E\Big(\|X_s(X_t(x_0))\|_V^2 \ \Big| \ \mathcal{F}_t^X\Big) = E\Big(\|X_{s+t}(x_0)\|_V^2 \ \Big| \ \mathcal{F}_t^X\Big). \qquad (3.4.11)$$

Therefore, we have

$$(\mathbf{L}\Lambda)(t_0, x_0) = \lim_{r\to 0} \frac{E\Lambda(t_0 + r, X_{t_0+r}(x_0)) - E\Lambda(t_0, x_0)}{r}$$

$$= \lim_{r\to 0} \frac{\int_{t_0+r}^{T+t_0+r} \int_{t_0+r}^u E\|X_s(X_{t_0+r}(x_0))\|_V^2 ds du - \int_{t_0}^{T+t_0} \int_{t_0}^u E\|X_s(x_0)\|_V^2 ds du}{r}$$

$$= \lim_{r\to 0} \frac{\int_{t_0}^{T+t_0} \int_{t_0}^u E\|X_{s+r}(X_{t_0+r}(x_0))\|_V^2 ds du - \int_{t_0}^{T+t_0} \int_{t_0}^u E\|X_s(x_0)\|_V^2 ds du}{r}$$

$$= \lim_{r\to 0} \frac{\int_{t_0}^{T+t_0} \int_0^{u-t_0} E\|X_{s+r+t_0}(x_0)\|_V^2 ds du - \int_{t_0}^{T+t_0} \int_{t_0}^u E\|X_s(x_0)\|_V^2 ds du}{r}$$

$$= \lim_{r\to 0} \int_{t_0}^{T+t_0} \frac{\int_r^{u-t_0+r} E\|X_{s+t_0}(x_0)\|_V^2 ds - \int_{t_0}^u E\|X_s(x_0)\|_V^2 ds}{r} du$$

$$= \lim_{r\to 0} \int_{t_0}^{T+t_0} \frac{\int_{u-t_0}^{u-t_0+r} E\|X_{s+t_0}(x_0)\|_V^2 ds - \int_0^r E\|X_{s+t_0}(x_0)\|_V^2 ds}{r} du$$

which, together with $\|x\|_H \le \beta\|x\|_V$, $\beta > 0$, for any $x \in V$, Lemma 3.4.1 and the continuity of $t \to E\|X_t(x_0)\|_H^2$, immediately implies

$$(\mathbf{L}\Lambda)(t_0, x_0) \le \int_{t_0}^{T+t_0} E\|X_u(x_0)\|_V^2 du - \frac{T}{\beta^2}\|x_0\|_H^2.$$

Now, substituting (3.4.8) into the above yields

$$(\mathbf{L}\Lambda)(t_0, x_0) \le \frac{1}{\alpha}\Big(1 + \frac{\alpha_1|\lambda|}{\beta_1}\Big)\|x_0\|_H^2 + \Big(\frac{|\lambda|\alpha_2}{\beta_2} + \frac{\gamma}{\mu}\Big) e^{-(\beta_2\wedge\mu)t_0} - \frac{T}{\beta^2}\|x_0\|_H^2.$$

Hence, noting (3.4.10) and choosing $T > 0$ large enough so that

$$T > \frac{\beta^2}{\alpha}\Big(1 + \frac{\alpha_1|\lambda|}{\beta_1}\Big) \vee \frac{\alpha_1}{\beta_1},$$

we are then in a position to obtain the desired results (b) and (c) in Theorem 3.4.1. If $\mu = 0$ or $\beta_2 = 0$, by a similar argument to the above, it can be also proved that Conditions (b) and (c) remain true. The proof is now complete.
$\square$

This result is powerful in studying stability and invariant measures of solutions by means of a first-order approximation argument (see Section 3.5 below). Before proceeding to this material, we first derive some corollaries which will play an important role in the subsequent investigation.

Consider the following nonlinear autonomous stochastic differential equation in V^*:

$$\begin{cases} dX_t = A(X_t)dt + B(X_t)dW_t, & t \in [0, \infty), \\ X_0 = x_0. \end{cases} \tag{3.4.12}$$

Suppose $A(\cdot) : V \to V^*$ and $B(\cdot) : V \to \mathcal{L}(K, H)$ satisfy

$$\|A(u)\|_{V^*} \le a_1 \|u\|_V \quad \text{and} \quad \|B(u)\| \le b_1 \|u\|_V \quad \text{for all} \quad u \in V, \tag{3.4.13}$$

for some constants $a_1, b_1 > 0$, with $A(0) = 0$ and $B(0) = 0$. Furthermore, assume the following coercive condition holds: there exist $\alpha > 0$ and $\lambda \in \mathbf{R}^1$ such that

$$2\langle u, A(u)\rangle_{V,V^*} + \|B(u)\|_{\mathcal{L}_2^0}^2 \le -\alpha\|u\|_V^2 + \lambda\|u\|_H^2, \quad \forall u \in V. \tag{3.4.14}$$

The following result is immediate from Theorem 3.4.1:

Corollary 3.4.1   *Assume the coercive condition (3.4.14) holds. Suppose X_t, $t \ge 0$, is the strong solution of (3.4.12). If there exists a function $\Lambda : H \to \mathbf{R}^1$ which satisfies the following:*

(a). $\Lambda(\cdot)$ satisfies all the conditions in Remark at the end of Subsection 1.3.1;
(b). $c_1\|x\|_H^2 \le \Lambda(x) \le c_2\|x\|_H^2, \quad \forall x \in V$;
(c). $(\mathbf{L}\Lambda)(x) \le -c_3\Lambda(x), \quad \forall x \in V$,

where $c_i > 0$, $i = 1, 2, 3$, are some constants and $\mathbf{L}$ is the infinitesimal generator of the Markov process X_t, i.e., $(\mathbf{L}\Lambda)(x) := \langle \Lambda'(x), A(x)\rangle_{V,V^*} + 1/2 \cdot \mathrm{tr}(\Lambda''(x)B(x)QB(x)^*)$, $x \in V$, then $X_t(x_0)$, $t \ge 0$, satisfies that

$$E\|X_t(x_0)\|_H^2 \le \alpha_1\|x_0\|_H^2 \cdot e^{-\beta_1 t}, \quad x_0 \in H, \quad t \ge 0, \quad X_0 = x_0, \tag{3.4.15}$$

for some $\alpha_1 > 0$ and $\beta_1 > 0$.
Conversely, suppose (3.4.15) holds and define

$$\Lambda(x_0) = \int_0^T \left(\int_0^u E\|X_s(x_0)\|_V^2 ds \right) du \quad \text{for all} \quad x_0 \in H, \quad t \ge 0, \tag{3.4.16}$$

where $X_0 = x_0$ and T is some proper positive constant and assume $\Lambda(\cdot)$ satisfies all the conditions in Remark at the end of Subsection 1.3.1, then there exist constants $c_i > 0$, $i = 1, 2, 3$, such that Conditions (b) and (c) above hold.

Generally speaking, it is not quite obvious whether $\Lambda(\cdot, \cdot)$ defined in (3.4.5) satisfies the conditions of using Itô's formula in Subsection 1.3.1. However, this is always the case if V is a real separable Hilbert space with inner product $\langle \cdot, \cdot \rangle_V$ and Equation (3.4.12) is linear. Precisely, consider the linear system given by

$$dX_t = AX_t dt + BX_t dW_t \quad \text{with} \quad X_0 = x_0, \qquad (3.4.17)$$

where $A \in \mathcal{L}(V, V^*)$, $B \in \mathcal{L}(V, \mathcal{L}(K, H))$ satisfy the coercive condition: there exist $\alpha > 0$ and $\lambda \in \mathbf{R}^1$ such that

$$2\langle u, Au \rangle_{V,V^*} + tr[(Bu)Q(Bu)^*] \leq -\alpha \|u\|_V^2 + \lambda \|u\|_H^2 \qquad (3.4.18)$$

for all $u \in V$. In particular, we have:

Corollary 3.4.2   *Assume the coercive condition (3.4.18) holds. Suppose X_t, $t \geq 0$, is the strong solution of (3.4.17). If there exists a function $\Lambda : H \to \mathbf{R}^1$ which satisfies the following:*

(a). $\Lambda(\cdot)$ satisfies all the conditions in Remark at the end of Subsection 1.3.1;
(b). $c_1\|x\|_H^2 \leq \Lambda(x) \leq c_2\|x\|_H^2$, $\forall x \in V$;
(c). $(\mathbf{L}\Lambda)(x) \leq -c_3\Lambda(x)$, $\forall x \in V$,

where $c_i > 0$, $i = 1, 2, 3$, are some constants and $\mathbf{L}$ is the infinitesimal generator of the Markov process X_t, i.e., $(\mathbf{L}\Lambda)(x) = \langle \Lambda'(x), Ax \rangle_{V,V^*} + 1/2 \cdot tr(\Lambda''(x)(Bx)Q(Bx)^*)$, $x \in V$, then $X_t(x_0)$, $t \geq 0$, satisfies

$$E\|X_t(x_0)\|_H^2 \leq \alpha_1 \|x_0\|_H^2 \cdot e^{-\beta_1 t}, \quad x_0 \in H, \quad t \geq 0, \quad X_0 = x_0, \qquad (3.4.19)$$

for some $\alpha_1 > 0$ and $\beta_1 > 0$.
    Conversely, suppose (3.4.19) holds and define

$$\Lambda(x_0) = \int_0^\infty E\|X_s(x_0)\|_V^2 ds \quad \text{for all} \quad x_0 \in H, \quad t \geq 0, \qquad (3.4.20)$$

where $X_0 = x_0$, then there exist constants $c_i > 0$, $i = 1, 2, 3$, such that $\Lambda(\cdot)$ defined above satisfies the conditions (a), (b) and (c). If, in addition, we assume $t \to E\|X_t(x_0)\|_V^2$ is continuous at zero for any $x_0 \in V$, then $(\mathbf{L}\Lambda)(x_0) = -\|x_0\|_V^2$, $x_0 \in V$.

Proof   All we need to prove is the second part. By a similar argument to Theorem 3.4.1, it is immediate that $\Lambda(\cdot) = \int_0^\infty E\|X_s(\cdot)\|_V^2 ds$ satisfies all the conditions in Corollary 3.4.2 except for Condition (a). To verify (a), note that $X_t(x)$ is linear in x and hence

$$T(x,y) := \int_0^\infty E\langle X_t(x), X_t(y)\rangle_V dt, \quad x,\ y \in V,$$

is a bilinear form on $V \times V$. By a straightforward computation, we can obtain from (3.4.18) and (3.4.19) that

$$\int_0^\infty E\|X_t(x)\|_V^2 dt \le M\|x\|_H^2,$$

for some constant $M > 0$. Using this and Schwartz inequality, we can get $|T(x,y)| \le M\|x\|_H\|y\|_H$ for all $x,\ y \in V$. Considering the fact that V is densely embedded into H, there exists a unique continuous extension $\tilde{T}$ of T to $H \times H$. Hence, there exists a continuous linear operator $\tilde{P}$ on $H \to H$ such that $\tilde{T}(x,y) = (\tilde{P}x,y)$. Now $\Lambda'(x) = 2\tilde{P}x$, $\Lambda''(x) = 2\tilde{P}$. Hence, $\Lambda, \Lambda', \Lambda''$ are locally bounded on H and Λ, Λ' are continuous on H as

$$|\Lambda(x)| \le \|\tilde{P}\|\|x\|_H^2$$

and

$$\|\Lambda'(x)\|_H = 2\|\tilde{P}x\|_H \le 2\|\tilde{P}\|\|x\|_H, \quad \forall x \in H.$$

For trace class Q, $Q\Lambda''(x) = 2Q\tilde{P}$; also, $tr(Q\tilde{P})$ being constant, it is continuous. To prove Condition (iv) of the Remark at the end of Subsection 1.3.1, we observe that $T(x,y)$ is bilinear on $V \times V$; the fact that $|T(x,y)| \le M\|x\|_H\|y\|_H$ and the continuity of the injection $V \to H$ imply that $T(x,y)$ is a continuous bilinear form on $V \times V$. It follows that there exists a continuous operator P on V such that $T(x,y) = \langle Px, y\rangle_V$ for $x,\ y \in V$. Hence $\Lambda'(x) = 2Px \in V$ for $x \in V$ and $x \to Px$ is continuous on $V \to V$. Since $\|\Lambda'(x)\|_V = 2\|Px\|_V \le 2\|P\|_{\mathcal{L}(V)}(1 + \|x\|_V)$ for all $x \in V$, we get the desired result.

Lastly, to conclude our proof, we notice that by (3.4.11) it follows that for any $x_0 \in V$,

$$\begin{aligned} E\Lambda(X_t(x_0)) - \Lambda(x_0) &= \int_0^\infty E\|X_{t+s}(x_0)\|_V^2 ds - \int_0^\infty E\|X_s(x_0)\|_V^2 ds \\ &= \int_t^\infty E\|X_s(x_0)\|_V^2 ds - \int_0^\infty E\|X_s(x_0)\|_V^2 ds \\ &= -\int_0^t E\|X_s(x_0)\|_V^2 ds \end{aligned}$$

which, using the continuity of the mapping: $t \to E\|X_t(x_0)\|_V^2$ at zero for $x_0 \in V$ and letting $t \to 0$, immediately yields

$$(\mathbf{L}\Lambda)(x_0) = -\|x_0\|_V^2$$

for any $x_0 \in V$. Now the proof is complete.   □

Remark   (1). Note that if A is bounded, it is certainly coercive with $V = H$ and hence the Lyapunov function is $\Lambda(x) = \int_0^\infty E\|X_t(x)\|_H^2 dt$ exactly as in the finite dimensional case. It is thus possible to extend the theory in Khas'minskii [1] to give analogues of results, for instance, in Daletskii and Krein [1] for the stochastic setting.

(2). In general, we don't know the exact conditions for the continuity of mapping: $t \to E\|X_t(x)\|_V^2$. However, if we choose a new Gelfand triplet (V_1, V, V_1^*) and operators $A : V_1 \to V_1^*$ and $B : V_1 \to \mathcal{L}(K, V)$ satisfying a coercive condition corresponding to the norms $\|\cdot\|_V$ and $\|\cdot\|_{V_1}$, then it can be proved (e.g., see Pardoux [1] or Rozovskii [1]) that for any $x \in V$, $t \to E\|X_t(x)\|_V^2$ is a continuous function. Typical examples of these are Sobolev spaces of higher order and A, B smooth operators. For instance, Krylov and Rozovskii [1] contains several interesting examples in this case.

The following result will play an important role in the study of invariant measures of strong solutions.

Corollary 3.4.3   *Assume the coercive condition (1.3.2) holds with $p = 2$. Let $X_t(x_0)$, $t \geq t_0$, denote the strong solution of (3.4.1). If there exists a function $\Lambda : \mathbf{R}_+ \times H \to \mathbf{R}^1$ which satisfies the following:*

(a). $\Lambda(\cdot, \cdot)$ satisfies all the conditions in Remark at the end of Subsection 1.3.1;
(b). $c_1\|x\|_H^2 - k_1 \leq \Lambda(t, x) \leq c_2\|x\|_H^2 + k_2$, $\forall x \in V$, $t \geq 0$;
(c). $(\mathbf{L}\Lambda)(t, x) \leq -c_3\Lambda(t, x) + k_3$, $\forall x \in V$, $t \geq 0$,

where $c_i > 0$, $k_i > 0$, $i = 1, 2, 3$, are some constants and operator $\mathbf{L}$ is the infinitesimal generator of the Markov process X_t, then X_t, $t \geq t_0 \geq 0$, satisfies

$$E\|X_t(x_0)\|_H^2 \leq \alpha_1\|x_0\|_H^2 \cdot e^{-\beta_1(t-t_0)} + \alpha_2,   x_0 \in H,   t \geq t_0,   X_{t_0} = x_0,$$
$$(3.4.21)$$

for some $\alpha_1 > 0$, $\beta_1 > 0$ and $\alpha_2 > 0$.
Conversely, suppose (3.4.21) holds and define

$$\Lambda(t_0, x_0) = \int_{t_0}^{T+t_0} \left(\int_{t_0}^u E\|X_s(x_0)\|_V^2 ds \right) du   \text{for all}   x_0 \in H,   t_0 \geq 0,$$
$$(3.4.22)$$

where $X_{t_0} = x_0$ and T is some proper positive constant. Assume $\Lambda(\cdot, \cdot)$ satisfies all the conditions in Remark at the end of Subsection 1.3.1, then there exist constants $c_i > 0$, $k_i > 0$, $i = 1, 2, 3$, such that Conditions (b) and (c) above hold.

In general, if (3.4.21) is true, we also call the process $X_t(x_0)$, $t \geq t_0$, is *exponentially ultimately bounded in mean square, or simply, ultimately bounded in mean square.*

3.5 Two Applications

In this section, we shall apply the results derived in the last section to investigate stability and invariant measures of strong solutions of nonlinear stochastic differential equations through the method of the so-called first order approximation by linear systems. In finite dimensional spaces, this method proves to be an effective tool in treating large time behavior of nonlinear deterministic systems (cf. Hahn [1]). Its generalization to finite dimensional stochastic differential equations has been presented in Miyahara [1], [2].

3.5.1 Stability in Probability

Consider the following nonlinear stochastic differential equation in V^*:

$$\begin{cases} dX_t = A(X_t)dt + B(X_t)dW_t, & t \in [0, \infty), \\ X_0 = x_0 \in H, \end{cases} \qquad (3.5.1)$$

where $A(\cdot) : V \to V^*$ is a measurable nonlinear mapping satisfying $\|A(u)\|_{V^*} \leq L_1 \|u\|_V$, $L_1 > 0$, and $B(u) \in \mathcal{L}(K, H)$, $\|B(u)\| \leq L_2 \|u\|_V$, $L_2 > 0$, for any $u \in V$. We first present a result whose proof can be derived by a similar argument to that in Theorem 3.3.3.

Theorem 3.5.1   *Let $\Lambda(x)$ be a function defined on the set $\{x \in H : \|x\|_H < \delta\}$ satisfying the following properties:*

(i). All the conditions in Remark at the end of Subsection 1.3.1 hold;
(ii). $\Lambda(x) \to 0$ as $\|x\|_H \to 0$;
(iii). $\inf_{\|x\|_H > \varepsilon} \Lambda(x) = \lambda_\varepsilon > 0$ for any $\varepsilon > 0$;
(iv). $(\mathbf{L}\Lambda)(x) \leq 0$ for $x \in V$ with $\|x\|_H < \delta$.

Suppose $\{X_t(x_0), t \geq 0\}$ is the strong solution of (3.5.1), then

$$\lim_{\|x_0\|_H \to 0} P\left\{ \sup_{t \geq 0} \|X_t(x_0)\|_H > \varepsilon \right\} = 0 \qquad \text{for each} \quad \varepsilon > 0,$$

i.e., the null solution is stable in probability. Indeed, it is strongly stable in probability (cf. Khas'minskii [1]).

Obviously, the function $\Lambda(\cdot)$ constructed in Corollary 3.4.2 of the linear system (3.4.17) under Condition (3.4.18) satisfies the conditions in Theorem

3.5.1. In particular, as an immediate consequence we can obtain the following result by using Corollary 3.4.2.

Theorem 3.5.2   *Consider the equation (3.4.17) with A, B satisfying (3.4.18) and let $\{X_t(x_0),\, t \geq 0\}$ be its strong solution. If the null solution of (3.4.17) is exponentially stable in mean square, then it is also (strongly) stable in probability.*

Now we may present theorems on stability of nonlinear systems through the first order approximation in the following manner.

Theorem 3.5.3   *Consider Equation (3.4.17) satisfying the condition (3.4.18) and let $\{X_t(x_0),\, t \geq 0\}$ be its strong solution with the property $t \to E\|X_t(x_0)\|_V^2$ continuous. Assume that the null solution of (3.4.17) is exponentially stable in mean square. Let $\{\tilde{X}_t(x_0),\, t \geq 0\}$ be the strong solution of*

$$\begin{cases} dX_t = \tilde{A}(X_t)dt + \tilde{B}(X_t)dW_t, & t \in [0, \infty), \\ X_0 = x_0 \in H, \end{cases} \tag{3.5.2}$$

where $\tilde{A} : V \to V^*$ with $\|\tilde{A}(v)\|_{V^*} \leq \tilde{L}_1\|v\|_V$, $\tilde{L}_1 > 0$, and $\tilde{B}(v) \in \mathcal{L}(K, H)$, $\|\tilde{B}(v)\| \leq \tilde{L}_2\|v\|_V$, $\tilde{L}_2 > 0$, for all $v \in V$. If, further,

$$\|\tilde{A}(v) - Av\|_{V^*}^2 + \left| tr\big[\tilde{B}(v)Q\tilde{B}(v)^* - (Bv)Q(Bv)^*\big] \right| \leq \varepsilon\|v\|_V^2 \tag{3.5.3}$$

for $\varepsilon > 0$ small enough in a sufficiently small neighborhood of $v = 0$ in $\|\cdot\|_V$. Then the null solution of (3.5.2) is stable in probability (indeed, the null solution is also exponentially stable in mean square).

Proof   For any twice Fréchet differentiable function Φ satisfying the conditions of using Itô's formula in Subsection 1.3.1, define for any $v \in V$,

$$(\tilde{\mathbf{L}}\Phi)(v) = \langle \Phi'(v), \tilde{A}(v)\rangle_{V,V^*} + 1/2 \cdot tr(\Phi''(v)\tilde{B}(v)Q\tilde{B}(v)^*).$$

Consider $\Lambda(\cdot)$ as in Corollary 3.4.2, then for any $v \in V$,

$$(\tilde{\mathbf{L}}\Lambda)(v) - (\mathbf{L}\Lambda)(v)$$
$$= \langle \Lambda'(v), \tilde{A}(v) - Av\rangle_{V,V^*} + 1/2 \cdot tr[\Lambda''(v)(\tilde{B}(v)Q\tilde{B}(v)^* - (Bv)Q(Bv)^*)].$$

By Corollary 3.4.2 and using the conditions of this theorem, we have that for any $v \in V$ and the operator P defined in the proof of Corollary 3.4.2,

$$(\tilde{\mathbf{L}}\Lambda)(v) \leq -\|v\|_V^2 + 4\varepsilon\|P\|_{\mathcal{L}(V)}\|v\|_V^2.$$

Since $\varepsilon > 0$ is sufficiently small, we get $(\tilde{\mathbf{L}}\Lambda)(v) \leq 0$ (indeed, $(\tilde{\mathbf{L}}\Lambda)(v) \leq -\beta^{-1}(1 - 4\varepsilon\|P\|_{\mathcal{L}(V)})\|v\|_H^2$ when $4\varepsilon\|P\|_{\mathcal{L}(V)} < 1$). Hence, by virtue of Theorem 3.5.1 or Corollary 3.4.1, we get the desired result.   □

Theorem 3.5.4 *Consider the equation (3.4.17) satisfying (3.4.18) and let $\{X_t(x_0), t \geq 0\}$ be its strong solution. Assume that the null solution of (3.4.17) is exponentially stable in mean square. Let $\{\tilde{X}_t(x_0), t \geq 0\}$ be the strong solution of (3.5.2). If, further,*

$$\|\tilde{A}(v) - Av\|_{V^*}^2 + \left| tr\left[\tilde{B}(v)Q\tilde{B}(v)^* - (Bv)Q(Bv)^*\right] \right| \leq \varepsilon \|v\|_H^2 \qquad (3.5.4)$$

for $\varepsilon > 0$ small enough in a sufficiently small neighborhood of $v = 0$ in $\| \cdot \|_H$. Then the null solution of (3.5.2) is stable in probability (indeed, the null solution is also exponentially stable in mean square).

Proof   We follow the ideas of the proof of Theorem 3.5.3, using Corollary 3.4.2 and the operator $\tilde{P}$ defined in the proof of Corollary 3.4.2, to obtain

$$(\tilde{\mathbf{L}}\Lambda)(x) \leq -M\|x\|_H^2 + 4\varepsilon\|\tilde{P}\|_{\mathcal{L}(H)}\|x\|_H^2$$

for some number $M > 0$. Since $\varepsilon > 0$ is sufficiently small, we can get $(\tilde{\mathbf{L}}\Lambda)(x) \leq 0$ (indeed, $(\tilde{\mathbf{L}}\Lambda)(x) \leq -(M - 4\varepsilon\|\tilde{P}\|_{\mathcal{L}(H)})\|x\|_H^2$ when $4\varepsilon\|\tilde{P}\|_{\mathcal{L}(H)} < M$). Hence, by virtue of Theorem 3.5.1 or Corollary 3.4.1, we get the desired result. $\qquad \Box$

Let us study two examples to illustrate how to apply the above results to deal with nonlinear stochastic systems.

Example 3.5.1   Let $\mathcal{O} \subset \mathbf{R}^n$ be a bounded open domain with smooth boundary $\partial\mathcal{O}$. Assume $H = L^2(\mathcal{O})$ and V is the Sobolev space $H_0^1(\mathcal{O})$. Also suppose

$$\{W_q(t, x); t \geq 0, x \in \mathcal{O}\}$$

is an H-valued Wiener process with associated covariance operator Q, $trQ < \infty$, given by a positively definite kernel

$$q(x, y) \in L^2(\mathcal{O} \times \mathcal{O}), \qquad q(x, x) \in L^2(\mathcal{O}).$$

Let A be the linear differential strictly elliptic operator of the second order on $\mathcal{O}$ and $B(u)$ be the operator of multiplication by u, i.e., $B(u)f(\cdot) = u(\cdot)f(\cdot)$ for $f \in L^2(\mathcal{O})$. It can be shown that the coercive condition (3.4.18) is fulfilled (Garding's inequality, e.g., see Pazy [1]). Then the problem (3.4.17) turns out to be

$$du(t, x) = Au(t, x)dt + u(t, x)dW_q(t, x).$$

In this case, let $\Lambda(u) = \|u\|_H^2$, then we get for any $u \in H_0^1(\mathcal{O})$,

$$\mathbf{L}(\|u\|_H^2) = 2\langle u, Au\rangle_{V,V^*} + tr\left[(Bu)Q(Bu)^*\right]$$

$$= 2\langle u, Au\rangle_{V,V^*} + \int_{\mathcal{O}} q(x, x)u(x)^2 dx.$$

On the other hand, let

$$\lambda_0 = \sup_{\substack{u \in H_0^1(\mathcal{O}) \\ \|u\|_H^2 \neq 0}} \frac{\mathbf{L}(\|u\|_H^2)}{\|u\|_H^2} = \sup_{\substack{u \in H_0^1(\mathcal{O}) \\ \|u\|_H^2 \neq 0}} \frac{2\langle u, Au \rangle_{V,V^*} + \langle Qu, u \rangle_H}{\|u\|_H^2}.$$

Then we have with the help of Corollary 3.4.2 that the null solution is exponentially stable in mean square if $\lambda_0 < 0$.

Example 3.5.2   Consider the nonlinear equation in $\mathcal{O}$ which is defined as in Example 3.5.1

$$du(t,x) = \tilde{A}(x, u(t,x))dt + \tilde{B}(x, u(t,x))dW_q(t,x) \qquad (3.5.5)$$

with the boundary condition

$$u(t,x)|_{\partial \mathcal{O}} = 0.$$

Assume that

$$\tilde{A}(x, u(t,x)) = Au(t,x) + \alpha_1(x, u(t,x)),$$
$$\tilde{B}(x, u(t,x)) = u(t,x) + \alpha_2(x, u(t,x)),$$

$\alpha_i(x, u) : \mathcal{O} \times \mathbf{R}^1 \to \mathbf{R}^1$, $i = 1, 2$, are two functions satisfying the following conditions: for $i = 1, 2$, there exists a constant $c > 0$ such that

$$\sup_{x \in \mathcal{O}} |\alpha_i(x, u_2) - \alpha_i(x, u_1)| \leq c|u_2 - u_1|, \qquad \forall u_1, u_2 \in \mathbf{R}^1,$$

$$\alpha_i(x, 0) = 0, \qquad x \in \mathcal{O},$$

$$\sup_{x \in \mathcal{O}} |\alpha_i(x, u)| = o(|u|) \quad \text{as} \quad |u| \to 0. \qquad (3.5.6)$$

Also assume that the operator A is strictly elliptic in $\mathcal{O}$ such that the conditions of Example 3.5.1 hold. These conditions guarantee the fulfillment of the assumptions of coercivity and monotonicity for (3.5.5). So there exists a unique strong solution of the problem. Furthermore, the last condition (3.5.6) is sufficient for (3.5.3). We can conclude, with the help of Theorem 3.5.4 and Example 3.5.1, that the null solution is stable in probability if $\lambda_0 < 0$.

3.5.2   Ultimate Boundedness and Invariant Measures

In this subsection, by using the results in Section 3.4, we will establish conditions for ultimate boundedness of strong solutions of nonlinear stochastic differential equations and study the associated problems of existence and uniqueness of invariant measures. As we did in the above subsection, we will deal with nonlinear systems by the first order linear approximation. In particular, by means of Corollary 3.4.3, the following results can be derived in a similar manner to Theorems 3.5.3 and 3.5.4.

Theorem 3.5.5 *Consider the equation (3.4.17) satisfying (3.4.18) and let $\{X_t(x_0), t \geq 0\}$ be its strong solution. Assume that the system (3.4.17) is exponentially ultimately bounded in mean square. Let $\{\tilde{X}_t(x_0), t \geq 0\}$ be the strong solution of (3.5.2). Moreover, we suppose $\tilde{A}(v) - Av \in H$ for all $v \in V$, and*

$$2\|v\|_H \|\tilde{A}(v) - Av\|_H + \left| tr\left[\tilde{B}(v)Q\tilde{B}(v)^* - (Bv)Q(Bv)^*\right]\right| \leq \varepsilon \|v\|_H^2 + k, \quad v \in V,$$

for $\varepsilon > 0$ small enough and some number $k > 0$. Then the strong solution of (3.5.2) is exponentially ultimately bounded in mean square.

Corollary 3.5.1 *Consider the equation (3.4.17) satisfying (3.4.18) and let $\{X_t(x_0), t \geq 0\}$ be its strong solution. Assume that this solution is exponentially ultimately bounded in mean square. Let $\{\tilde{X}_t(x_0), t \geq 0\}$ be the strong solution of (3.5.2). Furthermore, we suppose that for any $v \in V$, $\tilde{A}(v) - Av \in H$, and*

$$\|\tilde{A}(v) - Av\|_H^2 + \left| tr\left[\tilde{B}(v)Q\tilde{B}(v)^* - (Bv)Q(Bv)^*\right]\right| \leq L(1 + \|v\|_H^2)$$

for some $L > 0$. If for arbitrary $v \in V$, as $\|v\|_H \to \infty$,

$$\|\tilde{A}(v) - Av\|_H = o(\|v\|_H) \quad and \quad \left| tr\left[\tilde{B}(v)Q\tilde{B}(v)^* - (Bv)Q(Bv)^*\right]\right| = o(\|v\|_H^2).$$

Then the strong solution of (3.5.2) is exponentially ultimately bounded in mean square.

Example 3.5.3 Consider the following stochastic differential equation:

$$du(t) = Au(t)dt + F(u(t))dt + B(u(t))dW_t \qquad (3.5.7)$$

with initial condition $u(0) = u_0 \in H$. Suppose A, F and B satisfy the following conditions:

(i). $A : V \to V^*$ is coercive, that is, there exist constants $\alpha > 0$ and $\lambda \in \mathbf{R}^1$ such that for arbitrary $v \in V$,

$$2\langle v, Av \rangle_{V,V^*} \leq -\alpha \|v\|_V^2 + \lambda \|v\|_H^2;$$

(ii). $F : H \to H$ and $B : H \to \mathcal{L}(K, H)$ satisfy that for arbitrary $v \in H$,

$$\|F(v)\|_H^2 + \|B(v)\|^2 \leq L\left(1 + \|v\|_H^2\right)$$

for some $L > 0$;

(iii). For arbitrary $u, v \in H$,

$$\|F(u) - F(v)\|_H^2 + tr\left[(B(u) - B(v))Q(B(u)^* - B(v)^*)\right] \leq \lambda \|u - v\|_H^2.$$

If the solution $\{u^{u_0}(t), t \geq 0\}$ of $du(t) = Au(t)dt$ is exponentially decayable (or even exponentially ultimately bounded), and as $\|v\|_H \to \infty$,

$$\|F(v)\|_H = o(\|v\|_H), \qquad \|B(v)\| = o(\|v\|_H),$$

then the strong solution $\{u^{u_0}(t), t \geq 0\}$ of (3.5.7) is exponentially ultimately bounded in mean square.

Indeed, let $\tilde{A}(v) = Av + F(v)$ for $v \in V$. Since $F(v) \in H$,

$$2\langle v, \tilde{A}(v) \rangle_{V,V^*} + tr[B(v)QB(v)^*]$$
$$= 2\langle v, Av \rangle_{V,V^*} + 2\langle F(v), v \rangle_H + tr[B(v)QB(v)^*]$$
$$\leq -\alpha\|v\|_V^2 + \lambda\|v\|_H^2 + 2\|v\|_H\|F(v)\|_H + \|B(v)\|^2 trQ$$
$$\leq -\alpha\|v\|_V^2 + \lambda'\|v\|_H^2 + \gamma$$

for some constants $\lambda' \in \mathbf{R}^1$ and $\gamma \geq 0$. Hence, the equation (3.5.7) is coercive. Moreover, under the additional assumptions (ii) and (iii), the strong solution $\{u^{u_0}(t), t \geq 0\}$ of (3.5.7) exists. By the assumption (ii),

$$\|F(v)\|_H^2 + tr[B(v)QB(v)^*] \leq \|F(u)\|_H^2 + \|B(v)\|^2 trQ \leq (1+trQ)L(1+\|v\|_H^2)$$

and since

$$\|F(v)\|_H = o(\|v\|_H), \quad tr[B(v)QB(v)^*] \leq \|B(v)\|^2 trQ = o(\|v\|_H^2)$$

as $\|v\|_H \to \infty$, the assertion follows from Corollary 3.5.1.

Example 3.5.4   Let S^1 be the unit circle and $B(\cdot, \cdot)$ a standard Brownian sheet on $[0, \infty) \times S^1$. Consider the following stochastic heat equation:

$$\frac{\partial X_t}{\partial t}(\xi) = \frac{\partial^2 X_t}{\partial \xi^2}(\xi) - \alpha X_t(\xi) + f(X_t(\xi)) + g(X_t(\xi))\frac{\partial^2 B}{\partial t \partial \xi}, \qquad (3.5.8)$$

with initial condition

$$X_0(\cdot) = x_0(\cdot) \in L^2(S^1),$$

where α is a constant and $f(\cdot)$, $g(\cdot)$ are two real-valued functions with $f(0) = g(0) = 0$. Suppose

$$H = L^2(S^1), \quad V = W^{1,2}(S^1), \quad A(x) = \left(\frac{\partial^2}{\partial \xi^2} - \alpha\right)x,$$

and the mappings F and B are given for $\xi \in S^1$ and $x(\cdot), y(\cdot) \in L^2(S^1)$ by

$$F(x)(\xi) = f(x(\xi)), \quad B(x)[y](\xi) = g(x(\xi))y(\xi),$$

and let

$$\|x\|_H = \left(\int_{S^1} x^2(\xi)d\xi\right)^{1/2} \quad \text{for} \quad x \in H,$$

$$\|x\|_V = \left[\int_{S^1}\left(x^2(\xi) + \left(\frac{\partial x(\xi)}{\partial \xi}\right)^2\right)d\xi\right]^{1/2} \qquad \text{for} \quad x \in V,$$

then we have

$$2\langle x, Ax\rangle_{V,V^*} = -2\|x\|_V^2 + (-2\alpha + 2)\|x\|_H^2$$
$$\leq -2\|x\|_H^2 + (-2\alpha + 2)\|x\|_H^2 = -2\alpha\|x\|_H^2.$$

Therefore, the null solution of $dx(t) = Ax(t)dt$ is exponentially stable if $\alpha > 0$. Moreover, if we further assume $f(\cdot)$ and $g(\cdot)$ are both Lipschitz continuous and bounded, then from Example 3.5.3, the strong solution of (3.5.8) is exponentially ultimately bounded in mean square.

Suppose that X_t, $t \geq 0$, is a time-homogeneous, stochastically continuous Markov process (in norm $\|\cdot\|_H$) defined on some $(\Omega, \mathcal{F}, \mathcal{F}_t, P)$ with state space H. Denote the transition probability function by $P(x, t, \cdot)$ defined by the conditional probability:

$$P(x, t, \Gamma) = P\Big\{X_t \in \Gamma \ \Big| \ X_0 = x\Big\},$$

for $x \in H$ and $\Gamma \in \mathcal{B}(H)$, the Borel σ-field of H. A probability measure μ on $(H, \mathcal{B}(H))$ is called *invariant* for the given transition probability function if it satisfies the equation

$$\mu(\Gamma) = \int_H P(x, t, \Gamma)\mu(dx)$$

for any $t \geq 0$, or equivalently, the following relation holds:

$$\int_H \Psi(y)\mu(dy) = \int_H (P_t\Psi)(y)\mu(dy)$$

for any $\Psi \in C_b(H)$, $t \geq 0$, where the transition semigroup P_t, $t \geq 0$, is defined by

$$(P_t\Psi)(x) = \int_H \Psi(y)P(x, t, dy),$$

and $C_b(H)$ denotes the space of all bounded continuous functions on H. We also recall that $P(x, t, \cdot)$ has the *Feller property* if, for any bounded continuous function Ψ on H, $(P_t\Psi)(x)$ is continuous in x for any $t \geq 0$. It is clear that exponential decay of the process X_t, $t \geq 0$, implies that the system has invariant measure degenerate at zero. One of the remarkable consequences in the section is that we shall show the concept of exponentially ultimate boundedness has a close relationship with the existence of non trivial invariant measures.

Consider the following nonlinear stochastic differential equation in V^*:

$$\begin{cases} dX_t = A(X_t)dt + B(X_t)dW_t, & t \in [0, \infty), \\ X_0 = x_0 \in H, \end{cases} \qquad (3.5.9)$$

where $A(\cdot) : V \to V^*$ and $B(\cdot) : V \to \mathcal{L}(K, H)$ are both nonlinear measurable mappings which satisfy all the conditions in Theorem 1.3.1. In particular, it can be shown that the strong solution process X_t, $t \geq 0$, is a time-homogeneous, stochastically continuous Markov process and the associated transition probability function $P(x, t, \cdot)$ with its transition semigroup P_t, $t \geq 0$, has the Feller property. We now state and prove a theorem concerning the existence of invariant measures which extends to infinite dimensional spaces the corresponding finite dimensional results (see, e.g. Khas'minskii [1]).

Theorem 3.5.6   *Let $X_t(x_0)$, $t \geq 0$, be a strong solution of (3.5.9), and for some sequence $T_n \uparrow \infty$, define*

$$\mu_n(\Gamma) = \frac{1}{T_n} \int_0^{T_n} P(x_0, t, \Gamma) dt, \quad \Gamma \in \mathcal{B}(H). \tag{3.5.10}$$

If μ is the weak limit of a subsequence of $\{\mu_n\}$, then μ is an invariant measure of $P(x, t, \cdot)$. In other words, the existence of invariant measures is equivalent to the tightness of $\{\mu_n(\cdot)\}$.

Proof   Since the sequence $\{T_n\}$ was arbitrary, we may as well assume $\mu_n \xrightarrow{w} \mu$, i.e., μ_n converges weakly to μ on H. By the Feller property, for each $t \geq 0$ and $\Psi \in C_b(H)$, the function $(P_t \Psi)(\cdot)$ is in $C_b(H)$. It follows from the weak convergence of μ_n that

$$\int_H (P_t \Psi)(x) \mu(dx) = \lim_{n \to \infty} \int_H (P_t \Psi)(x) \mu_n(dx)$$

$$= \lim_{n \to \infty} \frac{1}{T_n} \int_0^{T_n} \int_H (P_t \Psi)(x) P(x_0, s, dx) ds \tag{3.5.11}$$

$$= \lim_{n \to \infty} \frac{1}{T_n} \int_0^{T_n} (P_{t+s} \Psi)(x_0) ds,$$

$$\tag{3.5.12}$$

where we have made use of the Fubini theorem and the Markov property. Now, for any fixed $t \geq 0$ and $x_0 \in H$, we can write

$$\lim_{n \to \infty} \frac{1}{T_n} \int_0^{T_n} (P_{t+s} \Psi)(x_0) ds$$

$$= \frac{1}{T_n} \left\{ \int_0^{T_n} (P_s \Psi)(x_0) ds + \int_{T_n}^{T_n+t} (P_s \Psi)(x_0) ds - \int_0^t (P_s \Psi)(x_0) ds \right\}. \tag{3.5.13}$$

Since $\|P_s \Psi\|_H \leq \|\Psi\|_H < \infty$ for any $s \geq 0$, by virtue of (3.5.11) and (3.5.12), we get

$$\int_H (P_t\Psi)(x)\mu(dx) = \lim_{n\to\infty} \frac{1}{T_n} \int_0^{T_n} (P_s\Psi)(x_0)ds = \lim_{n\to\infty} \int_H \Psi(x)\mu_n(dx)$$
$$= \int_H \Psi(x)\mu(dx)$$

by Fubini's theorem and weak convergence. Hence, μ is an invariant measure.

$\Box$

If H is compact, then the tightness of $\{\mu_n(\cdot)\}$ in Theorem 3.5.6 is true and existence follows. For non-compact H, one needs additional conditions which lead to the following corollary to get the existence of invariant measures.

Corollary 3.5.2 *Assume the embedding $V \hookrightarrow H$ is compact. Suppose that the strong solution $X_t(x_0)$ of the equation with $X_0 = x_0$ satisfies the condition that for some $x_0 \in H$, there exists a real number sequence $T_n \uparrow \infty$ such that*

$$\frac{1}{T_n} \int_0^{T_n} P\{\|X_t(x_0)\|_V > R\}dt \to 0 \quad \text{uniformly in} \quad n \quad \text{as} \quad R \to \infty.$$
(3.5.14)

Then Equation (3.5.9) has an invariant measure on $(H, \mathcal{B}(H))$.

Proof It suffices to show $\{\mu_n\}$ defined in (3.5.10) is weakly compact. For any $R > 0$, let $V_R = \{y \in V : \|y\|_V \le R\}$ and $V_R^{cH} = \{y \in H : y \notin V_R\}$. It is clear that V_R is compact in H and therefore by Condition (3.5.13), for any $\varepsilon > 0$, there exists a compact set $V_R \subset H$ such that

$$\mu_n\{H \setminus V_R\} = \mu_n(V_R^{cH}) < \varepsilon \quad \text{for all} \quad n \ge 1.$$

By the well-known Prokhorov theorem (see, e.g. Da Prato and Zabczyk [1]), the family $\{\mu_n\}$ is weakly compact.

$\Box$

As we pointed out before, there exists a close relationship between ultimate boundedness and invariant measures. As a matter of fact, we have the following result.

Theorem 3.5.7 *Assume the embedding $V \hookrightarrow H$ is compact. Suppose the strong solution $\{X_t; t \ge 0\}$ of (3.5.9) under the coercive condition (1.3.2) with $p = 2$ is ultimately bounded (in $\|\cdot\|_H$ norm), then there exists an invariant measure μ for $\{X_t; t \ge 0\}$.*

Proof Applying Itô's formula to $\Lambda(x) = \|x\|_H^2$, $x \in H$, taking expectation and using the coercive condition (1.3.2) with $p = 2$ there, we get

$$E\|X_t(x_0)\|_H^2 - \|x_0\|_H^2 = \int_0^t E\mathbf{L}\|X_s(x_0)\|_H^2 ds$$

$$= \lambda \int_0^t E\|X_s(x_0)\|_H^2 ds - \alpha \int_0^t E\|X_s(x_0)\|_V^2 ds + \gamma t.$$

Hence

$$\int_0^t E\|X_s(x_0)\|_V^2 ds \le \frac{1}{\alpha}\Big(\lambda \int_0^t E\|X_s(x_0)\|_H^2 ds + \|x_0\|_H^2 + \gamma t\Big),$$

which, by using Markov's inequality, immediately implies that for any $T > 0$,

$$\frac{1}{T}\int_0^T P\big\{\|X_t(x_0)\|_V > R\big\}dt \le \frac{1}{T}\int_0^T \frac{E\|X_t(x_0)\|_V^2}{R^2}dt$$

$$\le \frac{1}{\alpha R^2}\Big(\frac{|\lambda|}{T}\int_0^T E\|X_t(x_0)\|_H^2 dt + \frac{\|x_0\|_H^2}{T} + \gamma\Big).$$

Since $\{X_t(x_0); t \ge 0\}$ is ultimately bounded, there exist two constants $T_0 > 0$ and $M > 0$ such that

$$E\|X_t(x_0)\|_H^2 \le M \quad \text{for} \quad t \ge T_0,$$

which immediately yields

$$\lim_{R\to\infty} \liminf_{T\to\infty} \frac{1}{T}\int_0^T P\big\{\|X_t(x_0)\|_V > R\big\}dt$$

$$\le \lim_{R\to\infty} \liminf_{T\to\infty} \frac{|\lambda|}{\alpha R^2}\frac{1}{T}\int_0^T E\|X_t(x_0)\|_H^2 dt$$

$$= \lim_{R\to\infty} \liminf_{T\to\infty} \frac{|\lambda|}{\alpha R^2}\frac{1}{T}\Big(\int_0^{T_0} E\|X_t(x_0)\|_H^2 dt + \int_{T_0}^T E\|X_t(x_0)\|_H^2 dt\Big)$$

$$\le \lim_{R\to\infty} \liminf_{T\to\infty} \frac{|\lambda|}{\alpha R^2}\frac{1}{T}\Big(\int_0^{T_0} E\|X_t(x_0)\|_H^2 dt + M(T - T_0)\Big) = 0.$$

Therefore, the assertion of the theorem follows. $\qquad\qquad\qquad\qquad$ ▯

As a consequence of Theorem 3.5.7, we may easily get a result on the existence of invariant measures of the stochastic differential equations (3.5.7) and (3.5.8). For instance, as seen in Example 3.5.4, the solution of the stochastic heat equation is ultimately bounded in mean square of $\|\cdot\|_H$, and since $V \hookrightarrow H$ is compact by the Sobolev embedding theorem, the existence of an invariant measure follows.

3.6 Further Results on Invariant Measures

In the last section, we have investigated the existence of invariant measures. In infinite dimensional spaces, the uniqueness question is more subtle. Throughout this section, unless otherwise stated, we always suppose the embedding $V \hookrightarrow H$ is compact. The following condition for the uniqueness of invariant measures is somewhat stringent but seems to be natural.

Let $Y_t(y_0)$ denote the family of Markov processes such that $Y_0 = y_0$ and $P\{Y_t(y_0) \in \Gamma\} = P(y_0, t, \Gamma)$, $\forall t > 0$, $y_0 \in H$ and $\Gamma \in \mathcal{B}(H)$. Recall that $V_R = \{y \in V : \|y\|_V \leq R\}$ and $V_R^{cH} = \{y \in H : y \notin V_R\}$ for any $R > 0$.

Theorem 3.6.1 *Suppose that the following condition holds: for arbitrary $\varepsilon > 0$, $\delta > 0$ and $R > 0$, there exists $T_0(\varepsilon, \delta, R) > 0$ such that*

$$\frac{1}{T} \int_0^T P\Big\{ \|Y_t(\xi) - Y_t(\eta)\|_V \geq \delta \Big\} dt < \varepsilon \tag{3.6.1}$$

for any ξ, $\eta \in V_R$ and $T \geq T_0(\varepsilon, \delta, R)$. If there exists an invariant measure μ of $Y_t(y_0)$ with support in V, then it is unique.

Proof Suppose that μ and ν both are invariant measures with supports in V. It suffices to show that for any bounded, uniformly continuous function Ψ on H (see, e.g. Ikeda and Watanabe [1]),

$$\int_H \Psi(x) \mu(dx) = \int_H \Psi(x) \nu(dx).$$

Now define for arbitrary $\Gamma \in \mathcal{B}(H)$,

$$\mu_T^\xi(\Gamma) = \frac{1}{T} \int_0^T P(\xi, t, \Gamma) dt, \quad \xi \in H, \quad T > 0,$$

and, by the invariant properties of measures μ and ν, we have

$$\begin{aligned}
&\left| \int_H \Psi(x) \mu(dx) - \int_H \Psi(x) \nu(dx) \right| \\
&= \left| \int_H \int_H \Psi(x) \Big[\mu_T^\xi(dx) \mu(d\xi) - \mu_T^\eta(dx) \nu(d\eta) \Big] \right| \\
&\leq \int_{H \times H} \left| \int_H \Psi(x) \mu_T^\xi(dx) - \int_H \Psi(x) \mu_T^\eta(dx) \right| \mu(d\xi) \nu(d\eta) \\
&= \int_{V \times V} |F(\xi, \eta)| \mu(d\xi) \nu(d\eta),
\end{aligned} \tag{3.6.2}$$

where

$$F(\xi, \eta) = \int_H \Psi(x)\mu_T^\xi(dx) - \int_H \Psi(x)\mu_T^\eta(dx). \qquad (3.6.3)$$

Let $V_R^{cv} = V \backslash V_R$ and choose $R > 0$ such that

$$\mu(V_R^{cv}) + \nu(V_R^{cv}) < \varepsilon. \qquad (3.6.4)$$

Note (3.6.3) and (3.6.4), then the inequality (3.6.2) yields

$$\left| \int_H \Psi(x)\mu(dx) - \int_H \Psi(x)\nu(dx) \right| \leq \int_{V_R \times V_R} |F(\xi, \eta)|\mu(d\xi)\nu(d\eta) + 4b\varepsilon + 2b\varepsilon^2,$$

$$(3.6.5)$$

where $b = \sup_{x \in H} |\Psi(x)|$.

On the other hand, we have for some suitable real number $\delta > 0$,

$$\int_{V_R \times V_R} |F(\xi, \eta)|\mu(d\xi)\nu(d\eta)$$

$$\leq \int_{V_R \times V_R} \left\{ \frac{1}{T} \int_0^T E|\Psi(Y_t(\xi)) - \Psi(Y_t(\eta))|dt \right\} \mu(d\xi)\nu(d\eta)$$

$$\leq 2b \sup_{\xi, \eta \in V_R} \frac{1}{T} \int_0^T P\left\{ \|Y_t(\xi) - Y_t(\eta)\|_V \geq \delta \right\} dt + \sup_{\substack{\xi, \eta \in V_R \\ \|\xi - \eta\|_V < \delta}} |\Psi(\xi) - \Psi(\eta)|$$

$$< 2b\varepsilon + \varepsilon, \qquad (3.6.6)$$

for $T > 0$ sufficiently large, by Condition (3.6.1) and the uniform continuity of Ψ. In terms of (3.6.5) and (3.6.6), the uniqueness result follows. □

In the above proof, the property of having support in V of the measure $\mu(\cdot)$ is essential. Suppose that, for each n, $\mu_n(\cdot)$, defined in Theorem 3.5.6 has a support in the space V or $\mu_n(V) = 1$, $n = 1, 2, \cdots$. Then one expects the same is true for their weak limit μ. This is indeed the case as established by the following theorem.

Theorem 3.6.2   *Let all the conditions in Corollary 3.5.2 hold. If the family $\{\mu_n\}$ of probability measures is supported in V such that for any $\varepsilon > 0$, there exists $R_0 > 0$ such that*

$$\sup_n \mu_n(V_R^{cH}) < \varepsilon, \qquad \forall R > R_0, \qquad (3.6.7)$$

then any invariant measure, as the weak limit of a subsequence of $\{\mu_n\}$, has support in V.

Proof   Let $\{\mu_{n_k}\}$ be a subsequence converging weakly to μ so that $\mu_{n_k}(V) = 1$ for any $k \in \mathbf{N}_+$. Since $H_R = \{x \in H : \|x\|_V \leq R\}$ is compact, we have

$H_R = \overline{i(V_R)}_H$, where $\overline{i(V_R)}_H$ denotes the closure of $i(V_R) = \{x \in V : \|x\|_V < R\}$ in H. Let $\rho(x, i(V_R))$, $x \in H$, be the distance from x to set $i(V_R)$ given by

$$\rho(x, i(V_R)) = \inf\left\{\|y - x\|_V : y \in i(V_R)\right\},$$

and define function $\phi : \mathbf{R}_+ \to \mathbf{R}_+$ by $\phi(t) = t$ for $0 \le t \le 1$ and $\phi(t) = 1$ for $t > 1$. Now we introduce

$$\Psi_\delta^R(x) = \phi\left(\frac{1}{\delta}\rho(x, i(V_R))\right), \qquad \delta \in (0, 1).$$

Then $\Psi_\delta^R \in C_b(H)$ and

$$\int_H \Psi_\delta^R(x)\mu_{n_k}(dx) = \int_V \Psi_\delta^R(x)\mu_{n_k}(dx) \le \int_V \chi_{V_R^{c_H}}(x)\mu_{n_k}(dx) \qquad (3.6.8)$$
$$= \mu_{n_k}(V_R^{c_H}),$$

where $\chi_\Gamma(\cdot)$ denotes the indicator function of set $\Gamma \in \mathcal{B}(H)$. It follows from (3.6.7), (3.6.8) and the weak convergence that

$$\int_H \Psi_\delta^R(x)\mu(dx) = \lim_{k\to\infty}\int_H \Psi_\delta^R(x)\mu_{n_k}(dx) \le \lim_{k\to\infty}\mu_{n_k}(V_R^{c_H}) < \varepsilon, \quad \forall R > R_0.$$

Also, by the well-known dominated convergence theorem, we have

$$\lim_{\delta\downarrow 0}\int_H \Psi_\delta^R(x)\mu(dx) = \int_H \chi_{V_R^{c_H}}(x)\mu(dx) \le \varepsilon,$$

or

$$\mu(H\backslash V_R) = \mu(V_R^{c_H}) \le \varepsilon, \qquad \forall R > R_0.$$

Since $V = \bigcup_{n=1}^\infty V_n$, we deduce easily that

$$\mu(H\backslash V) = \mu\left(H \backslash \bigcup_{n=1}^\infty V_n\right) = 0$$

as required. $\square$

One of the most important consequences of the concept of invariant measure is that it is closely connected with some stationary properties of solution processes.

Proposition 3.6.1   *If μ is an invariant measure for (3.5.9) and x_0 is an H-valued, $\mathcal{F}_0$-measurable random variable such that the law of x_0 equals μ, then the solution process $X.(x_0)$ is stationary in the sense that for every finite sequence of numbers $t_1, \cdots, t_n$, the joint distribution of the random variables $X_{t_1+h}, \cdots, X_{t_n+h}$ is independent of $h \ge 0$.*

Proof   Let $\psi_1, \cdots, \psi_n$ be a family of bounded, measurable functions on H, and $0 \leq t_1 \leq t_2 \leq \cdots \leq t_n$. We have to show that the expectation

$$I = E\Big(\psi_1(X_{t_1+h}) \cdots \psi_n(X_{t_n+h})\Big)$$

is independent of $h \geq 0$. This is certainly true if $n = 1$. Note that

$$E\Big(\psi_1(X_{t_1+h}) \cdots \psi_n(X_{t_n+h})\Big)$$
$$= E\Big[\psi_1(X_{t_1+h}) \cdots \psi_{n-1}(X_{t_{n-1}+h}) E\Big(\psi_n(X_{t_n+h}) \;\Big|\; \mathcal{F}^X_{t_{n-1}+h}\Big)\Big].$$

By the Markov property,

$$I = E\Big(\psi_1(X_{t_1+h}) \cdots \psi_{n-1}(X_{t_{n-1}+h}) P_{t_n-t_{n-1}} \psi_n(X_{t_{n-1}+h})\Big).$$

So if the result is true for $n-1$, it is true for n, and consequently it holds by induction. □

The following theorem, which provides conditions to ensure the existence and uniqueness of invariant measures for Equation (3.5.9), is the main result of this section.

Theorem 3.6.3   *Let all the assumptions (1.3.2)–(1.3.5) on the strong solutions of the equation (3.5.9) hold. Further assume that the strong solution $X_t(x_0)$ with $X_0 = x_0$ satisfies the condition (3.5.13). Then the equation (3.5.9) has an invariant measure μ with its support in V, and the solution X_t with initial distribution μ is stationary. Moreover, if the condition (3.6.1) holds, the invariant measure μ is unique.*

Proof   Clearly, all we need to do is to show $\text{supp}\{\mu\} \subset V$. To this end, define

$$\mu_T(E) = \frac{1}{T} \int_0^T P\Big\{X_t(x_0) \in E\Big\} dt \qquad (3.6.9)$$

for arbitrary $T > 0$, $E \in \mathcal{B}(H)$, which is a probability measure on $(H, \mathcal{B}(H))$. In view of Theorem 3.6.2, it suffices to show that,

$$\mu_T\{H \setminus V\} = 0, \quad \forall T > 0.$$

First, by virtue of Theorem 1.3.1, we have for some $p > 1$,

$$E \int_0^T \|X_t(x_0)\|_V^p dt \leq M_T < \infty, \qquad x_0 \in H,$$

so that

$$X_t(x_0) \in V \quad \text{a.s.} \quad (t, \omega) \in \Omega_T := [0, T] \times \Omega.$$

Let

$$N = \left\{ (t,\omega) \in \Omega_T : X_t(x_0) \in H \setminus V \right\}.$$

Then

$$P\{N\} = \int_0^T \int_\Omega \chi_N(t,\omega) dt P(d\omega) = 0,$$

where $\chi_N(\cdot)$ is the indicator function of set N. Thus, by the Fubini theorem, we get for any $T > 0$,

$$P\{N^t\} = \int_\Omega \chi_N(t,\omega) P(d\omega) = 0, \quad t \in J \subset [0,T], \qquad (3.6.10)$$

where

$$N^t = \left\{ \omega \in \Omega : X_t(x_0) \in H \setminus V \right\},$$

and $J' = [0,T] \setminus J$ has Lebesgue measure zero. Now, by (3.6.9) and (3.6.10), we have for any $T > 0$,

$$\mu_T\{H \setminus V\} = \frac{1}{T} \int_0^T P\{N^t\} dt = \frac{1}{T} \int_0^T \chi_J(t) P\{N^t\} dt = 0$$

as was to be shown. $\qquad\qquad\qquad\qquad\qquad\qquad\qquad\qquad\qquad$ □

As an immediate consequence of the theorems derived above, we can obtain the following useful criterion which ensures a unique invariant measure of strong solutions by imposing some monotonicity conditions on Equation (3.5.9).

Corollary 3.6.1 *Let all the assumptions (1.3.2)–(1.3.5) on the strong solution of the equation (3.5.9) hold. If there exist positive constants $T_0 > 0$ and $M > 0$ such that*

$$\sup_{T > T_0} \left\{ \frac{1}{T} \int_0^T E\|X_t(x_0)\|_H^2 dt \right\} \leq M, \qquad (3.6.11)$$

then the equation (3.5.9) has a stationary solution with the initial invariant measure supported in V. This invariant measure is unique provided the following condition holds: there exist $c > 0$ and $\delta > 0$ such that

$$2\langle u - v, A(u) - A(v)\rangle_{V,V^*} + \|B(u) - B(v)\|_{\mathcal{L}_2^0}^2 \leq -c\|u - v\|_V^\delta, \quad \forall u, v \in V,$$
$$(3.6.12)$$

where $\|B(u) - B(v)\|_{\mathcal{L}_2^0}^2 = tr\left[(B(u) - B(v)) Q (B(u) - B(v))^* \right].$

Proof In view of Itô's formula, we have the following energy inequality

$$E\|X_t(x_0)\|_H^2 \leq \|x_0\|_H^2 + 2E \int_0^t \langle X_s(x_0), A(X_s(x_0))\rangle_{V,V^*} ds$$

$$+E \int_0^t \|B(X_s(x_0))\|_{\mathcal{L}_2^0}^2 ds$$

for any $t \geq 0$, which, in addition to the coercive condition (1.3.2), yields

$$E\|X_t(x_0)\|_H^2 + \alpha E \int_0^t \|X_s(x_0)\|_V^p ds \leq (\|x_0\|_H^2 + \gamma t) + \lambda E \int_0^t \|X_s(x_0)\|_H^2 ds.$$

It follows that for any $T_0 > 0$,

$$\sup_{T > T_0} \frac{1}{T} \int_0^T E\|X_s(x_0)\|_V^p ds \leq \frac{|\gamma| + \|x_0\|_H^2 / T_0}{\alpha}$$
$$+ \frac{|\lambda|}{\alpha} \sup_{T > T_0} \frac{1}{T} \int_0^T E\|X_s(x_0)\|_H^2 ds. \tag{3.6.13}$$

On the other hand, $\mu_T(\cdot)$ defined by (3.6.9) is supported in V. Hence, by the well-known Markov's inequality, we have for any $T > 0$, $R > 0$,

$$\frac{1}{T} \int_0^T P\Big\{\|X_s(x_0)\|_V > R\Big\} ds \leq \frac{1}{R^p} \cdot \Big\{\frac{1}{T} \int_0^T E\|X_s(x_0)\|_V^p ds\Big\},$$

which, together with (3.5.13) and (3.6.13), shows that the condition (3.6.11) is sufficient for the existence of an invariant measure μ on V by Theorem 3.6.3. By Proposition 3.6.1, the existence of a stationary solution follows.

To show uniqueness, let $X_t(x_1)$, $X_t(x_2)$ denote two solutions with the initial states x_1, $x_2 \in H$. We set $\Delta X_t = X_t(x_1) - X_t(x_2)$. Then, by Itô's formula, we have

$$E\|\Delta X_t\|_H^2 \leq \|x_1 - x_2\|_H^2 + 2E \int_0^t \langle \Delta X_s, A(X_s(x_1)) - A(X_s(x_2))\rangle_{V,V^*} ds$$

$$+E \int_0^t \|B(X_s(x_1)) - B(X_s(x_2))\|_{\mathcal{L}_2^0}^2 ds.$$

In view of Condition (3.6.12), we obtain

$$E\|\Delta X_t\|_H^2 \leq \|x_1 - x_2\|_H^2 - c \int_0^t E\|\Delta X_s\|_V^\delta ds,$$

or

$$\int_0^t E\|\Delta X_s\|_V^\delta ds \leq \frac{\|x_1 - x_2\|_H^2}{c}.$$

This implies that the condition (3.6.1) for uniqueness is satisfied by applying Markov's inequality:

$$P\Big\{\|X_t(x_1) - X_t(x_2)\|_V \geq \rho\Big\} \leq E\|X_t(x_1) - X_t(x_2)\|_V^\delta / \rho^\delta$$

for arbitrary $\rho > 0$. $\qquad\qquad\qquad\qquad\qquad\qquad\qquad\qquad\qquad\qquad\quad$ ☐

3.7   Stability, Ultimate Boundedness of Mild Solutions and Invariant Measures

In the previous sections, for instance, in the special case $A(v) = Av + f(v)$ of Equation (3.5.9) with A being coercive, Corollary 3.6.1 seems to have wider applications. However, the corresponding results for mild solutions are also useful and could be applied to a certain class of stochastic evolution equations without coercive and monotone conditions, such as stochastic wave equations. The existence, uniqueness and some associated problems for invariant measures of mild solutions recently received increasing attention. For instance, the reader may find a systematic presentation of recent developments in this area in the books Da Prato and Zabczyk [1], [2]. In this section, we first establish the Lyapunov function characteristic theorem of mild solutions for a class of semilinear stochastic evolution equations. By analogy with those in Section 3.5.2, we also study the existence and uniqueness of invariant measures, but concentrating on those results in connection with the ultimate boundedness of solutions in mean square and the use of Lyapunov function characteristic techniques.

3.7.1   Lyapunov Functions for Mild Solutions

An example was constructed in Section 3.4 to show that the usual Lyapunov functions in finite dimensional spaces are not strictly positive definite for strong solutions in the infinite dimensional setting. Examples may also be constructed to show that this is the case for mild solutions of semilinear stochastic evolution equations. In this section, we will construct Lyapunov functions for mild solutions and prove that the existence of such a Lyapunov function is a necessary and sufficient condition for mild solutions to be exponentially decayable in mean square. To this end, consider the semilinear stochastic differential equation (3.3.1). In particular, throughout this section we will impose the following condition: there exist constants $\gamma \geq 0$, $\lambda > 0$ and $\mu \geq 0$ such that for any $u \in \mathcal{D}(A)$,

$$2\langle Au + F(t, u), u\rangle_H + \|G(t, u)\|^2 \leq \lambda\|u\|_H^2 + \gamma \cdot e^{-\mu t}, \quad t \geq 0. \qquad (3.7.1)$$

Theorem 3.7.1   *Assume the condition (3.7.1) holds and $X_t(x_0)$, $t \geq t_0 \geq 0$, is a solution of (3.3.1). If there exists a function $\Lambda(\cdot, \cdot) \in C^{1,2}(\mathbf{R}_+ \times H; \mathbf{R}^1)$ which satisfies the following:*

(a) $\Lambda(\cdot, \cdot)$ satisfies all the conditions of using Itô's formula in Theorem 1.2.7;

(b) $|\Lambda_t'(t, x)| + \|x\|_H \|\Lambda_x'(t, x)\|_H + \|x\|_H^2 \|\Lambda_{xx}''(t, x)\| \leq c\|x\|_H^2$, $x \in H$, for some number $c = c(T) > 0$, $t \in [t_0, T]$, $T \geq 0$;

(c) $c_1\|x\|_H^2 - k_1 e^{-\mu_1 t} \leq \Lambda(t, x) \leq c_2\|x\|_H^2 + k_2 e^{-\mu_2 t}$, $\forall x \in H$, $t \geq 0$;

(d) $(\mathbf{L}\Lambda)(t, x) \leq -c_3\Lambda(t, x) + k_3 e^{-\mu_3 t}, \quad \forall x \in \mathcal{D}(A), \quad t \geq 0,$

where $c_i > 0$, $k_i \geq 0$, $\mu_i \geq 0$, $i = 1, 2, 3$, are some constants and $\mathbf{L}$ is defined by

$$(\mathbf{L}\Lambda)(t, x) = \Lambda'_t(t, x) + \langle Ax + F(t, x), \Lambda'_x(t, x)\rangle_H \\ + \tfrac{1}{2} \cdot tr(\Lambda''_{xx}(t, x)G(t, x)QG(t, x)^*), \quad (3.7.2)$$

where $x \in \mathcal{D}(A)$, $t \geq t_0$, then X_t, $t \geq t_0 \geq 0$, satisfies

$$E\|X_t(x_0)\|_H^2 \leq \alpha_1\|x_0\|_H^2 \cdot e^{-\beta_1(t-t_0)} + \alpha_2 \cdot e^{-\beta_2 t}, \quad x_0 \in H, \quad t \geq t_0, \quad X_{t_0} = x_0, \\ (3.7.3)$$

for some $\alpha_1 > 0$, $\alpha_2 \geq 0$, $\beta_1 > 0$ and $\beta_2 \geq 0$.
  Conversely, suppose (3.7.3) holds and define

$$\Lambda(t_0, x_0) = \int_{t_0}^{T+t_0} E\|X_s(x_0)\|_H^2 ds + \alpha\|x_0\|_H^2, \quad \text{for all} \quad x_0 \in H, \quad t_0 \geq 0, \\ (3.7.4)$$

where $X_{t_0} = x_0$ and T, α are two proper positive constants, independent on $t_0 \in \mathbf{R}_+$, $x_0 \in H$, and assume $\Lambda(\cdot, \cdot)$ lies in the domain of the infinitesimal generator of X_t and satisfies all the conditions of Theorem 1.2.7 and (b), then there exist constants $c_i > 0$, $k_i \geq 0$, $\mu_i \geq 0$, $i = 1, 2, 3$, such that Conditions (c) and (d) above hold.

Proof   To prove (3.7.3), first applying Itô's formula to $\Lambda(t, x)$ and X_t^n, the strong solution of (3.3.6), then taking expectations and using Condition (d), we can deduce that for any $t \geq s \geq t_0$,

$$E\Lambda(t, X_t^n) \leq E\Lambda(s, X_s^n) + \int_s^t \Big[-c_3 E\Lambda(u, X_u^n) + k_3 \cdot e^{-\mu_3 u}\Big] du \\ + \int_s^t E\Big\{ \langle\Lambda'_x(u, X_u^n), (R(n) - I)F(u, X_u^n)\rangle_H \\ + \tfrac{1}{2}tr\Big[R(n)G(u, X_u^n)Q\big(R(n)G(u, X_u^n)\big)^* \Lambda''_{xx}(u, X_u^n) \\ -G(u, X_u^n)QG(u, X_u^n)^*\Lambda''_{xx}(u, X_u^n)\Big]\Big\} du. \qquad (3.7.5)$$

Consequently, letting $n \to \infty$ in (3.7.5) and using the condition (b) above and Proposition 1.3.6, it follows

$$E\Lambda(t, X_t) \leq E\Lambda(s, X_s) + \int_s^t \Big[-c_3 E\Lambda(u, X_u) + k_3 \cdot e^{-\mu_3 u}\Big] du \qquad (3.7.6)$$

which, by carrying out a similar argument to the proof of Theorem 3.4.1 and using the condition (a), immediately yields that

$$E\|X_t(x_0)\|_H^2 \leq \alpha_1\|x_0\|_H^2 \cdot e^{-\beta_1(t-t_0)} + \alpha_2 \cdot e^{-\beta_2 t}, \quad x_0 \in H, \quad t \geq t_0, \quad X_{t_0} = x_0,$$

for some $\alpha_1 > 0$, $\alpha_2 \geq 0$, $\beta_1 > 0$ and $\beta_2 \geq 0$.

Now suppose (3.7.3) is true and let

$$\Lambda(t_0, x_0) = \int_{t_0}^{T+t_0} E\|X_s(x_0)\|_H^2 ds + \alpha\|x_0\|_H^2, \qquad (3.7.7)$$

where T and α are two proper positive constants to be determined later. Substituting (3.7.3) into the above equality yields

$$\Lambda(t_0, x_0) \leq \int_{t_0}^{T+t_0} \left\{\alpha_1\|x_0\|_H^2 \cdot e^{-\beta_1(s-t_0)} + \alpha_2 \cdot e^{-\beta_2 s}\right\} ds + \alpha\|x_0\|_H^2$$

$$\leq \int \left[\frac{\alpha_1}{\beta_1} + \alpha\right]\|x_0\|_H^2 + \alpha_2 T \cdot e^{-\beta_2 t_0}. \qquad (3.7.8)$$

Let

$$\psi(t_0, x_0) = \int_{t_0}^{T+t_0} E\|X_s(x_0)\|_H^2 ds$$

and

$$\psi^n(t_0, R(n)x_0) = \int_{t_0}^{T+t_0} E\|X_s^n(R(n)x_0)\|_H^2 ds$$

where $X_t^n(\cdot, \cdot)$ is the approximation solution of (3.3.1) in the form of (3.3.6), $R(n) = R(n, A)$, $R(n, A)$ is the resolvent of A, and by the same arguments as in the proofs of Theorem 3.4.1, we have

$$E\psi^n(r + t_0, X_{r+t_0}^n(R(n)x_0)) = \int_{r+t_0}^{T+r+t_0} E\|X_s^n(X_{r+t_0}^n(R(n)x_0))\|_H^2 ds$$

$$= \int_{r+t_0}^{T+r+t_0} E\|X_s^n(R(n)x_0)\|_H^2 ds.$$

Hence, by the continuity of $s \to E\|X_s^n(R(n)x_0)\|_H^2$, we get

$$(\mathbf{L}\psi^n)(t_0, x_0)$$

$$= \frac{d}{dr} E\psi^n(r + t_0, X_{r+t_0}^n(R(n)x_0))\Big|_{r=0} + \varepsilon(n)$$

$$= \lim_{r \to 0} \frac{E\psi^n(r + t_0, X_{r+t_0}^n(R(n)x_0)) - E\psi^n(t_0, x_0)}{r} + \varepsilon(n)$$

$$= \lim_{r \to 0} \frac{\int_{r+t_0}^{T+r+t_0} E\|X_s^n(R(n)x_0)\|_H^2 ds - \int_{t_0}^{T+t_0} E\|X_s^n(R(n)x_0)\|_H^2 ds}{r} + \varepsilon(n)$$

$$= \lim_{r \to 0}\left(-\frac{1}{r}\int_{t_0}^{r+t_0} E\|X_s^n(R(n)x_0)\|_H^2 ds + \frac{1}{r}\int_{T+t_0}^{T+r+t_0} E\|X_s^n(R(n)x_0)\|_H^2 ds\right)$$

$$+ \varepsilon(n)$$

where $\varepsilon(n) \to 0$ as $n \to \infty$. Letting $n \to \infty$ and using (3.7.3) immediately implies

$$(\mathbf{L}\psi)(t_0, x_0) = -\|x_0\|_H^2 + E\|X_{T+t_0}(x_0)\|_H^2$$

$$\leq -\left(1 - \alpha_1 \cdot e^{-\beta_1 T}\right)\|x_0\|_H^2 + \alpha_2 \cdot e^{-\beta_2 t_0}. \qquad (3.7.9)$$

On the other hand, by virtue of the condition (3.7.1), it is easy to deduce that for any $x \in \mathcal{D}(A)$,

$$\mathbf{L}\|x\|_H^2 = 2\langle Ax + F(t, x), x\rangle_H + \|G(t, x)\|_{\mathcal{L}_2^0}^2 \leq \lambda(1 + trQ)\|x\|_H^2 + \gamma(1 + trQ)e^{-\mu t}$$

which, together with (3.7.9), immediately yields

$$
\begin{aligned}
(\mathbf{L}\Lambda)(t_0, x_0) &= (\mathbf{L}\psi)(t_0, x_0) + \alpha(\mathbf{L}\|x\|_H^2)(t_0, x_0) \\
&\leq -\left[1 - \alpha_1 e^{-\beta_1 T} - \alpha\lambda(1 + trQ)\right]\|x_0\|_H^2 \qquad (3.7.10) \\
&\quad + \left[\alpha_2 + \gamma(1 + trQ)\right]e^{-(\beta_2 \wedge \mu)t_0}.
\end{aligned}
$$

Therefore, if $T > \ln \alpha_1/\beta_1$, then we can choose $\alpha > 0$ small enough such that $\Lambda(\cdot, \cdot)$ satisfies (d) above. By (3.7.8) and the very definition of $\Lambda(\cdot, \cdot)$, it is clear that $\Lambda(\cdot, \cdot)$ satisfies the condition (c) and the proof is complete.   □

As in Section 3.4, it is generally difficult to know whether or not $\Lambda(\cdot, \cdot)$ in (3.7.4) belongs to $C^{1,2}(\mathbf{R}_+ \times H; \mathbf{R}^1)$. However, this may be the case if Equation (3.3.1) is linear. Consider the following linear system

$$dX_t = AX_t dt + BX_t dW_t \quad \text{with} \quad X_0 = x_0, \qquad (3.7.11)$$

where $B \in \mathcal{L}(H, \mathcal{L}(K, H))$ and A is the infinitesimal generator of the C_0 semigroup $T(t)$, $t \geq 0$, on H satisfying $\|T(t)\| \leq e^{\mu t}$, $t \geq 0$, for some $\mu \in \mathbf{R}^1$. In particular, by similar arguments to the proof of Corollary 3.4.2, we can derive:

Corollary 3.7.1   *Suppose $X_t(x_0)$, $t \geq 0$, is the mild solution of (3.7.11). If there exists a function $\Lambda(\cdot) \in C^2(H; \mathbf{R}^1)$ which satisfies the following:*

(a) $\Lambda(\cdot)$ satisfies all the conditions of using Itô's formula in Theorem 1.2.7;
(b) $c_1\|x\|_H^2 \leq \Lambda(x) \leq c_2\|x\|_H^2$, $\forall x \in H$;
(c) $(\mathbf{L}\Lambda)(x) \leq -c_3\Lambda(x)$, $\forall x \in \mathcal{D}(A)$,

where $c_i > 0$, $i = 1, 2, 3$, are some constants and $\mathbf{L}$ is the infinitesimal generator of the Markov process $X_t(x_0)$, i.e., $(\mathbf{L}\Lambda)(x) = \langle Ax, \Lambda'(x)\rangle_H + 1/2 \cdot tr(\Lambda''(x)(Bx)Q(Bx)^*)$, $x \in \mathcal{D}(A)$, then X_t, $t \geq 0$, satisfies

$$E\|X_t(x_0)\|_H^2 \leq \alpha_1\|x_0\|_H^2 \cdot e^{-\beta_1 t}, \quad x_0 \in H, \quad t \geq 0, \quad X_0 = x_0, \qquad (3.7.12)$$

for some $\alpha_1 > 0$ and $\beta_1 > 0$.
   Conversely, suppose (3.7.12) holds and define

$$\Lambda(x_0) = \int_0^\infty E\|X_s(x_0)\|_H^2 ds + \alpha\|x_0\|_H^2 \quad \text{for all} \quad x_0 \in H, \qquad (3.7.13)$$

where $X_0 = x_0$ and α is a proper positive constant, independent on $x_0 \in H$, then $\Lambda(\cdot)$ defined by (3.7.13) has the property that there exist constants $c_i > 0$, $i = 1, 2, 3$, such that Conditions (a), (b) and (c) above hold.

Proof   The first part may be proved by carrying out a similar argument to Theorem 3.7.1.

Now we suppose (3.7.12) is true. Then it is easy to see that the validity of (3.7.1) is clear. Indeed, this is immediate by Proposition 2.1.3 since A is supposed to generate a C_0-semigroup $T(t)$ satisfying $\|T(t)\| \le e^{\mu t}$, $t \ge 0$, for some $\mu \in \mathbf{R}^1$. Since the null solution is exponentially stable in mean square, the term $\int_0^\infty E\|X_s(x_0)\|_H ds$ is well defined and there exists a symmetric and nonnegative operator $R \in \mathcal{L}(H)$ (see, e.g. Da Prato and Zabczyk [1]) such that for any $x \in H$,

$$\int_0^\infty E\|X_s(x)\|_H^2 ds = \langle Rx, x\rangle_H$$

and for any $x \in \mathcal{D}(A)$,

$$(\mathbf{L}\langle R\cdot, \cdot\rangle_H)(x) = -\|x\|_H^2.$$

Hence,

$$\Lambda(x) = \langle Rx, x\rangle_H + \alpha\|x\|_H^2$$

for arbitrary $x \in H$. It is obvious that $\Lambda(\cdot) \in C_b^2(H)$ and

$$\alpha\|x\|_H^2 \le \Lambda(x) \le (\|R\| + \alpha)\|x\|_H^2$$

for arbitrary $x \in H$. This proves (a) and (b). To prove (c), we note that A is the infinitesimal generator of the C_0-semigroup $T(t)$, $t \ge 0$, satisfying $\|T(t)\| \le e^{\mu t}$, which implies $\langle x, Ax\rangle_H \le \mu\|x\|_H^2$, $x \in \mathcal{D}(A)$, by Proposition 2.1.4. Hence we have for any $x \in \mathcal{D}(A)$,

$$(\mathbf{L}\|\cdot\|_H^2)(x) = 2\langle x, Ax\rangle_H + tr(BxQ(Bx)^*) \le (2\mu + \|B\|^2 trQ)\|x\|_H^2.$$

Therefore,

$$\begin{aligned}
(\mathbf{L}\Lambda)(x) &= (\mathbf{L}\langle R\cdot, \cdot\rangle_H)(x) + \alpha(\mathbf{L}\|\cdot\|_H^2)(x) \\
&\le -\|x\|_H^2 + \alpha(2\lambda + \|B\|^2 trQ)\|x\|_H^2 \\
&= \big[-1 + \alpha(2\lambda + \|B\|^2 trQ)\big]\|x\|_H^2
\end{aligned}$$

for any $x \in \mathcal{D}(A)$. Letting α be small enough and using (b) yield the condition (c). This proves the theorem. □

In a similar way to Theorem 3.7.1, we can also derive the following ultimate boundedness result which is quite useful in investigating invariant measures of mild solutions of the stochastic evolution equation (3.3.1).

Theorem 3.7.2   *Assume the condition (3.7.1) holds with $\mu = 0$. Suppose $X_t(x_0)$, $t \ge t_0 \ge 0$, is the solution of (3.3.1) with $X_{t_0} = x_0$. If there exists a function $\Lambda(\cdot, \cdot) \in C^{1,2}(\mathbf{R}_+ \times H; \mathbf{R}^1)$ satisfying the following:*

(a) $\Lambda(\cdot, \cdot)$ satisfies all the conditions of using Itô's formula in Theorem 1.2.7;

(b) $|\Lambda'_t(t, x)| + \|x\|_H \|\Lambda'_x(t, x)\|_H + \|x\|^2_H \|\Lambda''_{xx}(t, x)\| \le c\|x\|^2_H$, $x \in H$, for some number $c = c(T) \ge 0$, $t \in [t_0, T]$, $T \in \mathbf{R}_+$;

(c) $c_1\|x\|^2_H - k_1 \le \Lambda(t, x) \le c_2\|x\|^2_H + k_2$, $\forall x \in H$, $t \ge 0$;

(d) $(\mathbf{L}\Lambda)(t, x) \le -c_3\Lambda(t, x) + k_3$, $\forall x \in \mathcal{D}(A)$, $t \ge 0$,

where $c_i > 0$, $k_i \ge 0$, $i = 1, 2, 3$, are some constants and $\mathbf{L}$ is the infinitesimal generator defined as in Theorem 3.7.1 of the Markov process $X_t(x_0)$, then X_t, $t \ge t_0$, satisfies

$$E\|X_t(x_0)\|^2_H \le \alpha_1\|x_0\|^2_H \cdot e^{-\beta_1(t-t_0)} + \alpha_2, \quad t \ge t_0, \quad X_{t_0} = x_0 \in H, \quad (3.7.14)$$

for some $\alpha_1 > 0$, $\alpha_2 \ge 0$ and $\beta_1 > 0$.

Conversely, suppose (3.7.14) holds and define

$$\Lambda(t_0, x_0) = \int_{t_0}^{T+t_0} E\|X_s(x_0)\|^2_H ds + \alpha\|x_0\|^2_H \quad \text{for all} \quad x_0 \in H, \quad t_0 \ge 0,$$

where $X_{t_0} = x_0$ and T, α are two proper constants, independent on $t_0 \in \mathbf{R}_+$, $x_0 \in H$. Assume $\Lambda(\cdot, \cdot)$ lies in the domain of the infinitesimal generator of X_t and satisfies all the conditions of Theorem 1.2.7 and (b), then there exist constants $c_i > 0$, $k_i \ge 0$, $i = 1, 2, 3$, such that Conditions (c) and (d) above hold.

In nonlinear situations, generally speaking, we have difficulty in showing $\Lambda(t, x) \in C^{1,2}(\mathbf{R}^+ \times H; \mathbf{R}^1)$; it is therefore appropriate to use first order approximation technique as we did in Section 3.5 to study exponential decay of the solutions of Equation (3.3.1).

Proposition 3.7.1   *Suppose the null solution of Equation (3.7.11) is exponentially stable in mean square and the relation (3.7.12) holds. Then the solution $X_t(x_0)$ of (3.3.1) with $t_0 = 0$, $X_0 = x_0$ is exponentially decayable in mean square if*

$$2\|x\|_H \|F(t, x)\|_H + tr\left[G(t, x)QG(t, x)^* - (Bx)Q(Bx)^* \right] < \theta\|x\|^2_H + \gamma \cdot e^{-\mu t}$$

$$(3.7.15)$$

for some positive constants $\gamma > 0$, $\mu > 0$ and $0 < \theta < \beta_1/\alpha_1$.

Proof   Let

$$\Lambda_0(x) = \langle Rx, x \rangle_H + \alpha\|x\|^2_H$$

as defined in the proofs of Corollary 3.7.1 for arbitrary $x \in H$ and

$$(\mathbf{L}_0\langle R\cdot, \cdot\rangle_H)(x) = -\|x\|^2_H$$

for any $x \in \mathcal{D}(H)$ where $\mathbf{L}_0$ is the infinitesimal generator corresponding to the equation (3.7.11). Hence, $\|R\| \le \alpha_1/\beta_1$. On the other hand, since $\Lambda_0(x) \in$

$C^2(H; \mathbf{R}^+)$ by using Corollary 3.7.1, it suffices for our purposes to show that $\Lambda_0(x)$ satisfies the condition (d) in Theorem 3.7.1. Since

$$(\mathbf{L}\Lambda_0)(t,x) - (\mathbf{L}_0\Lambda_0)(x)$$
$$= \langle \Lambda_0'(x), F(t,x) \rangle_H + \frac{1}{2}tr\left[\Lambda_0''(x)\left(G(t,x)QG(t,x)^* - (Bx)Q(Bx)^*\right)\right]$$
$$= 2\langle (R+\alpha I)x, F(t,x) \rangle_H + tr\left[(R+\alpha I)\left(G(t,x)QG(t,x)^* - (Bx)Q(Bx)^*\right)\right]$$
$$\leq 2(\|R\| + \alpha)\|x\|_H\|F(t,x)\|_H$$
$$+ (\|R\| + \alpha)tr\left[G(t,x)QG(t,x)^* - (Bx)Q(Bx)^*\right]$$
$$= (\|R\| + \alpha)\left\{2\|x\|_H\|F(t,x)\|_H + tr\left[G(t,x)QG(t,x)^* - (Bx)Q(Bx)^*\right]\right\}$$

for arbitrary $x \in \mathcal{D}(A)$. By (3.7.10) and the assumption (3.7.15), $(\mathbf{L}\Lambda_0)(t,x)$ satisfies the condition (d) in Theorem 3.7.1 if we choose $\alpha > 0$ small enough and a suitable $T > 0$ in (3.7.10). This proves the proposition. □

Also, we use the first order approximation to derive some results about exponential ultimate boundedness in mean square of nonlinear equations, based on the same property of the corresponding linear equations. In fact, in a similar way to Theorem 3.5.5, we have the following results.

Proposition 3.7.2   *Consider the equation (3.7.11) and let $\{X_t^0(x_0), t \geq 0\}$ be its mild solution. Assume that this solution is exponentially ultimately bounded in mean square and satisfies the relation (3.7.14). Let $\{X_t(x_0), t \geq 0\}$ be a solution of the equation (3.3.1) and suppose*

$$2\|x\|_H\|F(t,x)\|_H + tr\left[G(t,x)QG(t,x)^* - (Bx)Q(Bx)^*\right] \leq \theta\|x\|_H^2 + M, \quad \forall x \in H,$$
$$(3.7.16)$$

for any constant $M > 0$ and some

$$0 < \theta < \max_{s > 0 \vee \frac{\ln \alpha_1}{\beta_1}} \frac{1 - \alpha_1 e^{-\beta_1 s}}{\alpha_1/\beta_1 + \alpha_2 s}. \qquad (3.7.17)$$

Then the solution of (3.3.1) is exponentially ultimately bounded in mean square.

Proof   The proofs are quite similar to those in Theorem 3.5.3 and are therefore omitted here (also, see Liu and Mandrekar [2]). □

3.7.2   Ultimate Boundedness and Invariant Measures

In this section, we shall consider the existence and uniqueness of invariant measures of mild solutions by means of boundedness of moments and by using

Lyapunov function characterization methods. Let $X_t(x_0)$, $t \geq 0$, be the mild solution of (3.3.10) whose coefficients $F(\cdot)$ and $G(\cdot)$ satisfy the usual Lipschitz and linear growth conditions.

Lemma 3.7.1   *Suppose the mild solution $X_t(x_0)$, $t \geq 0$, of (3.3.10) is ultimately bounded in mean square satisfying, for instance, (3.7.14). Then for any invariant measure ν of the Markov process $X_t(x_0)$,*

$$\int_H \|y\|_H^2 \nu(dy) \leq \alpha_2 < \infty, \tag{3.7.18}$$

where α_2 is the constant in (3.7.14).

Proof   Let $f(y) = \|y\|_H^2$ and $f_n(y) = \chi_{[0,n]}(f(y))f(y)$, $y \in H$, where $\chi_B(\cdot)$ is the indicator function of $B \subset \mathbf{R}^1$. We note that $f_n(\cdot) \in L^1(H, \mathcal{B}(H), \nu; \mathbf{R}^1)$. From the assumption of ultimate boundedness (3.7.14), there is a constant $\alpha_2 \geq 0$ such that

$$\limsup_{t \to \infty} E_{x_0} f(X_t) \leq \alpha_2 \quad \text{for any} \quad x_0 \in H, \tag{3.7.19}$$

where E_{x_0} means the conditional expectation under the condition $X_0 = x_0$. Using the Ergodic theorem for Markov processes with invariant measures (cf. Yosida [1]), there exists the following limit

$$\lim_{N \to \infty} \frac{1}{N} \sum_{k=1}^N P_k f_n(x_0) = f_n^*(x_0), \quad \nu - a.s. \tag{3.7.20}$$

and

$$E_\nu f_n^* = E_\nu f_n, \tag{3.7.21}$$

where $P_k f_n(x_0) = \int_H f_n(y) P(k, x_0, dy)$ and $E_\nu f_n = \int_H f_n(x) d\nu(x)$. ¿From the inequality $f_n(x) \leq f(x)$ and the assumption of ultimate boundedness of the solution (3.3.10), we have for any $x \in H$

$$\limsup_{N \to \infty} \frac{1}{N} \sum_{k=1}^N P_k f_n(x) \leq \limsup_{N \to \infty} \frac{1}{N} \sum_{k=1}^N P_k f(x) \leq \alpha_2,$$

which, together with (3.7.20), immediately yields $f_n^*(x) \leq \alpha_2$ (ν-a.s.). Therefore, from this inequality, we have

$$E_\nu f_n^* \leq \alpha_2. \tag{3.7.22}$$

The results (3.7.21), (3.7.22) and the fact $f_n(x) \uparrow f(x)$ ($n \to \infty$) imply that

$$E_\nu f = \lim_{n \to \infty} E_\nu f_n = \lim_{n \to \infty} E_\nu f_n^* \leq \alpha_2 < \infty.$$

$\square$

We are now in a position to obtain the main results in this subsection. First of all, we establish the uniqueness of invariant measures of mild solutions.

Theorem 3.7.3   *Suppose the solution $X_t(x_0)$, $t \geq 0$, of (3.3.10) is exponentially ultimately bounded in mean square. Suppose for each $R > 0$, $\delta > 0$ and $\varepsilon > 0$, there exists $T_0 = T_0(R, \delta, \varepsilon) > 0$ such that*

$$P\left\{\left\|X_t(x_1) - X_t(x_2)\right\|_H > \delta\right\} < \varepsilon \quad \text{for any} \quad x_1, \, x_2 \in H_R \quad \text{whenever} \quad t \geq T_0,$$
$$(3.7.23)$$

where $H_R = \{x \in H : \|x\|_H \leq R\}$. Then there exists at most one invariant measure.

Proof   Let μ_i, $i = 1, \, 2$, be invariant measures of (3.3.10). Then by Lemma 3.7.1, for each $\varepsilon > 0$, there exists $R > 0$ such that $\mu_i(H \backslash H_R) < \varepsilon$. Let $\psi \in C_w^b(H)$, the space of all bounded weakly continuous functions on H (cf. Kantorovich and Akilov [1]). We firstly claim that there exists $T = T(\varepsilon, R, \psi) > 0$ such that

$$\left\|(P_t\psi)(x_1) - (P_t\psi)(x_2)\right\|_H \leq \varepsilon \quad \text{for} \quad x_1, \, x_2 \in H_R \quad \text{if} \quad t \geq T. \quad (3.7.24)$$

Indeed, to this end, let L be a weakly compact set in H and recall that the weak topology on L is equivalent to the topology defined by the metric

$$d(x, y) = \sum_{k=1}^{\infty} (1/2^k)|\langle e_k, x - y\rangle_H|, \quad x, \, y \in L,$$

for any orthonormal set $\{e_k\}$ of H. We shall first prove that for each $\delta > 0$ and $\varepsilon > 0$, there exists a number $T_1 = T_1(\varepsilon, R, \delta, \psi) > 0$ such that $t \geq T_1$ implies

$$P\left\{\left|\psi(X_t(x_1)) - \psi(X_t(x_2))\right| \leq \delta\right\} \geq 1 - \varepsilon \quad \text{for all} \quad x_1, \, x_2 \in H_R.$$

By the ultimate boundedness, there exists $T_2 = T_2(\varepsilon, R) > 0$ such that $t \geq T_2$ implies

$$P\left\{X_t(x) \in H_R\right\} \geq 1 - \varepsilon/3 \quad \text{for any} \quad x \in H_R.$$

Note that $\psi(\cdot)$ on H_R is uniformly continuous with respect to the metric d. Hence, there exists a $\delta' > 0$ such that $x, \, y \in H_R$ and $d(x, y) \leq \delta'$ implies $|\psi(x) - \psi(y)| \leq \delta$. Note also that there exists an integer $J > 0$ such that

$$\sum_{k=J+1}^{\infty} (1/2^k)|\langle e_k, x - y\rangle_H| \leq \delta'/2 \quad \text{for all} \quad x, \, y \in H_R.$$

Now choose $T_1 \geq T_2$ such that $t \geq T_1$ implies

$$\sum_{k=1}^{J} P\left\{\left|\langle e_k, X_t(x_1) - X_t(x_2)\rangle_H\right| \leq \delta'/2\right\} \geq 1 - \varepsilon/3 \quad \text{for all} \quad x_1, x_2 \in H_R,$$

a case which is possible by using the condition (3.7.23). Hence, for $t \geq T_1$ we have

$$P\left\{\left|\psi(X_t(x_1)) - \psi(X_t(x_2))\right| \leq \delta\right\}$$

$$\geq P\left\{X_t(x_1), X_t(x_2) \in H_R, \ d(X_t(x_1), X_t(x_2)) \leq \delta'\right\}$$

$$\geq P\left\{X_t(x_1), X_t(x_2) \in H_R, \ \sum_{k=1}^{J}(1/2^k)\left|\langle e_k, X_t(x_1) - X_t(x_2)\rangle_H\right| \leq \delta'/2\right\}$$

$$\geq P\left\{X_t(x_1), X_t(x_2) \in H_R, \ \left|\langle e_k, X_t(x_1) - X_t(x_2)\rangle_H\right| \leq \delta'/2, \ k = 1, 2, \cdots J\right\}$$

$$\geq 1 - \varepsilon/3 - \varepsilon/3 - \varepsilon/3 = 1 - \varepsilon.$$

Now for the given $\varepsilon > 0$, choosing $T > 0$ such that $t \geq T$ implies

$$P\left\{\left|\psi(X_t(x_1)) - \psi(X_t(x_2))\right| \leq \varepsilon/2\right\} \geq 1 - \frac{\varepsilon}{4M_0}$$

where $M_0 = \sup|\psi(x)| < \infty$. Then

$$E\left|\psi(X_t(x_1)) - \psi(X_t(x_2))\right| \leq \varepsilon/2 + 2M_0(\varepsilon/4M_0) = \varepsilon$$

as required. In order to conclude our proof, first note that

$$\int_H \psi(x)m_i(dx) = \int_H [P_t\psi](x)m_i(dx), \quad i = 1, 2.$$

On the other hand, we have for any $t \geq T$,

$$\left|\int_H \psi(x_1)m_1(dx_1) - \int_H \psi(x_2)m_2(dx_2)\right|$$

$$= \left|\int_H \int_H [\psi(x_1) - \psi(x_2)]m_1(dx_1)m_2(dx_2)\right|$$

$$= \left|\int_H \int_H \left\{[P_t\psi](x_1) - [P_t\psi](x_2)\right\}m_1(dx_1)m_2(dx_2)\right|$$

$$\leq \int_H \int_H \left|[P_t\psi](x_1) - [P_t\psi](x_2)\right|m_1(dx_1)m_2(dx_2)$$

$$= \left(\int_{H_R} + \int_{H\backslash H_R}\right)\left(\int_{H_R} + \int_{H\backslash H_R}\right)\left|[P_t\psi](x_1) - [P_t\psi](x_2)\right|m_1(dx_1)m_2(dx_2)$$

$$\leq \varepsilon + 2(2M_0)\varepsilon + 2M_0\varepsilon^2.$$

Since $\varepsilon > 0$ is arbitrary, we deduce

$$\int_H \psi(x_1) m_1(dx_1) = \int_H \psi(x_2) m_2(dx_2)$$

which immediately implies $m_1(\cdot) = m_2(\cdot)$.   □

Corollary 3.7.2   *Assume the solution $X_t(x_0)$, $t \geq 0$, of (3.3.10) is exponentially ultimately bounded in mean square. Suppose the conditions of Propositions 3.3.1 and 3.3.2 hold, then (3.7.23) is satisfied. In other words, in this case there exists at most one invariant measure.*

To establish existence results similar to those in Section 3.5 for mild solutions, first notice that in this situation the condition (3.5.13), for instance, does not make sense any more. A suitable version however is possible with the V-norm replaced by a H-norm plus some additional assumptions on the associated infinitesimal generator A. For instance, we impose the following assumption:

(H6) A is self-adjoint and has eigenvectors $\{e_k\}$, $k = 1, 2, \cdots$, which form an orthonormal basis of H and eigenvalues $\{-\lambda_k\} \downarrow -\infty$ as $k \to \infty$.

Theorem 3.7.4   *Suppose the condition (H6) holds and the solution of (3.3.10) satisfies*

$$\frac{1}{T} \int_0^T E\|X_s(x_0)\|_H^2 ds \leq M(1 + \|x_0\|_H^2) \qquad (3.7.25)$$

for some $M > 0$ and any $T \geq 0$. Then there exists an invariant measure for the equation (3.3.10).

To obtain this result, we first show the following lemma.

Lemma 3.7.2   *Suppose the conditions in Theorem 3.7.4 hold. Then*

$$m_t(\cdot) = \frac{1}{t} \int_0^t P(s, x_0, \cdot) ds \quad \text{for} \ \ t \geq 0$$

is weakly compact. Here $P(t, x_0, \Gamma)$, $\Gamma \in \mathcal{B}(H)$, is the transition probability of the solution process $X_t(x_0)$, $t \geq 0$.

Proof   By a well-known result (cf. Gihman and Skorohod [1]), it suffices to show

$$\frac{1}{T} \int_0^T \sum_{k=1}^\infty Ex_k^2(t) dt \quad \text{is uniformly convergent in } T \geq 0,$$

where $x_k(t) = \langle X_t(x_0), e_k \rangle_H$ and $\{e_k\}$ is the orthonormal basis given in (H6). Since the assumption (H6) holds, the semigroup $T(t)$ has the representation

$$T(t)h = \sum_{k=1}^{\infty} e^{-\lambda_k t} \langle e_k, h \rangle_H e_k, \quad h \in H,$$

which, by the definition of mild solutions, immediately yields

$$x_k(t) = e^{-\lambda_k t} x_k + \int_0^t e^{-\lambda_k(t-r)} \langle e_k, F(X_r(x_0)) \rangle_H dr$$

$$+ \int_0^t e^{-\lambda_k(t-r)} \langle e_k, G(X_r(x_0)) dW_r \rangle_H$$

where $x_k = \langle x_0, e_k \rangle_H$, $k = 1, 2, \cdots$. Hence,

$$Ex_k^2(t) = 3e^{-2\lambda_k t} x_k^2 + 3E \left| \int_0^t e^{-\lambda_k(t-r)} \langle e_k, F(X_r(x_0)) \rangle_H dr \right|^2$$

$$+ 3E \left| \int_0^t e^{-\lambda_k(t-r)} \langle e_k, G(X_r(x_0)) dW_r \rangle_H \right|^2.$$

Letting N be large enough so that $\lambda_N > 0$ and using Hölder's inequality, we can deduce that for any integer $m > 0$,

$$\sum_{k=N}^{N+m} \frac{1}{T} \int_0^T E \left| \int_0^t e^{-\lambda_k(t-r)} \langle e_k, F(X_r(x_0)) \rangle_H dr \right|^2 dt$$

$$\leq \frac{\int_0^T E\|F(X_r(x_0))\|_H^2 dr}{4\delta(\lambda_N - \delta)T} \leq \frac{C_1(1 + \|x_0\|_H^2)}{\delta(\lambda_N - \delta)}$$

for a small $\delta > 0$ and some $C_1 > 0$. In a similar way, it can be deduced that

$$\sum_{k=N}^{N+m} \frac{1}{T} \int_0^T E \left| \int_0^t e^{-\lambda_k(t-r)} \langle e_k, G(X_r(x_0)) dW_r \rangle_H \right|^2 dt$$

$$\leq \frac{trQ \int_0^T E\|G(X_t(x_0))\|^2 dt}{2\lambda_N T} \leq \frac{C_2 trQ(1 + \|x_0\|_H^2)}{\lambda_N}$$

for some $C_2 > 0$. Thus,

$$\sum_{k=N}^{N+m} \frac{1}{T} \int_0^T Ex_k(t)^2 dt$$

$$\leq 3\|x_0\|_H^2/2\lambda_N + 3(C_1 + C_2)(1 + \|x_0\|_H^2)[1/\delta(\lambda_N - \delta) + trQ/\lambda_N]$$

which tends to zero uniformly in $T \geq 0$ as $N \to \infty$. $\square$

Proof of Theorem 3.7.4   For integer $n \geq 0$, define

$$m_n(\Gamma) = \frac{1}{n} \int_0^n P(s, x_0, \Gamma)ds, \quad \Gamma \in \mathcal{B}(H).$$

Then $m_n(\cdot)$ is a probability measure and

$$\int_H \|y\|_H^2 m_n(dy) \leq M(1 + \|x_0\|_H^2).$$

Hence, for each $\varepsilon > 0$, there exists $R > 0$ such that

$$m_n(H_R) > 1 - \varepsilon, \quad H_R = \{x \in H : \|x\|_H \leq R\}.$$

By Lemma 3.7.2, $m_n(\cdot)$, $n \geq 0$, is weakly compact and there exists a subsequence, still denoted by $m_n(\cdot)$, which is weakly convergent to some probability measure $m(\cdot)$. By a similar argument to the proof of Theorem 3.5.6, it can be deduced that $m(\cdot)$ is an invariant measure of $P(t, x_0, \cdot)$. The proof is now complete.   □

Remark   Nowadays, there exists an extensive literature on the topic of invariant measures of the stochastic evolution equations (3.3.10). In particular, various extensions of Theorems 3.7.3 and 3.7.4 in one way or another have been made by some researchers. The reader is referred to the existing references such as Da Prato and Zabczyk [2] for more material in this respect.

Example 3.7.1   Consider the following stochastic heat equation

$$\begin{cases} dy(x,t) = \partial^2/\partial x^2 y(x,t)dt + \left[cy^2(x,t)/(1+|y(x,t)|)\right]dB_t^1 + g(x)dB_t^2 \\ \quad\quad 0 < x < 1, \quad c > 0, \\ y(0,t) = y(1,t) = 0, \quad y(x,0) = y_0(x), \quad y_0(\cdot), \quad g(\cdot) \in L^\infty(0,1), \end{cases}$$

$$(3.7.26)$$

where B_t^i, $i = 1, 2$, are mutually independent real Brownian motions. In this situation, we take $H = L^2(0,1)$ and $A = d^2/dx^2$ with $\mathcal{D}(A) = \{y \in H : y', y'' \in H, y(0) = y(1) = 0\}$. Then it is easy to deduce $\langle Ay, y \rangle_H \leq -\pi^2\|y\|_H^2$. Also note that the condition (H6) is satisfied since A has eigenvectors $\{\sqrt{2}\sin n\pi x\}$ and eigenvalues $\{-n^2\pi^2\}$, $n = 1, 2, \cdots$. Now let $\Lambda(y) = \|y\|_H^2$, then

$$(\mathbf{L}\| \cdot \|_H^2)(y) \leq -2[\pi^2 - c^2/2]\|y\|_H^2 + \|g\|_{L^\infty(0,1)}^2, \quad y \in \mathcal{D}(A).$$

If $2\pi^2 > c^2$, then the solution is exponentially ultimately bounded in mean square. Moreover, by virtue of Theorem 3.7.4 there exists an invariant measure. We also have

$$(\mathbf{L}_d\| \cdot \|_H^2)(y) \leq -[2\pi^2 - c^2]\|y\|_H^2, \quad y \in \mathcal{D}(A),$$

where $\mathbf{L}_d$ is defined exactly as that in Proposition 3.3.2. Hence, if $2\pi^2 > c^2$, by virtue of Theorem 3.7.3, the invariant measure is unique. Furthermore, if $g(\cdot) = 0$, the null solution is exponentially stable in mean square.

3.8   Decay Rates of Systems

The stochastic differential equation (3.2.1) or (3.3.1) could be commonly illustrated as the perturbed stochastic system by white noise sources of a deterministic differential equation. For instance, consider the system (3.3.1) which could be regarded as the perturbed stochastic system of the corresponding deterministic one

$$dX_t = T(t)x_0 + \int_0^t T(t-s)F(s, X_s)ds, \qquad X_0 = x_0 \in H. \qquad (3.8.1)$$

We have shown in Example 3.2.1 that when the solution of (3.8.1) is exponentially decayable (for instance, suppose $F(t, x) = 0$ and A generates a stable C_0-semigroup $T(t)$, $t \geq 0$, on H), it cannot be generally deduced that its perturbed stochastic system (3.3.1) remains exponentially decayable, despite the fact that the solution might remain decayable with a slower decay. To put this more precisely, suppose $H = L^2(0, 1)$ and

$$\left\{ W_q(t, x); \ t \geq 0, \ x \in [0, 1] \right\}$$

is an H-valued Wiener process with associated covariance operator Q, $trQ < \infty$, determined by a positive definite kernel

$$q(x, y) \in L^2\Big([0, 1] \times [0, 1]; \mathbf{R}^1\Big), \qquad q(x, x) \in L^2(0, 1; \mathbf{R}^1).$$

Let A be a linear strictly elliptic differential operator of the second order, for instance, the Laplace operator on $[0, 1]$, $F(t, \cdot) = 0$ and $G(t, \cdot)$ in (3.3.1) is the operator defined by $G(t, \cdot)h(\cdot) = (1 + t)^{-\mu}h(\cdot)$ for some constant $\mu > 0$ and arbitrary $h(\cdot) \in L^2(0, 1)$. Then we may formulate the following problem with null initial data

$$
\begin{aligned}
dX_t(x) &= \frac{\partial^2 X_t(x)}{\partial x^2}dt + (1 + t)^{-\mu}dW_q(t, x), \quad t \geq 0, \quad x \in [0, 1], \\
X_0(x) &= 0, \quad x \in [0, 1]; \quad X_t(0) = X_t(1) = 0, \quad t \geq 0.
\end{aligned}
\qquad (3.8.2)
$$

It is easy to obtain the explicit mild solution:

$$X_t(x) = \int_0^t T(t-s)(1+s)^{-\mu}dW_q(s, x), \qquad t \geq 0,$$

where $T(t)$, $t \geq 0$, is the strongly continuous semigroup generated by the Laplace operator $\frac{\partial^2}{\partial x^2}$. It may be shown by a direct computation that for arbitrarily given $\mu > 0$, the Lyapunov exponents

$$\limsup_{t\to\infty} \frac{\log E\|X_t(\cdot)\|_H^2}{t} = 0 \quad \text{and} \quad \limsup_{t\to\infty} \frac{\log \|X_t(\cdot)\|_H}{t} = 0 \quad a.s.$$

However, we want to point out that although the mild solution is not exponentially decayable, we show below that the solution is decayable with a slower, polynomial, decay rate.

3.8.1 Decay in the p-th Moment

Consider the semilinear stochastic differential equation (3.3.1) in the separable, real Hilbert space H. Assume $\Lambda(t,x) : \mathbf{R}_+ \times H \to \mathbf{R}_+$ is any function differentiable of the first order in $\mathbf{R}_+$ and twice Fréchet differentiable in H. Denote it by $\Lambda(t,x) \in C^{1,2}(\mathbf{R}_+ \times H; \mathbf{R}_+)$. Assume that $\Lambda(t,x)$ satisfies all the conditions of using Itô's formula in Theorem 1.2.7. For our purpose, we introduce the following operator $\mathbf{L}$:

$$(\mathbf{L}\Lambda)(t,x) := \Lambda_t'(t,x) + \langle \Lambda_x'(t,x),\, Ax + F(t,x) \rangle_H$$
$$+ \frac{1}{2} \cdot tr\Big[\Lambda_{xx}''(t,x)G(t,x)QG(t,x)^*\Big], \quad t \geq 0, \quad x \in \mathcal{D}(A).$$

$$(3.8.3)$$

Definition 3.8.1 Let $p \geq 2$ and $\lambda(t) \uparrow \infty$, as $t \to \infty$, which is some non-decreasing, continuous function defined for sufficiently large $t > 0$. The mild solution of Equation (3.3.1) is said to be *decayable with rate* $\lambda(t)$ *in the p-th moment* if there exist a positive constant $\gamma > 0$ and function $O(\lambda(t))$, $t \geq 0$, such that

$$\limsup_{t \to \infty} \frac{\log E\|X_t(x_0)\|_H^p}{\log O(\lambda(t))} \leq -\gamma, \quad (3.8.4)$$

holds for any $X_0 = x_0 \in H$, where $O(\lambda(t))$ is the big oh of $\lambda(t)$.

Theorem 3.8.1 Let $\Lambda(t,x) \in C^{1,2}(\mathbf{R}_+ \times H; \mathbf{R}_+)$ and $\psi_1(t)$, $\psi_2(t)$ be two non-negative, continuous functions. Assume that there exist a positive constant $m > 0$ and real numbers ν, $\theta \in \mathbf{R}^1$ such that for some $p \geq 2$,

(1). $\|x\|_H^p \lambda(t)^m \leq \Lambda(t,x)$, $(t,x) \in \mathbf{R}_+ \times H$;
(2). $|\Lambda_t'(t,x)| + |\Lambda(t,x)| + \|x\|_H\|\Lambda_x'(t,x)\|_H + \|x\|_H^2\|\Lambda_{xx}''(t,x)\| \leq c\|x\|_H^p$, $x \in H$, *for some constant $c = c(T) > 0$, $t \in [0,T]$, $T \in \mathbf{R}_+$;*
(3). $(\mathbf{L}\Lambda)(t,x) \leq \psi_1(t) + \psi_2(t)\Lambda(t,x)$, $(t,x) \in \mathbf{R}_+ \times \mathcal{D}(A)$;
(4).

$$\limsup_{t \to \infty} \frac{\log \int_0^t \psi_1(s)ds}{\log \lambda(t)} \leq \nu, \qquad \limsup_{t \to \infty} \frac{\int_0^t \psi_2(s)ds}{\log \lambda(t)} \leq \theta. \quad (3.8.5)$$

Then, if $\gamma = m - \theta - \nu > 0$, we have

$$\limsup_{t \to \infty} \frac{\log E\|X_t(x_0)\|_H^p}{\log \lambda(t)} \leq -\gamma. \quad (3.8.6)$$

In other words, the solution of Equation (3.3.1) is decayable with rate $\lambda(t)$ in the p-th moment sense.

Proof   In order to show (3.8.6), let us carry out an approximating solution procedure exactly as we did in Section 3.3. To this end, applying Itô's formula to $\Lambda(t, x)$ and the strong solution X_t^n of (3.3.6) yields

$$\Lambda(t, X_t^n) = \Lambda(0, X_0^n) + \int_0^t (\mathbf{L}^n \Lambda)(s, X_s^n)ds + \int_0^t \langle \Lambda_x'(s, X_s^n), G(s, X_s^n)dW_s \rangle_H,$$

$$(3.8.7)$$

where

$$(\mathbf{L}^n \Lambda)(t, x) = \Lambda_t'(t, x) + \langle \Lambda_x'(t, x), Ax + R(n)F(t, x) \rangle_H$$
$$+ \frac{1}{2} tr\Big[\Lambda_{xx}''(t, x)R(n)G(t, x)Q\big(R(n)G(t, x)\big)^* \Big], \qquad x \in \mathcal{D}(A),$$

and $n \in \rho(A)$, the resolvent set of A. Now, for fixed $n \in \mathbf{R}_+$ and arbitrary $m \in \mathbf{R}_+$, define an increasing sequence of stopping times

$$\tau_m^n = \begin{cases} \inf\Big\{ t > 0 : \Big| \int_0^t \langle \Lambda_x'(s, X_s^n), G(s, X_s^n)dW_s \rangle_H \Big| > m \Big\}, \\ \infty \qquad \text{if the set is empty.} \end{cases}$$

Clearly, for any fixed n, $\tau_m^n \uparrow \infty$, as $m \to \infty$, and since

$$\int_0^t \langle \Lambda_x'(s, X_s^n), G(s, X_s^n)dW_s \rangle_H, \qquad t \in \mathbf{R}_+,$$

is a continuous localmartingale, it follows that for fixed $n \geq 1$, and any $m \geq 1$,

$$E\Big(\int_0^{t \wedge \tau_m^n} \langle \Lambda_x'(s, X_s^n), G(s, X_s^n)dW_s \rangle_H \Big) = 0, \qquad t \in \mathbf{R}_+.$$

Therefore, taking expectations in (3.8.7) and using Conditions (2) and (3) in Theorem 3.8.1 yield

$$E\Lambda(t \wedge \tau_m^n, X_{t \wedge \tau_m^n}^n) \leq E\Lambda(0, X_0^n) + \int_0^{t \wedge \tau_m^n} \Big[\psi_1(s) + \psi_2(s)E\Lambda(s, X_s^n) \Big] ds$$

$$+ \int_0^{t \wedge \tau_m^n} E\Big\{ \langle \Lambda_x'(s, X_s^n), (R(n) - I)F(s, X_s^n) \rangle_H$$

$$+ \frac{1}{2} tr\Big[R(n)G(s, X_s^n)Q\big(R(n)G(s, X_s^n)\big)^* \Lambda_{xx}''(s, X_s^n)$$

$$- G(s, X_s^n)QG(s, X_s^n)^* \Lambda_{xx}''(s, X_s^n) \Big] \Big\} ds$$

$$(3.8.8)$$

where $R(n) = nR(n, A)$, $R(n, A)$ is the resolvent of A.

Firstly, letting $m \to \infty$ and using the Fatou's lemma in (3.8.8), then taking $n \to \infty$, together with Condition (2) above and using Proposition 1.3.6, it follows that

$$E\Lambda(t, X_t) \le E\Lambda(0, x_0) + \int_0^t \Big(\psi_1(s) + \psi_2(s) E\Lambda(s, X_s) \Big) ds.$$

Hence, by virtue of the Gronwall's lemma we derive

$$E\Lambda(t, X_t) \le \Big[E\Lambda(0, x_0) + \int_0^t \psi_1(s) ds \Big] \exp \Big(\int_0^t \psi_2(s) ds \Big).$$

Hence, for arbitrary $\varepsilon > 0$, in view of Condition (4) we have that for any $t \in \mathbf{R}_+$ sufficiently large,

$$\log E\Lambda(t, X_t) \le \log \Big[E\Lambda(0, x_0) + \lambda(t)^{\nu + \varepsilon} \Big] + \log \lambda(t)^{\theta + \varepsilon},$$

which, letting $\varepsilon \to 0$ and using Condition (1), immediately implies

$$\limsup_{t \to \infty} \frac{\log E \| X_t(x_0) \|_H^p}{\log \lambda(t)} \le \limsup_{t \to \infty} \frac{\log \big[\lambda(t)^{-m} E\Lambda(t, X_t) \big]}{\log \lambda(t)}$$
$$\le -[m - (\nu + \theta)]$$

as required. □

Next, let us investigate a class of stochastic differential equations which includes as a special case (3.8.2) to illustrate the theory derived above.

Example 3.8.1 Let $\mu > 1/2$ be a positive constant. Consider the following stochastic partial differential equation

$$\begin{cases} dY_t(x) = \frac{\partial^2}{\partial x^2} Y_t(x) dt + (1+t)^{-\mu} \alpha(Y_t(x)) dW_q(t, x), & t > 0, \quad x \in (0, 1), \\ Y_0(x) = y_0(x), \quad Y_t(0) = Y_t(1) = 0, & t \ge 0, \end{cases}$$
$$(3.8.9)$$

where $W_q(t, x)$ is the Q-Wiener process defined exactly as in the beginning of the section, and $\alpha(\cdot) : \mathbf{R}^1 \to \mathbf{R}^1$ is a certain bounded, Lipschitz continuous function with $|\alpha(y)| \le C$, $C > 0$, for each $y \in \mathbf{R}^1$. Let $H = L^2[0, 1]$. Clearly, on this occasion $F(t, u) = 0$, $G(t, u) = (1+t)^{-\mu} \alpha(u)$ and $Au = \frac{d^2}{dx^2} u(x)$ in (3.8.3) with $\mathcal{D}(A) = \big\{ u(x) \in H : u(x), u'(x) \text{ are absolutely continuous with } u'(x), u''(x) \in H, u(0) = u'(0) = 0 \big\}$.

Introduce a Lyapunov function on H in the following way

$$\Lambda(t, u) = (1+t)^{2(\pi \wedge \mu)} \int_0^1 u(x)^2 dx, \quad \forall u \in H.$$

A simple computation gives that for any $u \in \mathcal{D}(A)$, $t \geq 0$,

$$(\mathbf{L}\Lambda)(t, u) \leq C^2 \int_0^1 q(x, x)dx < \infty.$$

In view of Theorem 3.8.1, we obtain the result that the mild solution of (3.8.9) is polynomially decayable in mean square. Moreover,

$$\limsup_{t \to \infty} \frac{\log E\|X_t(x_0)\|_H^2}{\log t} \leq -\big[2(\pi \wedge \mu) - 1\big].$$

3.8.2   Almost Sure Pathwise Decay

When one tries to investigate almost sure decay rate, the following difficulty might be encountered. The usual method, i.e., consider the decay of solution processes in p-th moment and then (sometimes additional conditions are imposed) deduce their pathwise decay, may not work any more. This is because in the usual investigation of exponential decay such as Theorem 3.2.2, a Borel-Cantelli lemma type of technique is used. This method usually involves an argument of convergence of a certain series with terms in the form (3.2.12). For exponential decay, this series which consists of exponential decay terms is always convergent. But, for general decay rates, this series could be divergent. In this section, we shall present a different approach to handle this difficulty.

Definition 3.8.2   Suppose $\lambda(t) \uparrow \infty$, as $t \to \infty$, is some non-decreasing, continuous function defined for sufficiently large $t > 0$. The solution of Equation (3.3.1) is said to be *almost surely decayable with rate* $\lambda(t)$ if there exist a positive constant $\gamma > 0$ and function $O(\lambda(t))$, $t \geq 0$, such that

$$\limsup_{t \to \infty} \frac{\log \|X_t(x_0)\|_H}{\log O(\lambda(t))} \leq -\gamma \qquad a.s.$$

holds for any $x_0 \in H$, where $O(\lambda(t))$ is the big oh of $\lambda(t)$.

In order to obtain the main results, we need the following exponential martingale inequality. Recall that $\mathcal{W}^2([0, T]; \mathcal{L})$ denotes the family of all $\mathcal{L}(K, H)$-valued predictable $\{\mathcal{F}_t\}$-adapted processes $\Phi(t, \omega)$, $t \geq 0$, such that

$$E \int_0^T tr\Big(\Phi(t, \omega)Q\Phi(t, \omega)^*\Big)dt < \infty \quad \text{for} \quad T \geq 0.$$

Lemma 3.8.1   *Assume $J(t, \omega) \in H$ is an arbitrary H-valued continuous process with $\int_0^T E\|J(t, \omega)\|_H^2 dt < \infty$. Let $\Phi(t) \in \mathcal{W}^2([0, T]; \mathcal{L})$ and T, a, b be* *any positive numbers. Then*

$$P\Big\{ \sup_{0 \leq t \leq T} \Big[\int_0^t \langle J(s), \Phi(s)dW_s \rangle_H$$
$$-\frac{a}{2} \int_0^t tr\Big(J(s) \otimes J(s)(\Phi(s)Q\Phi(s)^*)\Big)ds \Big] > b \Big\} \leq e^{-ab},$$

(3.8.10)

where W_t is some Q-Wiener process and $J(t) \otimes J(t)$ is the linear operator defined by $(J(t) \otimes J(t))h = \langle J(t), h \rangle_H J(t)$ for any $h \in H$, $t \geq 0$.

Proof   For every integer $n \geq 1$, define the stopping time

$$\tau_n = \begin{cases} \inf \Big\{ t \geq 0 : \Big| \int_0^t \langle J(s), \Phi(s)dW_s \rangle_H \\ \qquad\qquad + \int_0^t tr\Big(J(s) \otimes J(s)(\Phi(s)Q\Phi(s)^*)\Big)ds > n \Big\}, \\ \infty \qquad\qquad \text{if the set is empty}, \end{cases}$$

and the process

$$x_n(t) = a \int_0^t \chi_{[[0,\tau_n]]}(s)\langle J(s), \Phi(s)dW_s \rangle_H$$
$$-\frac{a^2}{2} \int_0^t \chi_{[[0,\tau_n]]}(s) tr\Big(J(s) \otimes J(s)(\Phi(s)Q\Phi(s)^*)\Big)ds$$

where $[[\cdot,\cdot]]$ is the stochastic interval (cf. Métivier [1]). Clearly, $x_n(t)$ is bounded and $\tau_n \uparrow \infty$ almost surely as $n \to \infty$. Applying Itô's formula to $\exp[x_n(t)]$, we obtain

$$\exp[x_n(t)] = 1 + \int_0^t \exp[x_n(s)]dx_n(s)$$
$$+ \frac{a^2}{2} \int_0^t \exp[x_n(s)]\chi_{[[0,\tau_n]]}(s) tr\Big(J(s) \otimes J(s)(\Phi(s)Q\Phi(s)^*)\Big)ds$$
$$= 1 + a \int_0^t \exp[x_n(s)]\chi_{[[0,\tau_n]]}(s)\langle J(s), \Phi(s)dW_s \rangle_H.$$

Therefore, one can easily see that $\exp[x_n(t)]$ is a nonnegative martingale over $\mathbf{R}_+$ with $E\Big(\exp[x_n(t)] \Big) = 1$, $t \geq 0$. Hence, by Doob's maximal inequality, we get that

$$P\Big\{ \sup_{0 \leq t \leq T} \exp[x_n(t)] \geq e^{ab} \Big\} \leq e^{-ab} E\Big(\exp[x_n(T)] \Big) = e^{-ab}.$$

That is,

$$P\Big\{ \sup_{0 \leq t \leq T} \Big[\int_0^t \chi_{[[0,\tau_n]]}(s)\langle J(s), \Phi(s)dW_s \rangle_H$$

$$-\frac{a}{2}\int_0^t \chi_{[[0,\tau_n]]}(s)tr\Big(J(s)\otimes J(s)(\Phi(s)Q\Phi(s)^*)\Big)ds\Big] > b\Big\} \le e^{-ab}.$$

Now the required result follows by letting $n \to \infty$.   □

For arbitrary $\Lambda(t,x) \in C^{1,2}(\mathbf{R}_+ \times H; \mathbf{R}_+)$, consider the equation (3.3.1) and define the operator

$$(\mathbf{Q}\Lambda)(t,x) := tr\Big[\Lambda_x' \otimes \Lambda_x'(t,x)G(t,x)QG(t,x)^*\Big], \quad t \ge 0, \quad x \in H. \quad (3.8.11)$$

Theorem 3.8.2   *Let $\lambda(t) \uparrow \infty$, as $t \to \infty$, be some non-decreasing, continuous function satisfying the properties that there exists a $T \ge 0$ such that*

(i). *$\log \lambda(t)$ is uniformly continuous over $[T, \infty)$;*
(ii). *There exists a non-negative constant $\tau \ge 0$ such that*
    $\limsup_{t\to\infty} \frac{\log\log t}{\log \lambda(t)} \le \tau;$
(iii). *For all s, $t \ge T$, $\lambda(s)\lambda(t) \ge \lambda(s+t)$.*

Let $\Lambda(t,x) \in C^{2,1}(\mathbf{R}_+ \times H \to \mathbf{R}_+)$ and assume $\psi_1(t)$, $\psi_2(t)$ are two non-negative continuous functions on $\mathbf{R}_+$. Suppose that there exist a positive constant $m > 0$, and real numbers ν, $\theta \in \mathbf{R}^1$ such that for some $p \ge 2$,

(1). *$\|x\|_H^p \lambda(t)^m \le \Lambda(t,x)$, $(t,x) \in \mathbf{R}_+ \times H$;*
(2). *$|\Lambda_t'(t,x)| + |\Lambda(t,x)| + \|x\|_H\|\Lambda_x'(t,x)\|_H + \|x\|_H^2\|\Lambda_{xx}''(t,x)\| \le c\|x\|_H^p$, $x \in H$, for some constant $c = c(T) > 0$, $t \in [0,T]$, $T \in \mathbf{R}_+$;*
(3). *$(\mathbf{L}\Lambda)(t,x) + (\mathbf{Q}\Lambda)(t,x) \le \psi_1(t) + \psi_2(t)\Lambda(t,x)$, $x \in \mathcal{D}(A)$, $t \in \mathbf{R}_+$;*
(4).

$$\limsup_{t\to\infty} \frac{\log\big(\int_0^t \psi_1(s)ds\big)}{\log \lambda(t)} \le \nu, \qquad \limsup_{t\to\infty} \frac{\int_0^t \psi_2(s)ds}{\log \lambda(t)} \le \theta.$$

Then, if $\gamma = m - \theta - \tau - \nu > 0$, we have almost surely

$$\limsup_{t\to\infty} \frac{\log\|X_t(x_0)\|_H}{\log \lambda(t)} \le -\frac{\gamma}{p}.$$

In other words, the solution of Equation (3.3.1) is almost surely decayable with rate $\lambda(t)$.

Proof   This is similar to the proof of Theorem 3.8.1: applying Itô's formula to $\Lambda(t,x)$ and the strong solution X_t^n of (3.3.6), we can derive that for any $t \ge 0$,

$\Lambda(t, X_t^n)$

$$= \Lambda(0, X_0^n) + \int_0^t (\mathbf{L}^n\Lambda)(s, X_s^n)ds + \int_0^t \langle\Lambda_x'(s, X_s^n), R(n)G(s, X_s^n)dW_s\rangle_H$$

$$= \Lambda(0, X_0^n) + \int_0^t (\mathbf{L}\Lambda)(s, X_s^n)ds + \int_0^t \langle \Lambda_x'(s, X_s), G(s, X_s)dW_s \rangle_H$$
$$+ I_1(t, n) + I_2(t, n) \qquad\qquad (3.8.12)$$

where

$$I_1(t, n) = \int_0^t (\mathbf{L}^n\Lambda)(s, X_s^n)ds - \int_0^t (\mathbf{L}\Lambda)(s, X_s^n)ds,$$

$$I_2(t, n) = \int_0^t \langle \Lambda_x'(s, X_s^n), [R(n) - I]G(s, X_s^n)dW_s \rangle_H,$$

and

$$(\mathbf{L}^n\Lambda)(t, x) = \Lambda_t'(t, x) + \langle \Lambda_x'(t, x), Ax + R(n)F(t, x) \rangle_H$$
$$+ \frac{1}{2}tr\Big[\Lambda_{xx}''(t, x)R(n)G(t, x)Q(R(n)G(t, x))^*\Big], \qquad x \in \mathcal{D}(A),$$

for any $t \geq 0$, $x \in \mathcal{D}(A)$. By virtue of the uniform continuity of $\log \lambda(t)$, for any $\varepsilon > 0$ there exist two positive integers $N = N(\varepsilon)$ and $k_1 = k_1(\varepsilon)$ such that if $\frac{k-1}{2^N} \leq t \leq \frac{k}{2^N}$, $k \geq k_1(\varepsilon)$, we have

$$\left| \log \lambda\Big(\frac{k}{2^N}\Big) - \log \lambda(t) \right| \leq \epsilon.$$

On the other hand, owing to Lemma 3.8.1 we have

$$P\Big\{\omega : \sup_{0 \leq t \leq w} \Big[\int_0^t \langle \Lambda_x'(s, X_s), G(s, X_s)dW_s \rangle_H - \int_0^t \frac{u}{2}(\mathbf{Q}\Lambda)(s, X_s)ds\Big] > v \Big\}$$
$$\leq e^{-uv}$$

for any positive constants u, v and w. In particular, taking

$$u = 2, \quad v = \log\Big(\frac{k-1}{2^N}\Big), \quad w = \frac{k}{2^N}, \quad k = 2, 3, \dots,$$

we then apply the well-known Borel-Cantelli lemma to obtain the existence of a random integer $k_0(\varepsilon, \omega) > 0$ such that almost surely

$$\int_0^t \langle \Lambda_x'(s, X_s), G(s, X_s)dW_s \rangle_H \leq \log\Big(\frac{k-1}{2^N}\Big) + \int_0^t (\mathbf{Q}\Lambda)(s, X_s)ds$$

for all $0 \leq t \leq \frac{k}{2^N}$, $k(\omega) \geq k_0(\varepsilon, \omega)$. Substituting this into (3.8.12) and using Condition (3), we see that for almost all $\omega \in \Omega$,

$$\Lambda(t, X_t^n) \leq \log\Big(\frac{k-1}{2^N}\Big) + \Lambda(0, X_0^n) + \int_0^t (\mathbf{L}\Lambda)(s, X_s^n)ds + \int_0^t (\mathbf{Q}\Lambda)(s, X_s^n)ds$$

$$+ I_1(t, n) + I_2(t, n) + I_3(t, n)$$
$$\leq \log\left(\frac{k-1}{2^N}\right) + \Lambda(0, X_0^n) + \int_0^t \left(\psi_1(s) + \psi_2(s)\Lambda(s, X_s^n)\right)ds$$
$$+ I_1(t, n) + I_2(t, n) + I_3(t, n),$$

$$(3.8.13)$$

where

$$I_3(t, n) = \int_0^t \left((\mathbf{Q}\Lambda)(s, X_s) - (\mathbf{Q}\Lambda)(s, X_s^n)\right)ds$$

for all $t \in [0, \frac{k}{2^N}]$, $k \geq k_0(\varepsilon, \omega) \vee k_1(\varepsilon)$. So by the well-known Gronwall's lemma, we derive that almost surely

$$\Lambda(t, X_t^n) \leq \left[\Lambda(0, X_0^n) + \log\left(\frac{k-1}{2^N}\right) + |I_1(t, n)| + |I_2(t, n)|\right.$$
$$\left. + |I_3(t, n)| + \int_0^t \psi_1(s)ds\right] \cdot \exp\left(\int_0^t \psi_2(s)ds\right)$$

$$(3.8.14)$$

for all $0 \leq t \leq \frac{k}{2^N}$, $k \geq k_0(\varepsilon, \omega) \vee k_1(\varepsilon)$. By virtue of Proposition 1.3.6, there exists a subsequence $X_t^{n_k}$ of X_t^n, denote it still by X_t^n, such that $X_t^n \to X_t$ almost surely, as $n \to \infty$ uniformly with respect to $t \in [0, \frac{k}{2^N}]$, i.e., there exists $\Omega_k \subset \Omega$ with $P(\Omega_k) = 0$ such that for any $\omega \in \Omega\backslash\Omega_k$, $X_t^n \to X_t$, as $n \to \infty$, uniformly with respect to $t \in [0, \frac{k}{2^N}]$. Hence, we have, by using Condition (2), that for any $\omega \in \Omega\backslash\bigcup_{k=1}^\infty \Omega_k$, $I_i(t, n) \to 0$, $i = 1, 2, 3$, as $n \to \infty$ for all $0 \leq t \leq \frac{k}{2^N}$, $k \geq k_0(\varepsilon, \omega) \vee k_1(\varepsilon)$. Hence, letting $n \to \infty$ in (3.8.14), we may deduce that almost surely

$$\Lambda(t, X_t) \leq \left[\Lambda(0, x_0) + \log\left(\frac{k-1}{2^N}\right) + \int_0^t \psi_1(s)ds\right] \exp\left(\int_0^t \psi_2(s)ds\right)$$

for all $0 \leq t \leq \frac{k}{2^N}$, $k \geq k_0(\varepsilon, \omega) \vee k_1(\varepsilon)$. Using Condition (4) and the uniform continuity of $\log \lambda(t)$, for the preceding $\varepsilon > 0$ there exists a positive integer $k_2(\varepsilon, \omega) > 0$ such that

$$\log \Lambda(t, X_t) \leq \log\left[\Lambda(0, x_0) + \lambda(t)^{(\nu+\epsilon)}\right] + \log\log\left(\frac{k-1}{2^N}\right) + (\theta + \epsilon)\log \lambda(t)$$

for all $\frac{k-1}{2^N} \leq t \leq \frac{k}{2^N}$, $k \geq k_0(\varepsilon, \omega) \vee k_1(\varepsilon) \vee k_2(\epsilon, \omega)$, which, letting $\varepsilon \to 0$ and using Condition (1), immediately implies that almost surely

$$\limsup_{t\to\infty} \frac{\log \|X_t(x_0)\|_H}{\log \lambda(t)} \leq \limsup_{t\to\infty} \frac{1}{p} \frac{\log\left[\lambda(t)^{-m}\Lambda(t, X_t)\right]}{\log \lambda(t)} \leq -\frac{m - [\nu + \tau + \theta]}{p}$$

as required.   □

Clearly, replacing $\lambda(t)$ in Theorem 3.8.2 by e^t, $1 + t$ or $\log t$ leads to exponential, polynomial or logarithmic decay, respectively. In particular, as an immediate consequence of Theorem 3.8.2, we may apply it to the equation (3.8.9) to obtain its almost sure decay rate. Indeed, to this end we may introduce the following Lyapunov function on H,

$$\Lambda(t, u) = (1 + t)^{2(\pi \wedge \mu - \delta)} \int_0^1 u(x)^2 dx, \quad \forall u \in H, \quad t \geq 0,$$

where $2\mu > \delta > 1$ is some constant to be determined later on. A direct computation easily yields that for any $u \in \mathcal{D}(A)$, $t \geq 0$,

$$(\mathbf{L}\Lambda)(t, u) \leq \left[2(\pi \wedge \mu - \delta)(1 + t)^{2(\pi \wedge \mu) - 1 - \delta} - 2\pi (1 + t)^{2\pi \wedge \nu - \delta}\right] \int_0^1 u(x)^2 dx$$

$$+ (1 + t)^{-\delta} C^2 \int_0^1 q(x, x) dx$$

$$\leq (1 + t)^{-\delta} C^2 \int_0^1 q(x, x) dx < \infty,$$

and

$$(\mathbf{Q}\Lambda)(t, u) \leq \left[4(1 + t)^{-\delta} \int_0^1 q(x, x) dx\right] \Lambda(t, u).$$

Using Theorem 3.8.2, we can obtain that if $2(\pi \wedge \mu) - \delta - (1 - \delta) > 0$, i.e., $\mu > 1/2$ (letting $\delta \to 1$), the mild solution of (3.8.9) is polynomially decayable in the almost sure sense. Moreover,

$$\limsup_{t \to \infty} \frac{\log \|X_t(x_0)\|_H}{\log t} \leq -[(\pi \wedge \mu) - 1/2] \quad a.s.$$

3.9 Stabilization of Systems by Noise

In Example 3.2.1, we have shown that although the deterministic system $dX_t = AX_t dt$, where A generates an exponentially stable C_0-semigroup, is exponentially stable, its stochastically perturbed system might fail to be exponentially stable if the perturbation does not "decay fast". However, in some situations we find that even though the deterministic system is not exponentially stable, it is still possible to choose some proper noise sources so as to make its perturbed system exponentially stable. To put this clearly, let us go back to Example 2.4.1:

$$\begin{cases} \frac{\partial X_t(x)}{\partial t} = \frac{\partial^2 X_t(x)}{\partial x^2} + r X_t(x), & t > 0, \quad 0 < x < \pi, \\ X_t(0) = X_t(\pi) = 0, & t \geq 0, \\ X_0(x) = x_0(x), & 0 \leq x \leq \pi. \end{cases} \tag{3.9.1}$$

Recall that if we assume $r = r_0$, a constant, then this equation has an explicit solution

$$X_t(x) = \sum_{n=1}^{\infty} a_n e^{-(n^2 - r_0)t} \sin nx$$

with $x_0(x) = \sum_{n=1}^{\infty} a_n \sin nx$. Therefore, it is immediate to obtain exponential stability if $r_0 < 1$ despite the fact that for $r_0 \geq 1$, the null solution is generally not stable.

Now suppose that r is random, and it is modelled as $r_0 + r_1 \dot{B}_t$ so that the equation (3.9.1) becomes

$$dY_t(x) = \left(\frac{\partial^2}{\partial x^2} + r_0 \right) Y_t(x) \, dt + r_1 Y_t(x) dB_t,$$

$$t > 0, \quad 0 < x < \pi, \tag{3.9.2}$$

$$Y_t(0) = Y_t(\pi) = 0, \quad t \geq 0; \quad Y_0(x) = y_0(x), \quad 0 \leq x \leq \pi,$$

where B_t is a one-dimensional standard Brownian motion. In Example 2.4.1, it was shown that when $r_0 < 1$, i.e., the unperturbed system (3.9.1) is stable, the perturbed system (3.9.2) remains pathwise exponentially stable if $r_1^2 < 2(1 - r_0)$. On the other hand, since a new noise term is included in the perturbed system (3.9.2), a natural problem now arises: as $r_0 \geq 1$, is it possible to obtain any stability results for the perturbed system (3.9.2)? In other words, is it possible to stabilize the unperturbed (3.9.1) by a suitable noise source? In this chapter, we shall provide sufficient conditions for exponential stabilization of a class of (stochastic) differential equations in infinite dimensions.

3.9.1   Nonlinear Deterministic Equations

First of all, recall that $V \hookrightarrow H \equiv H^* \hookrightarrow V^*$, and β is a positive constant such that $\|u\|_H \leq \beta \|u\|_V$, $\forall u \in V$. Consider the following deterministic nonlinear equation

$$\begin{cases} dX_t = A(t, X_t)dt, & t \geq 0, \\ X_0 = x_0 \in H, \end{cases} \tag{3.9.3}$$

where $A(t, \cdot) : V \to V^*$, $A(t, 0) = 0$, $t \in \mathbf{R}_+$, is a measurable family of nonlinear operators satisfying the following hypothesis:

(H7). There exist a continuous function $\nu(t)$, $t \geq 0$, and a real number $\nu_0 \in \mathbf{R}^1$ such that

$$2\langle u, A(t, u) \rangle_{V, V^*} \leq \nu(t) \|u\|_H^2 \quad \text{for all} \quad t \in \mathbf{R}_+, \quad u \in V,$$

where

$$\limsup_{t \to \infty} \frac{1}{t} \int_0^t \nu(s) \, ds \leq \nu_0.$$

Now we may present the following question. If the system (3.9.3) is not exponentially stable, is it possible to stabilize it by using a stochastic perturbation, for simplicity of the type $B(t, X_t)\dot{B}_t$? Here B_t, $t \geq 0$, is some standard real Brownian motion, and $B(t, \cdot) : H \to H$ satisfies $B(t, 0) = 0$ and the following condition:

(H8).

$$\|B(t, u) - B(t, v)\|_H^2 \leq \lambda(t)\|u - v\|_H^2, \quad t \in \mathbf{R}_+, \quad u, \ v \in V,$$

where $\lambda(t)$, $t \geq 0$, is a nonnegative continuous function such that

$$\limsup_{t \to \infty} \frac{1}{t} \int_0^t \lambda(s)ds \leq \lambda_0$$

for some number $\lambda_0 \geq 0$.

The answer to this question will be affirmative if we choose a suitable $B(\cdot, \cdot)$. Indeed, consider the following stochastic perturbed differential equation on H

$$\begin{cases} dY_t = A(t, Y_t)dt + B(t, Y_t)dB_t, \quad t \geq 0, \\ Y_0 = y_0 \in H. \end{cases} \tag{3.9.4}$$

The following result can be used to handle this problem.

Theorem 3.9.1 *Assume the strong solution $Y_t(y_0)$, $t \geq 0$, of (3.9.4) satisfies the condition that $\|Y_t(y_0)\|_H \neq 0$ almost surely for all $t \geq 0$ provided $\|y_0\|_H \neq 0$. In addition to hypotheses (H7) and (H8), assume the following:*

$$\langle B(t, u), u \rangle_H^2 \geq \rho(t)\|u\|_H^4 \quad \text{for all} \quad t \in \mathbf{R}_+, \quad u \in H,$$

where $\rho(t)$, $t \geq 0$, is a nonnegative continuous function such that

$$\liminf_{t \to \infty} \frac{1}{t} \int_0^t \rho(s)\, ds \geq \rho_0 \quad \text{for some} \quad \rho_0 \in \mathbf{R}_+.$$

Then, the following relation

$$\limsup_{t \to \infty} \frac{1}{t} \log \|Y_t(y_0)\|_H \leq -\left(\rho_0 - \frac{\nu_0 + \lambda_0}{2}\right)$$

holds almost surely for any $y_0 \in H$ with $\|y_0\|_H \neq 0$. In particular, if $2\rho_0 > \nu_0 + \lambda_0$ the null solution of Equation (3.9.4) is almost surely exponentially stable.

Proof Fix $\|y_0\|_H \neq 0$ and then it is easy to deduce by Itô's formula that

$$\log \|Y_t\|_H^2 \leq \log \|y_0\|_H^2 + \int_0^t \frac{2}{\|Y_s\|_H^2} \langle Y_s, B(s, Y_s) \rangle_H dB_s$$

$$+ \int_0^t \left(\frac{2\langle Y_s, A(s, Y_s) \rangle_{V, V^*} + \|B(s, Y_s)\|_H^2}{\|Y_s\|_H^2} - \frac{2\langle Y_s, B(s, Y_s) \rangle_H^2}{\|Y_s\|_H^4} \right) ds.$$

$$\tag{3.9.5}$$

Due to the exponential martingale inequality, i.e., Lemma 3.8.1, we have

$$P\left\{\omega: \sup_{0\le t\le w}\left[\int_0^t \frac{2}{\|Y_s\|_H^2}\langle Y_s,\,B(s,Y_s)\rangle_H dB_s\right.\right.$$
$$\left.\left.-\int_0^t \frac{2u}{\|Y_s\|_H^4}\langle Y_s,\,B(s,Y_s)\rangle_H^2 ds\right] > v\right\} \le e^{-uv}$$

for any positive constants u, v and w. Assigning $\varepsilon > 0$ arbitrarily and taking

$$u = \alpha, \quad v = 2\alpha^{-1}\log k, \quad w = k\varepsilon, \quad k = 1, 2, 3, \ldots,$$

where $0 < \alpha < 1$, we then apply the well-known Borel-Cantelli lemma to show that there exists an integer $k_0(\varepsilon,\omega) > 0$ for almost all $\omega \in \Omega$ such that

$$\int_0^t \frac{2}{\|Y_s\|_H^2}\langle Y_s,\,B(s,Y_s)\rangle_H dB_s \le 2\alpha^{-1}\log k + \alpha\int_0^t \frac{2}{\|Y_s\|_H^4}\langle Y_s,\,B(s,Y_s)\rangle_H^2 ds$$

for all $0 \le t \le k\varepsilon$, $k \ge k_0(\varepsilon,\omega)$. Substituting this into (3.9.5) and using the conditions in Theorem 3.9.1, we see that for the preceding $\varepsilon > 0$, there exists a positive integer $k_1(\varepsilon) > 0$ such that almost surely

$$\frac{\log\|Y_t\|_H}{t}$$
$$\le \frac{1}{2t}\left(\log\|y_0\|_H^2 + 2\alpha^{-1}\log k + \int_0^t (\nu(s)+\lambda(s))ds - (1-\alpha)\int_0^t \rho(s)ds\right)$$
$$\le \frac{1}{2t}\left(\log\|y_0\|_H^2 + 2\alpha^{-1}\log k + (\nu_0+\lambda_0+\varepsilon)t - (1-\alpha)(2\rho_0+\varepsilon)t\right).$$

for all $(k-1)\varepsilon \le t \le k\varepsilon$, $k \ge k_0(\varepsilon,\omega) \vee k_1(\varepsilon)$, which immediately implies

$$\limsup_{t\to\infty}\frac{\log\|Y_t\|_H}{t} \le \frac{1}{2}\left[(\nu_0+\lambda_0+\varepsilon) - (1-\alpha)(2\rho_0+\varepsilon)\right] \qquad a.s.$$

Therefore, letting $\alpha \to 0$ and $\varepsilon \to 0$ yields

$$\limsup_{t\to\infty}\frac{\log\|Y_t(y_0)\|_H}{t} \le -\left(\rho_0 - \frac{\nu_0+\lambda_0}{2}\right) \qquad a.s.$$

as required. $\square$

As an immediate consequence, let us apply Theorem 3.9.1 to the simple linear case described at the beginning of this section. Indeed, we may formulate this problem by setting $V = H_0^1([0,\pi])$, $H = L^2([0,\pi])$, $A(t,u) = \frac{\partial^2 u}{\partial x^2} + r_0 u$, $u \in V$, $B(t,u) = r_1 u$, $u \in H$. It is obvious that this equation has a unique trivial solution if $\|y_0\|_H = 0$ by the uniqueness of solutions. Since $A(t,0) = 0$, $B(t,0) = 0$, it is easy to conclude, for instance, by an argument of backward stochastic differential equations, that $\|Y_t(y_0)\|_H \ne 0$ for all $t \ge 0$ almost surely if $\|y_0\|_H \neq 0$. Now it is easy to check that Theorem 3.9.1 can be applied to this problem with $\rho_0 = \lambda_0 = r_1^2$ and $\nu_0 = 2(r_0 - 1)$. Thus, it can be deduced that the strong solution of (3.9.4) satisfies

$$\limsup_{t \to \infty} \frac{\log \|Y_t(y_0)\|_H}{t} \leq -\left[\frac{r_1^2}{2} - (r_0 - 1)\right] \qquad a.s.$$

for any $\|y_0\|_H \neq 0$. That is, we get pathwise exponential stability with probability one if $r_1^2 > 2(r_0 - 1)$, a case which actually improves the corresponding results in Example 2.4.1. In particular, for any $r_0 \in \mathbf{R}^1$, the above result shows that it is always possible to choose a suitable multiplicative noise source to stabilize the unperturbed system (3.9.1) when $r = r_0$.

To close this subsection, let us investigate a nonlinear example.

Example 3.9.1   Let $\mathcal{O}$ be an open, bounded subset in $\mathbf{R}^n$ with regular boundary and $2 \leq p < \infty$. Consider the Sobolev spaces $V = W_0^{1,p}(\mathcal{O})$, $H = L^2(\mathcal{O})$ with its usual inner product, and the monotone operator $A: V \to V^*$ defined as

$$\langle v, Au \rangle_{V,V^*} = -\sum_{i=1}^n \int_{\mathcal{O}} \left|\frac{\partial u(x)}{\partial x_i}\right|^{p-2} \frac{\partial u(x)}{\partial x_i} \frac{\partial v(x)}{\partial x_i} dx$$

$$+ \int_{\mathcal{O}} au(x)v(x)dx, \quad \forall u, v \in V,$$

where $a \in \mathbf{R}^1$. We also let $B(t, u) = bu$, $u \in H$, where $b \in \mathbf{R}^1$ and B_t be a standard real Brownian motion.

Then, the condition (H5) in Section 3.2 turns out to be

$$2\langle u, A(t, u) \rangle_{V,V^*} + \|B(t, u)\|^2 = -2\|u\|_V^p + 2a\|u\|_H^2 + b^2\|u\|_H^2, \quad u \in V.$$

Condition (i) of Theorem 3.2.1 requires $2a + b^2 < 0$, i.e., $a < 0$ and $b^2 < -2a$. Alternatively, (ii) will hold whenever $(2a + b^2)\beta^2 - 2 < 0$, i.e., $b^2 < 2\beta^{-2} - 2a$. Therefore, Theorem 3.2.1 guarantees almost surely exponential stability of the null solution only for these values of a and b for which the deterministic system $dX_t = A(t, X_t)dt$ is exponentially stable and the random perturbation is small enough. However, it is easy to prove by an argument similar to the above that Theorem 3.9.1 can be applied in this situation to ensure exponential stability of the null solution for sufficiently large perturbations although the deterministic system is unstable. Note that, in this case, it is not difficult to see that

$$2\langle u, A(t, u) \rangle_{V,V^*} = -2\|u\|_V^p + 2a\|u\|_H^2 \leq \begin{cases} 2a\|u\|_H^2 & \text{if } p > 2, \\ (2a - 2\beta^{-2})\|u\|_H^2 & \text{if } p = 2. \end{cases}$$

Therefore, by virtue of Theorem 3.9.1 it immediately follows that the strong solution of (3.9.4) satisfies

$$\limsup_{t \to \infty} \frac{1}{t} \log \|Y_t(y_0)\|_H \leq \begin{cases} -(b^2/2 - a) & \text{if } p > 2, \\ -(b^2/2 - a + \beta^{-2}) & \text{if } p = 2, \end{cases}$$

for any $\|y_0\|_H \neq 0$. Consequently, we get almost surely exponential stability if

$$b^2 > \begin{cases} 2a & \text{if } p > 2, \\ 2a - 2\beta^{-2} & \text{if } p = 2. \end{cases}$$

3.9.2 Nonlinear Stochastic Equations

It is interesting that a noise source can also be used to stabilize stochastic differential equations. For instance, let us go back to Equation (3.9.2) once again. When $r_0 > 1$, $r_1 \in \mathbf{R}^1$ with $r_1^2 \leq 2(r_0 - 1)$, we do not know whether or not the null solution is exponentially stable. Indeed, we may show below that if the system is perturbed by another multiplicative noise source of the same type, say $r_2 Y_t \dot{\bar{B}}_t$, where $\bar{B}_t$ is another one-dimensional Brownian motion independent of B_t, it can be specifically deduced that the system

$$dY_t(x) = \left(\frac{\partial^2}{\partial x^2} + r_0 \right) Y_t(x)\, dt + r_1 Y_t(x)\, dB_t + r_2 Y_t(x)\, d\bar{B}_t,$$

$$t \geq 0, \quad 0 \leq x \leq \pi, \tag{3.9.6}$$

$$Y_t(0) = Y_t(\pi) = 0, \quad t \geq 0; \quad Y_0(x) = y_0(x), \quad x \in [0, \pi],$$

becomes pathwise exponentially stable once again if $r_2 \in \mathbf{R}^1$ is chosen large enough.

Consider the following stochastic system

$$\begin{cases} dY_t = A(t, Y_t)dt + B(t, Y_t)dB_t + H(t, Y_t)d\bar{B}_t, & t > 0, \\ Y_0 = y_0 \in H, \end{cases} \tag{3.9.7}$$

where B_t and $\bar{B}_t$ are two independent standard real Brownian motions on the same probability space $(\Omega, \mathcal{F}, P)$. $A(t, \cdot) : V \to V^*$, $A(t, 0) = 0$, is a family of nonlinear measurable operators, and $B(t, \cdot)$, $H(t, \cdot) : H \to H$ are both Lipschitz continuous, $B(t, 0) = H(t, 0) = 0$, and satisfy the following assumptions:

(H9). There exists a continuous function $\tilde{\nu}(t)$, $t \geq 0$, and $\tilde{\nu}_0 \in \mathbf{R}^1$ such that

$$2\langle u, A(t, u) \rangle_{V, V^*} + \|B(t, u)\|_H^2 \leq \tilde{\nu}(t)\|u\|_H^2, \quad t \geq 0, \quad u \in V,$$

and

$$\limsup_{t \to \infty} \frac{1}{t} \int_0^t \tilde{\nu}(s)\, ds \leq \tilde{\nu}_0.$$

There exist nonnegative continuous functions $\tilde{\lambda}(t)$, $\tilde{\rho}(t)$, $t \geq 0$, and $\tilde{\lambda}_0$, $\tilde{\rho}_0 \in \mathbf{R}_+$ such that

$$\|H(t, u)\|_H^2 \leq \tilde{\lambda}(t)\|u\|_H^2, \quad t \geq 0, \quad u \in H,$$

$$\langle u, H(t, u) \rangle_H^2 \geq \tilde{\rho}(t)\|u\|_H^4, \quad t \geq 0, \quad u \in H,$$

where

$$\limsup_{t\to\infty}\frac{1}{t}\int_0^t\tilde{\lambda}(s)ds\le\tilde{\lambda}_0,\qquad\liminf_{t\to\infty}\frac{1}{t}\int_0^t\tilde{\rho}(s)\,ds\ge\tilde{\rho}_0.$$

Theorem 3.9.2   *Assume the strong solution $Y_t(y_0)$, $t\ge 0$, of (3.9.7) satisfies that $\|Y_t(y_0)\|_H\ne 0$ for all $t\ge 0$ almost surely provided $\|y_0\|_H\ne 0$. Suppose the hypothesis (H9) holds. Then the strong solution of Equation (3.9.7) satisfies*

$$\limsup_{t\to\infty}\frac{1}{t}\log\|Y_t(y_0)\|_H\le-\left(\tilde{\rho}_0-\frac{\tilde{\nu}_0+\tilde{\lambda}_0}{2}\right)\qquad a.s.\qquad(3.9.8)$$

for any $y_0\in H$ with $\|y_0\|_H\ne 0$. In particular, if $2\tilde{\rho}_0>\tilde{\nu}_0+\tilde{\lambda}_0$, the null solution of (3.9.7) is exponentially stable in the almost sure sense.

Proof   Fix arbitrarily $y_0\in H$ such that $\|y_0\|_H\ne 0$. Then, by virtue of Itô's formula, it follows that

$$\begin{aligned}
\log\|Y_t\|_H^2\\
=\ &\log\|y_0\|_H^2+\int_0^t\frac{2}{\|Y_s\|_H^2}\langle Y_s,A(s,Y_s)\rangle_{V,V^*}ds\\
&+\frac{1}{2}\int_0^t\left[\frac{2\|B(s,Y_s)\|_H^2}{\|Y_s\|_H^2}-\frac{4\langle Y_s,B(s,Y_s)\rangle_H^2}{\|Y_s\|_H^4}\right]ds\\
&+\frac{1}{2}\int_0^t\left[\frac{2\|H(s,Y_s)\|_H^2}{\|Y_s\|_H^2}-\frac{4\langle Y_s,H(s,X_s)\rangle_H^2}{\|X_s\|_H^4}\right]ds\\
&+\int_0^t\frac{2\langle Y_s,B(s,Y_s)\rangle_H}{\|Y_s\|_H^2}dB_s+\int_0^t\frac{2\langle Y_s,H(s,Y_s)\rangle_H}{\|Y_s\|_H^2}d\bar{B}_s.\\
\le\ &\log\|y_0\|_H^2+\int_0^t\frac{2}{\|Y_s\|_H^2}\langle Y_s,A(s,Y_s)\rangle_{V,V^*}ds+\frac{1}{2}\int_0^t\frac{2\|B(s,Y_s)\|_H^2}{\|Y_s\|_H^2}ds\\
&+\frac{1}{2}\int_0^t\left[\frac{2\|H(s,Y_s)\|_H^2}{\|Y_s\|_H^2}-\frac{4\langle Y_s,H(s,X_s)\rangle_H^2}{\|X_s\|_H^4}\right]ds\\
&+\int_0^t\frac{2}{\|Y_s\|_H^2}\langle Y_s,B(s,Y_s)\rangle_H dB_s+\int_0^t\frac{2}{\|Y_s\|_H^2}\langle Y_s,H(s,Y_s)\rangle_H d\bar{B}_s.
\end{aligned}$$

$$(3.9.9)$$

Taking into account the hypotheses of the present theorem, applying Lemma 3.8.1 to the last two terms of the right hand side in (3.9.9) and carrying out a similar argument to that in the proof of Theorem 3.9.1, it is easy to deduce that

$$\limsup_{t\to\infty}\frac{1}{t}\log\|Y_t(y_0)\|_H\le-\left(\tilde{\rho}_0-\frac{\tilde{\nu}_0+\tilde{\lambda}_0}{2}\right)\qquad a.s.$$

$\square$

As an immediate consequence, we can apply this result to stabilize (3.9.2). Recall that we had no idea whether (3.9.2) is exponentially stable or not when $r_1^2 \leq 2(r_0 - 1)$, $r_0 > 1$, $r_1 \in \mathbf{R}^1$. Indeed, we can handle this problem in the form of (3.9.6) by introducing another multiplicative noise source. More precisely, for the example modelled by (3.9.6), one can easily check that

$$\tilde{\nu}_0 = 2(r_0 - 1) + r_1^2, \qquad \tilde{\lambda}_0 = \tilde{\rho}_0 = r_2^2.$$

Therefore, the relation (3.9.8) becomes

$$\limsup_{t \to \infty} \frac{1}{t} \log \|Y_t(y_0)\|_H \leq -\left[\frac{r_2^2 - r_1^2}{2} - (r_0 - 1)\right] \qquad a.s.$$

In other words, when $r_2^2 > 2(r_0 - 1) + r_1^2$, the null solution is exponentially stable in the almost sure sense.

3.10   Lyapunov Exponents and Stabilization

In the history of the study of asymptotic properties (mainly in finite dimensional spaces), one of the most important methods for handling stochastic stability of finite dimensional stochastic systems was the development of the so-called Lyapunov exponent approach. This is certainly the stochastic counterpart in some sense of the notion of characteristic exponents introduced in Lyapunov's classic work on asymptotic (exponential) stability of deterministic systems. Under some circumstances, this approach provides necessary and sufficient conditions for stability, but significant computational problems must be solved. In infinite dimensional cases, the Lyapunov exponent method, especially for nonlinear stochastic systems, needs to use sophisticated mathematics and is far from being a mature subject area.

In this section, we would like to inaugurate the study of Lyapunov exponent methods by presenting some elementary results, and then applying them to the stabilization problem for a class of linear stochastic evolution equations. Consider the following deterministic equation in H,

$$\begin{cases} dX_t = AX_t dt, & t \geq 0, \\ X_0 = x_0 \in H. \end{cases} \tag{3.10.1}$$

We shall prove that if A generates a C_0-semigroup, then this equation can be stabilized, in terms of Lyapunov exponents, by suitable noise sources.

To this end, we assume that A generates a strongly continuous semigroup of bounded operators $T(t)$, $t \geq 0$, on H. It is then well known that $X_t(x_0) = T(t)x_0$, $t \geq 0$, is the unique mild solution of (3.10.1). For arbitrary $x_0 \in H$

with $\|x_0\|_H \neq 0$, we define the Lyapunov exponent of the equation (3.10.1) as follows:

$$\lambda_X(x_0) = \limsup_{t \to \infty} \frac{\log \|X_t(x_0)\|_H}{t}.$$

Lemma 3.10.1   *If the operator A generates a C_0-semigroup $T(t)$, $t \geq 0$, then for any initial value $x_0 \in H$,*

$$\lambda_X(x_0) \leq \lambda < \infty,$$

where $\lambda \in \mathbf{R}^1$ is independent of $x_0 \in H$.

Proof   This may be deduced from the fact that since $T(t)$, $t \geq 0$, is a C_0-semigroup, there exist constants $\lambda \in \mathbf{R}^1$ and $M \geq 1$ such that

$$\|T(t)\| \leq M \cdot \exp(\lambda t), \qquad \forall t \geq 0.$$

$\square$

Let us next consider the following stochastic system

$$\begin{cases} dY_t = AY_t dt + GY_t dB_t, & t \geq 0, \\ Y_0 = y_0 \in H, \end{cases} \tag{3.10.2}$$

where $A : \mathcal{D}(A) \subset H \to H$, generates a C_0-semigroup $T(t)$, $t \geq 0$, $G \in \mathcal{L}(H, H)$. Assume that $T(t)$ and G commute for any $t \geq 0$, and B_t, $t \geq 0$, is a real standard Brownian motion. It is easy to deduce that there exists a unique mild solution to the problem (3.10.2) and the solution is given by the following formula

$$Y_t(y_0) = \exp\left(B_t G - \frac{1}{2} t G^2\right) T(t) y_0. \tag{3.10.3}$$

For arbitrary $y_0 \in H$ with $\|y_0\|_H \neq 0$, the Lyapunov exponent of the system (3.10.2) is defined as:

$$\lambda_Y(y_0, \omega) = \limsup_{t \to \infty} \frac{\log \|Y_t(y_0, \omega)\|_H}{t}.$$

From Lemma 3.10.1 and the next theorem, it will follow that $\lambda_Y(y_0, \omega) < \infty$ almost surely.

Theorem 3.10.1   *Let $y_0 \in H$ with $\|y_0\|_H \neq 0$. Suppose $Y_t(y_0)$, $t \geq 0$, is the mild solution of (3.10.2) given by (3.10.3); then the following inequality holds:*

$$\lambda_Y(y_0, \omega) \leq \lambda_X(y_0) - \alpha \qquad a.s.$$

where $\lambda_X(y_0)$ is the Lyapunov exponent of (3.10.1) with initial value $X_0 = y_0$ and

$$\alpha = -\limsup_{t \to \infty} \frac{1}{t} \log \left\| \exp[-(1/2)tG^2] \right\| > 0.$$

Proof   We have by virtue of (3.10.3) that

$$\lambda_Y(y_0, \omega)$$
$$= \limsup_{t \to \infty} \frac{1}{t} \log \left\| \exp\left(B_t G - \frac{1}{2}tG^2\right)T(t)y_0 \right\|_H$$
$$\leq \limsup_{t \to \infty} \frac{1}{t} \log \left\| \exp[(-1/2)tG^2] \right\| \left\| \exp(B_t G) \right\| + \limsup_{t \to \infty} \frac{1}{t} \log \|T(t)y_0\|_H$$
$$\leq \lambda_X(y_0) + \limsup_{t \to \infty} \frac{1}{t} \log \left\| \exp[(-1/2)tG^2] \right\| + \limsup_{t \to \infty} \frac{1}{t} \log \left\| \exp(B_t G) \right\|.$$

$$(3.10.4)$$

Note that

$$\limsup_{t \to \infty} \frac{1}{t} \log \left\| \exp(B_t G) \right\| \leq \limsup_{t \to \infty} \frac{1}{t} \log \exp(\|G\| |B_t|) = 0,$$

since by the strong law of large numbers $\limsup_{t \to \infty} |B_t|/t = 0$ with probability one. Therefore, substituting this into (3.10.4) immediately yields the desired result.   □

A similar argument to Theorem 3.10.1 can be applied to the following stochastic system driven by multiplicative white noise sources,

$$\begin{cases} dZ_t = AZ_t dt + \sum_{k=1}^{n} b_k Z_t dB_t^k, & t \geq 0, \\ Z_0 = z_0 \in H, \end{cases}$$

$$(3.10.5)$$

where $b_k \in \mathbf{R}^1$, $k = 1, \cdots, n$, and $B_t^1, \cdots, B_t^n$ are mutually independent, real Brownian motions. Then we have:

Theorem 3.10.2   Let $z_0 \in H$ with $\|z_0\|_H \neq 0$. Suppose $Z_t(z_0)$, $t \geq 0$, is the mild solution of (3.10.5) and $\lambda_Z(z_0, \omega)$ is the corresponding Lyapunov exponent, then the following inequality holds:

$$\lambda_Z(z_0, \omega) \leq \lambda_X(z_0) - \frac{1}{2} \sum_{k=1}^{n} b_k^2 \quad a.s.$$

Corollary 3.10.1   Suppose that the operator A generates a C_0-semigroup; then the deterministic system (3.10.1) can be stabilized by noise.

Proof   Let us fix $z_0 \in H$ with $\|z_0\|_H \neq 0$. Lemma 3.10.1 ensures the existence of a constant $\lambda > 0$ such that $\lambda_X(z_0) \leq \lambda$. Choosing $b_k \in \mathbf{R}^1$, $k = 1, \cdots, n$, properly such that

$$\frac{1}{2}\sum_{k=1}^{n} b_k^2 > \lambda,$$

then we have $\lambda_Z(z_0, \omega) < 0$ almost surely. The proof is complete by means of Theorem 3.10.2. $\qquad\square$

Next, we shall apply the theory derived above to some partial differential equations of parabolic type. Let $\mathcal{O}$ be a bounded domain in $\mathbf{R}^n$ with smooth boundary $\partial\mathcal{O}$. Consider the differential operator of the form

$$A = \sum_{i,j}^{n} \frac{\partial}{\partial x_i}\left(a_{ij}(x)\frac{\partial}{\partial x_j}\right) + a(x),$$

where the coefficients satisfy the following assumptions:

(1). $a_{ij}(\cdot)$, $a(\cdot)$ are sufficiently smooth real-valued functions on $\bar{\mathcal{O}}$, the closure of $\mathcal{O}$;

(2). $a_{ij}(\cdot) = a_{ji}(\cdot)$ for any $i, j \in \{1, \cdots, n\}$;

(3). $\mu_1\|\xi\|_{\mathbf{R}^n}^2 \leq \sum_{i,j} a_{ij}\xi_i\xi_j \leq \mu_2\|\xi\|_{\mathbf{R}^n}^2$, $\|\xi\|_{\mathbf{R}^n}^2 = \xi_1^2 + \cdots + \xi_n^2$, $\xi_i \in \mathbf{R}^1$, $i = 1, \cdots, n$;

(4). $\left|\frac{\partial a_{ij}(\cdot)}{\partial x_k}\right| \leq \mu_3$ for any $1 \leq i, j, k \leq n$;

(5). $|a(\cdot)| \leq \mu_4$,

where $\mu_i > 0$, $i = 1, \cdots, 4$. It is well known (Pazy [1]) that there exists an orthonormal basis of $L^2(\mathcal{O})$, $\{e_j\}$, $j = 1, 2, \cdots$, consisting of eigenvectors of the operator A such that

$$Ae_j = \lambda_j e_j, \quad \text{where} \quad \lambda_j \downarrow -\infty \quad \text{as} \quad j \to \infty.$$

Moreover, the operator A generates an analytic semigroup of bounded operators $T(t)$, $t \geq 0$, on the space $L^2(\mathcal{O})$. We will study the following parabolic equation with Dirichlet boundary conditions:

$$\begin{cases} \frac{\partial X_t(x)}{\partial t} = AX_t(x), & \forall t \geq 0, \ x \in \mathcal{O}; \\ X_0(x) = x_0(x) \in L^2(\mathcal{O}), & \forall x \in \mathcal{O}; \ X_t(x) = 0, \ \forall t \geq 0, \ x \in \partial\mathcal{O}. \end{cases}$$
$$(3.10.6)$$

It is easy to deduce that the unique mild solution of (3.10.6) is given by the following formula

$$X_t(x) = T(t)x_0(x) = \sum_{j=1}^{\infty} \exp(t\lambda_j)x_0^j e_j(x)$$

where

$$x_0(x) = \sum_{j=1}^{\infty} x_0^j e_j(x), \qquad x_0^j = \langle x_0, e_j \rangle_{L^2(\mathcal{O})}. \tag{3.10.7}$$

Proposition 3.10.1   *Let* $\|x_0(\cdot)\|_{L^2(\mathcal{O})} \neq 0$ *and* j_0 *be the smallest integer* $j \geq 1$ *in the expansion (3.10.7) of* x_0 *such that* $x_0^{j_0} \neq 0$. *Then the Lyapunov exponent of (3.10.6) exists and is given by*

$$\lambda_X(x_0) = \lambda_{j_0}.$$

Proof   It is immediate to deduce

$$\frac{1}{t} \log \Big\| \sum_{j=0}^{\infty} \exp(t\lambda_j) x_0^j e_j \Big\|_{L^2(\mathcal{O})} \leq \frac{1}{t} \log \Big(\sum_{j=j_0}^{\infty} \Big| \exp(t\lambda_{j_0}) x_0^j \Big|^2 \Big)^{1/2}$$

$$= \lambda_{j_0} + \frac{1}{t} \log \|x_0\|_{L^2(\mathcal{O})}$$

and

$$\frac{1}{t} \log \Big\| \sum_{j=0}^{\infty} \exp(t\lambda_j) x_0^j e_j \Big\|_{L^2(\mathcal{O})} \geq \frac{1}{t} \log \Big| \exp(t\lambda_{j_0}) x_0^{j_0} \Big| = \lambda_{j_0} + \frac{1}{t} \log |x_0^{j_0}|.$$

The desired result follows.   ☐

Now let us consider the following stochastic perturbation of the deterministic problem (3.10.6)

$$\frac{\partial Y_t(x)}{\partial t} = AY_t(x) + \sum_{k=1}^{n} b_k Y_t(x) dB_t^k, \quad \forall t \geq 0, \quad x \in \mathcal{O};$$

$$Y_0(x) = y_0(x) \in L^2(\mathcal{O}), \quad \forall x \in \mathcal{O}, \text{ and } Y_t(x) = 0, \quad \forall t \geq 0, \ x \in \partial\mathcal{O},$$

$$\tag{3.10.8}$$

where $b_k \neq 0$, $k = 1, \cdots, n$, and $B_t^1, \cdots, B_t^n$ are mutually independent, real Brownian motions.

It is easy to show that the mild solution of the equation (3.10.8) is equal to

$$Y_t(x) = \exp \Big(\sum_{k=1}^{n} b_k B_t^k \Big) \exp \Big[\Big(-\frac{1}{2} \sum_{k=1}^{n} b_k^2 \Big) t \Big] X_t(x),$$

where $X_t(x)$ is the mild solution of the equation (3.10.6). It is now easy to prove the following theorem using the above formula and the fact that $\lim_{t\to\infty} B_t^k / t = 0$ almost surely for $k = 1, \cdots, n$.

Proposition 3.10.2 *The Lyapunov exponent of the system (3.10.8) almost surely exists as a limit which is non-random and the following formula holds:*

$$\lambda_Y(y_0, \omega) = \lambda_X(y_0) - \frac{1}{2} \sum_{k=1}^{n} b_k^2 = \lambda_{j_0} - \frac{1}{2} \sum_{k=1}^{n} b_k^2 \quad a.s.$$

In particular, we have the following: if $y_0^1 = \langle y_0, e_1 \rangle_{L^2(\mathcal{O})} \neq 0$, then the stochastic system (3.10.8) is exponentially stable in the almost sure sense if and only if its top Lyapunov exponent satisfies

$$\lambda_1 - \frac{1}{2} \sum_{k=1}^{n} b_k^2 < 0.$$

In conjunction with Proposition 3.10.2, the stabilization strategy by noise above can be stated in a slightly different way. We may say that the Lyapunov exponents of the linear equation (3.10.1) are essentially different from those of its perturbed Itô equation (3.10.2) because of the inclusion of multiplicative white noise sources. However, we may show that this is not the case if we consider the Stratonovich version of (3.10.2). Precisely, we will show that the Lyapunov exponents of (3.10.2) in the Stratonovich interpretation are as same as those of its unperturbed deterministic model for a wide range of stochastic perturbations.

Let us consider that a linear noise is added to the problem (3.10.1) in a Stratonovich sense

$$\begin{cases} dY_t = AY_t dt + GY_t \circ dB_t, & t \geq 0, \\ Y_0 = y_0 \in H, \end{cases} \tag{3.10.9}$$

where A generates a C_0-semigroup $T(t)$, $t \geq 0$, and $G : \mathcal{D}(G) \subset H \to H$ is assumed to be the generator of a C_0-group, denoted by $S(t)$, $t \in \mathbf{R}^1$, satisfying $\mathcal{D}(A) \subset \mathcal{D}(G)$. We assume that for each $y_0 \in H$, there exists a unique mild solution of (3.10.9) (see Da Prato and Zabczyk [1] or Kunita [2] for suitable conditions).

Theorem 3.10.3 *Assume that A and $S(t)$, $t \geq 0$, commute. Then the null solution of Equation (3.10.1) is exponentially stable if and only if the null solution of (3.10.9) is exponentially stable in the almost sure sense.*

Proof Let $y_0 \in \mathcal{D}(A)$ and $Y_t(y_0)$ be a solution of (3.10.9). Define the transformation Z_t by the following

$$Z_t = Z_t(y_0) = S^{-1}(B_t(\omega))Y_t(y_0). \tag{3.10.10}$$

Now, it is not difficult to check that

$$dZ_t = S^{-1}(B_t(\omega))dY_t - S^{-1}(B_t(\omega))GY_t \circ dB_t$$
$$= S^{-1}(B_t(\omega))AY_t dt = AZ_t dt,$$

and $Z_0 = y_0$. Consequently, the process $Z_t = Z_t(y_0) = X_t(y_0) = T(t)y_0$ is the unique mild solution of Equation (3.10.1). Since $\mathcal{D}(A)$ is dense in H and Z_t, $T(t)$ both are linear, $Z_t(y_0) = X_t(y_0)$ for any $y_0 \in H$.

Assume that the null solution of (3.10.1) is exponentially stable. This means that there exist constants $M \geq 1$, $\mu > 0$ such that $\|T(t)\| \leq Me^{-\mu t}$ for all $t \geq 0$. On the other hand, as the operator G is the generator of the C_0-group $S(t)$, it is well known (see Pazy [1]) that there exist $C \geq 1$, $\gamma \in \mathbf{R}^1$ such that $\|S(t)\| \leq Ce^{\gamma|t|}$ for all $t \in \mathbf{R}^1$. Then, since

$$\lim_{t \to \infty} \frac{|B_t|}{t} = 0 \quad a.s.$$

there exists $\Omega_0 \subset \Omega$, $P(\Omega_0) = 0$ such that if $\omega \in \Omega \backslash \Omega_0$,

$$\lim_{t \to \infty} \left(\mu - \gamma \frac{|B_t|}{t} \right) = \mu,$$

and there exists $T(\omega) \geq 0$ such that for all $t \geq T(\omega)$, $\mu - \gamma |B_t|/t \geq \mu/2$. Thus, for arbitrarily given $y_0 \in \mathcal{D}(A)$, $\omega \in \Omega \backslash \Omega_0$, and taking into account the fact that $X_t(y_0) = T(t)y_0$,

$$\|Y_t(y_0)\|_H = \|S(B_t(\omega))X_t(y_0)\|_H \leq MCe^{\gamma|B_t(\omega)|}e^{-\mu t}\|y_0\|_H$$

$$\leq MC\|y_0\|_H e^{-(\mu - \frac{\gamma|B_t(\omega)|}{t})t} \leq MC\|y_0\|_H e^{-\mu_0 t}, \quad \forall t \geq T(\omega),$$

where $\mu_0 = \mu/2$. Therefore, the null solution of (3.10.9) is exponentially stable in the almost sure sense.

Conversely, if G is an operator such that the null solution of (3.10.9) is almost surely exponentially stable, there exist $\Omega_0' \subset \Omega$, $P(\Omega_0') = 0$ and constants $\alpha > 0$, $\theta > 0$ such that if $\omega \in \Omega \backslash \Omega_0'$, then

$$\|Y_t(y_0)\|_H \leq \alpha\|y_0\|_H e^{-\theta t}, \quad \forall t \geq T_0(\omega)$$

for some random variable $T_0(\omega) \geq 0$. Now, for any fixed $\omega \in \Omega \backslash \Omega_0'$, the equation (3.10.10) implies

$$\|T(t)y_0\|_H \leq \|S(-B_t(\omega))Y_t(y_0)\|_H \leq \alpha C\|y_0\|_H e^{-(\theta - \frac{\gamma|B_t(\omega)|}{t})t}, \quad \forall t \geq T_0(\omega).$$

On the other hand, we can assure the existence of $\Omega_0'' \subset \Omega$, $P(\Omega_0'') = 0$ satisfying that for all $\omega \in \Omega \backslash \Omega_0''$, there exists $T_1(\omega) \geq 0$ such that for all $t \geq T_1(\omega)$, we have

$$\theta - \frac{\gamma|B_t|}{t} \geq \frac{\theta}{2}.$$

Denote $\bar{\Omega}_0 = \Omega_0' \cup \Omega_0''$ and take any fixed $\omega \in \Omega \backslash \bar{\Omega}_0$, it is easy to deduce that

$$\|T(t)y_0\|_H \leq Me^{-\mu t}\|y_0\|_H, \quad \forall t \geq \bar{T}(\omega),$$

where $\bar{T}(\omega) = \max\{T_0(\omega), T_1(\omega)\}$. The proof is now complete. $\square$

Note that if we consider the particular case $Gy = \sigma y$ for some $\sigma \in \mathbf{R}^1$, the C_0-group $S(t)$ is given by $S(t) = e^{\sigma t}I$, and the hypotheses in Theorem 3.10.3 are fulfilled. Therefore, the null solution of (3.10.1) is exponentially stable if and only if the null solution of Equation (3.10.9) is exponentially stable in the almost sure sense.

3.11   Notes and Comments

Important progress was made on the stability of nonlinear stochastic differential equations in infinite dimensional spaces over the last two decades. The main results in Subsection 3.1.1 are taken from Ichikawa [6]. Absolute stability in the class L_k of finite dimensional differential equations in Subsection 3.1.2 has been extensively studied since it appears in some feedback control systems (Hahn [1]). The systems (3.1.22), (3.1.23) are often called stochastic differential equations of Lue's type. In finite dimensional cases, Morozan [1] showed that the system (3.1.23) is absolutely stable in the class L_k defined in Subsection 3.1.2 and that the null solution is exponentially stable in mean square for each $g \in L_k$ if and only if the linear equation (3.1.22) is stable. The material in Subsection 3.1.2 is mainly due to Ichikawa [4]. Following the classic work of Pardoux [1] who established the fundamental results on existence and uniqueness of solutions of stochastic nonlinear partial differential equations of monotone type in which a coercive condition plays an important role, Chow [2] pointed out that under some circumstances the coercive condition actually takes the role of a stability criterion. The main results in Section 3.2 which are due to Caraballo and Liu [1] improved and generalized those in Chow [2] to non-autonomous cases. Stochastic stability of partial differential equations using finite dimensional approximations can be found in Yavin [1], [2].

In the history of the study of asymptotic stability of stochastic systems, the Lyapunov function method is probably the most influential tool, for instance, see Khas'minskii [1]. The basic technique involved is to construct, firstly, a proper Lyapunov function and then deal with the stability of the nonlinear case by means of the first order approximation procedure. In infinite dimensional cases, the first investigation in this respect was due to Khas'minskii and Mandrekar [1]. The general non-autonomous version presented in Theorem 3.4.1 is taken from Liu [5]. The relation between ultimate boundedness in the mean square sense and invariant measures of stochastic differential equations has been pointed out by Miyahara [1], [2] in finite dimensional cases. But the basic ideas go back at least to Wonham [2] and Zakai [1], [2]. The corresponding investigation in infinite dimensional cases was carried out in Liu and Mandrekar [1] for strong solutions and Ichikawa [5], Liu and Mandrekar [2] for mild solutions. The infinite dimensional generalizations, Theorem 3.5.6, Corollary 3.5.2 and Theorem 3.6.1, of finite dimensional results are presented in Chow and Khas'minskii [1]. The reader can also find one approach in Ethier and Kurtz [1] to deal with the existence of invariant measures by constructing a Lyapunov function and solving the relevant martingale problem.

The research about decay rates of finite dimensional Itô's stochastic differential equations goes back at least to Mao [1] in which polynomial decay was investigated. The material in Section 3.8.2 is closely related to Liu [1]. One of the most interesting topics in stability theory is the so-called stabilization by white noise sources of deterministic (stochastic) systems. In this respect there is a systematic statement in finite dimensions in Khas'minskii [1]. There exists some related work, for instance, in Arnold, Crauel and Wihstutz [1], Mao [2] and Scheutzow [1] among others. The presentation in Section 3.9 is mainly taken from Caraballo, Liu and Mao [1]. Some material about Lyapunov exponents such as Theorem 3.10.1 in Section 3.10 is taken from Kwiecińska [2]. But the theorem 3.10.3 is obtained in Caraballo and Langa [1] in spite of the fact that in finite dimensional cases, Arnold [4] has proved that the deterministic system (3.10.1) can be stabilized by a suitable Stratonovich linear noise if and only if $trA < 0$.

Chapter 4

Stability of Stochastic Functional Differential Equations

In this chapter, we shall investigate stability properties of stochastic functional differential equations in infinite dimensions. We begin with an argument of reducing the stability problem of retarded functional linear deterministic equations to a class of C_0-semigroups of bounded linear operators so as to find exact regions of stability. The characteristic conditions of mean square exponential stability established in Chapter 2 for linear equations are extended to a class of stochastic linear functional equations with time lag. A kind of coercive condition is formulated to secure desired decay behavior of strong solutions for nonlinear stochastic functional differential equations. The methods of Lyapunov and Razumikhin functionals are emphasized and contrasted to describe stability properties of mild solutions for semilinear stochastic evolution equations with memory.

4.1 Linear Deterministic Equations

Recall that S denotes a separable Banach space with norm $\|\cdot\|_S$ over the real field $\mathbf{R}^1$ and suppose $r \geq 0$ is a given constant. Let $C_r = C([-r, 0]; S)$ denote the Banach space of all continuous S-valued functions on $[-r, 0]$ with the usual supremum norm $\|\cdot\|_{C_r}$, which is defined by $\|\phi\|_{C_r} = \max_{-r \leq \theta \leq 0} \|\phi(\theta)\|_S$ for any $\phi \in C_r$. For arbitrary real numbers $a \leq b$, $t \in [a, b]$ and any continuous function $u(\cdot) : [a - r, b] \to S$, u_t denotes the element of C_r given by $u_t(\theta) = u(t + \theta)$ for $\theta \in [-r, 0]$.

In this section, we wish to consider the following abstract integral equation on S

$$u(t) = T(t)u(0) + \int_0^t T(t - s)Fu_s ds, \qquad t \geq 0,$$
$$u_0 = \phi \in C_r, \tag{4.1.1}$$

where $F : C_r \to S$ is a bounded linear operator with norm $\|F\|$ and $\{T(t)\}$, $t \geq 0$, is a strongly continuous semigroup of bounded linear operators over S with its infinitesimal generator A_T. It can be proved by the standard Picard iteration procedure that for arbitrarily given $\phi \in C_r$, there exists a unique continuous function $u(\cdot)(\phi) : [-r, \infty) \to S$, which solves the equation (4.1.1). Clearly, all the stability definitions in Section 1.4 can be applied to Equation (4.1.1) in an obvious manner. For instance, we say that the null solution of (4.1.1) is *stable* if for each $\varepsilon > 0$, there exists $\delta > 0$ such that the solution $u(t)(\phi)$ satisfies that for any $\|\phi\|_{C_r} < \delta$,

$$\|u(t)(\phi)\|_S < \varepsilon$$

for any $t \geq 0$.

4.1.1   Stable Semigroups (Finite Delays)

In order to investigate the stability of Equation (4.1.1), we define a family of operators $U(t) : C_r \to C_r$ given by

$$U(t)\phi = u_t(\phi), \qquad t \geq 0,$$

where $u_t(\phi) = u(t + \cdot)(\phi)$, $t \geq 0$, denotes the solution of (4.1.1). As an immediate consequence, all the stability definitions of Equation (4.1.1) can be restated by means of $U(t)$, $t \geq 0$, in a straightforward way. For instance, we say that the null solution of (4.1.1) is *asymptotically stable* if it is stable and there exists $\delta > 0$ such that the relation $\|\phi\|_{C_r} < \delta$ implies

$$\lim_{t \to \infty} \|U(t)\phi\|_{C_r} = 0.$$

By a standard argument, we may deduce immediately from the solution of (4.1.1) that:

Proposition 4.1.1 $U(t)$, $t \geq 0$, *is a strongly continuous semigroup of bounded linear operators on C_r satisfying the condition that for any $\phi \in C_r$, $t \geq 0$,*

$$\begin{aligned} \|U(t)\phi\|_{C_r} &\leq M\|\phi\|_{C_r} \cdot e^{(\mu + M\|F\|)t}, & if & \quad \mu \geq 0, \\ \|U(t)\phi\|_{C_r} &\leq Me^{-\mu r}\|\phi\|_{C_r} \cdot e^{(\mu + M\|F\|e^{-\mu r})t}, & if & \quad \mu < 0, \end{aligned} \qquad (4.1.2)$$

where $\|T(t)\| \leq M \cdot e^{\mu t}$, $M \geq 1$, $\mu \in \mathbf{R}^1$, for all $t \geq 0$.

Proof   The linearity of $U(t)$ is immediate and its strong continuity follows from the fact that solutions of (4.1.1) are continuous. The semigroup property follows from that for arbitrary $t, \tilde{t} \geq 0$, $\phi \in C_r$,

$$u(t + \tilde{t})(\phi)$$

$$= T(t + \tilde{t})\phi(0) + \int_0^t T(t + \tilde{t} - s)Fu_s(\phi)ds + \int_t^{t+\tilde{t}} T(t + \tilde{t} - s)Fu_s(\phi)ds$$

$$= T(\tilde{t})\left(T(t)\phi(0) + \int_0^t T(t - s)Fu_s(\phi)ds\right) + \int_0^{\tilde{t}} T(\tilde{t} - s)Fu_{t+s}(\phi)ds$$

$$= T(\tilde{t})u(t)(\phi) + \int_0^{\tilde{t}} T(\tilde{t} - s)Fu_{t+s}(\phi)ds.$$

By the uniqueness of solutions of Equation (4.1.1), this implies that $u_{t+\tilde{t}}(\phi) = u_{\tilde{t}}(u_t(\phi))$. On the other hand, from (4.1.1) we have that for arbitrary $t \geq -r$,

$$\|u(t)(\phi)\|_S \leq Me^{\mu t}\|\phi(0)\|_S + M\|F\| \int_0^t e^{\mu(t-s)}\|u_s(\phi)\|_{C_r}ds.$$

If $\mu \geq 0$, then for $t \geq 0$,

$$\|u_t(\phi)\|_{C_r} \leq Me^{\mu t}\|\phi\|_{C_r} + M\|F\| \int_0^t e^{\mu(t-s)}\|u_s(\phi)\|_{C_r}ds,$$

and if $\mu < 0$, then for $t \geq 0$,

$$\|u_t(\phi)\|_{C_r} \leq Me^{-\mu r}e^{\mu t}\|\phi\|_{C_r} + M\|F\|e^{-\mu r} \int_0^t e^{\mu(t-s)}\|u_s(\phi)\|_{C_r}ds.$$

By virtue of the well-known Gronwall lemma, we have the desired (4.1.2). ⬜

Define the following operator $A_U : \mathcal{D}(A_U) \subset C_r \to C_r$ by

$$\mathcal{D}(A_U) = \left\{\phi \in C_r : \phi' \in C_r, \ \phi(0) \in \mathcal{D}(A_T), \ \phi'_-(0) = A_T\phi(0) + F\phi\right\},$$
$$A_U\phi(\theta) = \phi'(\theta), \quad -r \leq \theta \leq 0, \quad \phi \in \mathcal{D}(A_U).$$

It can be shown (cf. see Travis and Webb [1]) that A_U is the infinitesimal generator of the C_0-semigroup $\{U(t)\}_{t\geq 0}$. In addition, Propositon 4.1.1 actually provides a sufficient condition for the null solution of (4.1.1) to be stable or asymptotically stable in terms of the growth rate of the semigroup $\{T(t)\}_{t\geq 0}$ and the norm $\|F\|$.

Corollary 4.1.1 *Suppose $u(t)(\phi)$, $t \geq 0$, is the solution of the equation (4.1.1). Assume*

$$\mu + M\|F\|e^{-\mu r} \leq 0; \tag{4.1.3}$$

then the null solution is stable. Moreover, assume

$$\mu + M\|F\|e^{-\mu r} < 0; \tag{4.1.4}$$

then the null solution is globally asymptotically stable in the sense that it is stable and for any $\phi \in C_r$, the relation $\lim_{t\to\infty} \|u_t(\phi)\|_{C_r} = 0$ holds.

However, we would also like to point out that the stability condition (4.1.3) or (4.1.4) is not easy to apply to many practical situations. For instance, in a variety of situations we find that the stability of (4.1.1) still remains true for some time retarded parameters $r > 0$, which do not satisfy (4.1.3) or (4.1.4). In the sequel, we shall carry out a detailed investigation of this problem so as to establish some more effective stability criteria. In particular, for the sake of simplicity we shall always assume in the remainder of this subsection that $\{T(t)\}_{t\geq 0}$ is a strongly continuous semigroup of bounded linear operators over S satisfying

$$\|T(t)\| \leq e^{\mu t} \quad \text{for all} \quad t \geq 0,$$

where $\mu \in \mathbf{R}^1$ is some real number.

Remark   It is obvious that Proposition 4.1.1 remains true even for certain nonlinear operators, for instance, if the term F is supposed to be a class of nonlinear operators $F(\cdot) : C_r \to S$ with global Lipschitz constants, denoted still by $\|F\|$. That is, the term F satisfies

$$\|F(\phi) - F(\tilde{\phi})\|_S \leq \|F\| \cdot \|\phi - \tilde{\phi}\|_{C_r} \quad \text{for any} \quad \phi, \ \tilde{\phi} \in C_r.$$

Proposition 4.1.2   *Let $\|F\|$ denote the operator norm of linear operator F. If $\mu \geq -\|F\|$ and $\operatorname{Re}\lambda > \|F\| + \mu$ where $\operatorname{Re}\lambda$ denotes the real part of the complex number λ, then $(A_U - \lambda I)^{-1}$ exists and has domain all of C_r.*

Proof   Given $\psi \in C_r$, we must solve

$$(A_U - \lambda I)\phi = \psi.$$

That is,

$$\phi' - \lambda\phi = \psi, \qquad \phi'(0) = \lambda\phi(0) + \psi(0) = A_T\phi(0) + F\phi. \tag{4.1.5}$$

This means that

$$\begin{aligned} \phi(\theta) &= e^{\lambda\theta}\phi(0) + \int_0^\theta e^{\lambda(\theta-s)}\psi(s)ds, \qquad \theta \in [-r, 0], \\ \phi(0) &= (A_T - \lambda I)^{-1}(\psi(0) - F\phi). \end{aligned} \tag{4.1.6}$$

The mapping

$$x \to (A_T - \lambda I)^{-1}\left[\psi(0) - F\left(e^{\lambda\cdot}x + \int_0^{\cdot} e^{\lambda(\cdot-s)}\psi(s)ds\right)\right]$$

is a strict contraction from S to S since by the well-known Hille-Yosida theorem

$$\left\|(A_T - \lambda I)^{-1}F(e^{\lambda\cdot}x)\right\|_S \leq \frac{\|F\|}{\operatorname{Re}\lambda - \mu}\|x\|_S$$

for all $x \in S$. Then (4.1.6) and hence (4.1.5) has a unique solution. But this means that $(A_U - \lambda I)$ is onto and injective and the proof is complete. $\Box$

Proposition 4.1.3   *If $\mu \geq -\|F\|$ and $\operatorname{Re}\lambda > \|F\|+\mu$, then $\|(A_U-\lambda I)^{-1}\| \leq 1/(\operatorname{Re}\lambda - \|F\| - \mu)$.*

Proof   Let $\phi = (A_U - \lambda I)^{-1}\psi$ for $\psi \in C_r$. Suppose $\varepsilon > 0$ and $\theta \in [-r, 0]$ has the property that $\|\phi(\theta)\|_S \geq \|\phi\|_{C_r} - \varepsilon$. Using (4.1.6) and Hille-Yosida theorem, we have

$$\|\phi(\theta)\|_S \leq \left\|e^{\lambda\theta}(A_T - \lambda I)^{-1}[\psi(0) - F\phi]\right\|_S + \left\|\int_0^\theta e^{\lambda(\theta-s)}\psi(s)ds\right\|_S$$

$$\leq \frac{e^{Re\,\lambda\theta}}{Re\,\lambda - \mu}(\|\psi\|_{C_r} + \|F\|\|\phi\|_{C_r}) + \frac{1 - e^{Re\,\lambda\theta}}{Re\,\lambda}\|\psi\|_{C_r}$$

$$= \frac{\|F\|e^{Re\,\lambda\theta}}{Re\,\lambda - \mu}\|\phi\|_{C_r} + \frac{1 - (\frac{\mu}{Re\,\lambda})(1 - e^{Re\,\lambda\theta})}{Re\,\lambda - \mu}\|\psi\|_{C_r}.$$

But this implies

$$\frac{Re\,\lambda - \mu - \|F\|e^{Re\,\lambda\theta}}{Re\,\lambda - \mu}\|\phi\|_{C_r} \leq \varepsilon + \frac{1 - (\frac{\mu}{Re\,\lambda})(1 - e^{Re\,\lambda\theta})}{Re\,\lambda - \mu}\|\psi\|_{C_r}.$$

Since

$$\frac{1 - (\frac{\mu}{Re\,\lambda})(1 - e^{Re\,\lambda\theta})}{Re\,\lambda - \mu - e^{Re\,\lambda\theta}\|F\|} \leq \frac{1}{Re\,\lambda - \mu - \|F\|},$$

the assertion follows.   ⬜

As an immediate by-product of Propositions 4.1.2 and 4.1.3, we may deduce the following results which will help us to formulate more refined stability conditions than those in Corollary 4.1.1.

Corollary 4.1.2   *If* $-\|F\| = \mu$*, then for arbitrary* $\phi \in C_r$*,* $t \geq 0$*,*

$$\|U(t)\phi\|_{C_r} \leq \|\phi\|_{C_r}. \tag{4.1.7}$$

Proof   By virtue of Propositions 4.1.2 and 4.1.3 and the fact that

$$(I - \lambda A_U)^{-1} = (A_U - (1/\lambda)I)^{-1}(-1/\lambda),$$

it is easy to deduce (choosing $\mu \geq -\|F\|$ if necessary) that $(I - \lambda A_U)^{-1}$ exists with domain C_r and for all real λ with $0 < \lambda < 1/(\|F\| + \mu)$,

$$\|(I - \lambda A_U)^{-1}\| \leq \frac{1}{1 - \lambda(\|F\| + \mu)}.$$

Then it is easy to get the desired result by using the following result of Crandall and Liggett [1],

$$\lim_{n\to\infty}(I - (t/n)A_U)^{-n}\phi = U(t)\phi \quad \text{for all} \quad \phi \in C_r.$$

⬜

Corollary 4.1.3   *If* $-\|F\| > \mu$, *then for arbitrary* $\phi \in C_r$, $t \geq 0$, *and each positive integer* n,

$$\|U(t)\phi\|_{C_r}$$
$$\leq \left[(-\|F\|/\mu)^n + \left(\sum_{k=0}^{n-1} \|F\|^k(1 - (-\|F\|/\mu)^{n-k})e^{\mu(t-(k+1)r)}t^k/k!\right)\right]\|\phi\|_{C_r}.$$

$$(4.1.8)$$

Furthermore, there exists a unique $\phi_0 \in C_r$ *such that* $U(t)\phi_0 = \phi_0$ *for arbitrary* $t > r$ *and* $\lim_{t\to\infty} U(t)\phi = \phi_0$ *for all* $\phi \in C_r$.

Proof   Since $-\|F\| > \mu$, it is easy to show that (4.1.7) remains true by virtue of Corollary 4.1.2 (μ can always be chosen larger than any given μ). Then, for $t \geq 0$

$$\begin{aligned}\|u(t)(\phi)\|_S &\leq e^{\mu t}\|\phi(0)\|_S + \|F\| \int_0^t e^{\mu(t-s)}\|u_s(\phi)\|_{C_r}ds \\ &\leq \left[(-\|F\|/\mu) + (1 - (-\|F\|/\mu))e^{\mu t}\right]\|\phi\|_{C_r}.\end{aligned} \qquad (4.1.9)$$

Then, for arbitrary $t \geq r$,

$$\|U(t)\phi\|_{C_r} \leq \left[(-\|F\|/\mu) + (1 - (-\|F\|/\mu))e^{\mu(t-r)}\right]\|\phi\|_{C_r}. \qquad (4.1.10)$$

However, since $e^{\mu(t-r)} \geq 1$ for $0 \leq t \leq r$, (4.1.10) holds for all $t \geq 0$. In a similar manner, we can substitute the inequality (4.1.10) into (4.1.9) and integrate to obtain

$$\begin{aligned}\|U(t)\phi\|_{C_r} \leq \Big[(-\|F\|/\mu)^2 + (1 - (-\|F\|/\mu)^2)e^{\mu(t-r)} \\ + \|F\|(1 - (-\|F\|/\mu))e^{\mu(t-2r)}t\Big]\|\phi\|_{C_r}\end{aligned}$$

for all $t \geq 0$. An induction argument yields (4.1.8). By virtue of (4.1.10), $U(t)$, $t > r$, is a commutative family of strict contractions on C_r and therefore has a unique fixed point. That is, for arbitrary $t > r$, there exists $\phi_t \in C_r$ such that $U(t)\phi_t = \phi_t$. Thus, $U(s)\phi_t = U(s)U(t)\phi_t = U(t)U(s)\phi_t$ for any $s > r$ which implies $\phi_t = U(s)\phi_t$, i.e., $\phi_s = \phi_t$. The last statement now follows from (4.1.8). $\qquad \square$

As a direct consequence of Corollaries 4.1.2 and 4.1.3, we can immediately deduce that the null solution of (4.1.1) is stable if $-\|F\| = \mu$ and asymptotically stable if $-\|F\| > \mu$.

Example 4.1.1 Let $S = C_0[0, \pi]$, the space of all continuous real-valued functions over $[0, \pi]$ which are zero at the ends 0 and π, and has the usual supremum norm. Let $A_T : S \to S$ be defined by

$$A_T y(x) = \partial^2 y(x)/\partial x^2, \quad \mathcal{D}(A_T) = \{y(\cdot) \in S : \partial^2 y(x)/\partial x^2 \in S\}.$$

Then A_T is the infinitesimal generator of a strongly continuous semigroup of bounded linear operators $T(t)$ satisfying $\|T(t)\| \leq e^{-t}$, $t \geq 0$, and for an arbitrarily given real number θ, there exists a unique mild solution of the following equation

$$\begin{cases} u'_t(x, t) = u''_{xx}(x, t) + \theta \cdot u(x, t - r), & 0 \leq x \leq \pi, \quad t \geq 0, \\ u(0, t) = u(\pi, t) = 0, & t \geq 0, \\ u(\cdot, t) = \phi(t)(\cdot) \in C_r, & -r \leq t \leq 0. \end{cases} \quad (4.1.11)$$

Moreover, the arguments above apply and the null solution of (4.1.11) is stable if $\theta = 1$ and asymptotically stable if $\theta < 1$.

Under some circumstances, some more refined results of stability than Corollaries 4.1.2 and 4.1.3 can be derived. In fact, it is possible to determine the exact regions of stability for Equation (4.1.1) by a different argument. To illustrate this, let us first explore some compact properties of solutions of (4.1.1).

Lemma 4.1.1 *Let $\{T(t)\}_{t \geq 0}$ be a strongly continuous semigroup of bounded linear operators on S. Assume also that $T(t) : S \to S$ is compact for each $t \geq 0$. Let B be a bounded subset of S and $\{f_\gamma; \gamma \in \Gamma\}$ a family of continuous functions from some finite interval $[c, d] \subset (0, \infty)$ to B. Then $W = \{\int_c^d T(s) f_\gamma(s) ds; \gamma \in \Gamma\}$ is a precompact subset of S.*

Proof Let $L = \{T(t)x; t \in [c, d], x \in B\}$. We will show that L is totally bounded. For any $\varepsilon > 0$, by virtue of the uniform continuity of the mapping $T(\cdot) : [c, d] \to \mathcal{L}(S, S)$, there exist $c = t_1 < t_2 < \cdots < t_n = d$ such that

$$\|T(t_i) - T(t)\| \leq \frac{\varepsilon}{2C} \quad \text{for} \quad t \in [t_{i-1}, t_i], \quad i = 2, \cdots, n, \quad (4.1.12)$$

where $C > 0$ is some bound of B. Since for each t_i, $T(t_i)B$ is totally bounded, there exist $\{x_1^i, x_2^i, \cdots, x_{k(i)}^i\} \subset B$ such that if $x \in B$, then

$$\|T(t_i)x_j^i - T(t_i)x\|_S \leq \varepsilon/2 \quad \text{for some} \quad x_j^i. \quad (4.1.13)$$

The total boundedness of L now follows from (4.1.12) and (4.1.13). Since L is precompact, so is the closed convex hull of L. The lemma follows since W is contained in the closed convex hull of $(d - c)L$. $\qquad \square$

Proposition 4.1.4   *Suppose that $T(t)$, $t \geq 0$, defined in (4.1.1) is compact for each $t > 0$. Then the mapping $(t, \phi) \to u_t = u_t(\phi)$ defined by the solution of (4.1.1) is compact in ϕ for each fixed $t > r$.*

Proof   Let $\{\phi_\gamma : \gamma \in \Gamma\}$ be a bounded subset of C_r and $t > r$. For each $\gamma \in \Gamma$, define $f_\gamma \in C_r$ by $f_\gamma = u_t(\phi_\gamma)$. Then, for $\theta \in [-r, 0]$ we have $t + \theta > 0$, and so

$$f_\gamma(\theta) = u_t(\phi_\gamma)(\theta) = u(\phi_\gamma)(t + \theta) = T(t + \theta)\phi_\gamma(0)$$
$$+ \int_0^{t+\theta} T(t + \theta - s)Fu_s(\phi_\gamma)ds.$$

We first show that this family is equicontinuous. By virtue of (4.1.2), one may show that $\{Fu_s(\phi_\gamma); s \in [0, t], \gamma \in \Gamma\}$ is bounded by a constant, say $M > 0$. Let $\gamma \in \Gamma$, $0 < c < t - r$, $-r \leq \tilde{\theta} < \theta \leq 0$, and observe that

$$\begin{aligned}
\|f_\gamma(\theta) &- f_\gamma(\tilde{\theta})\|_S \\
&\leq \|T(t + \theta)\phi_\gamma(0) - T(t + \tilde{\theta})\phi_\gamma(0)\|_S \\
&+ \left\| \int_0^{t+\theta} T(t + \theta - s)Fu_s(\phi_\gamma)ds - \int_0^{t+\tilde{\theta}} T(t + \theta - s)Fu_s(\phi_\gamma)ds \right\|_S \\
&+ \left\| \int_{t+\tilde{\theta}-c}^{t+\tilde{\theta}} \left[T(t + \theta - s) - T(t + \tilde{\theta} - s) \right] Fu_s(\phi_\gamma)ds \right\|_S \\
&+ \left\| \int_0^{t+\tilde{\theta}-c} \left[T(t + \theta - s) - T(t + \tilde{\theta} - s) \right] Fu_s(\phi_\gamma)ds \right\|_S \\
&\leq \|T(t + \theta) - T(t + \tilde{\theta})\| \|\phi_\gamma(0)\|_S + |\theta - \tilde{\theta}| Me^{\mu t} + 2cMe^{\mu t} \\
&+ Mt \sup_{s \in [0, t+\tilde{\theta}-c]} \|T(t + \theta - s) - T(t + \tilde{\theta} - s)\|.
\end{aligned}$$

One can now use the uniform continuity of $T(s)$, $s \in [c, t]$, in $\mathcal{L}(S, S)$ to demonstrate the claimed equicontinuity.

Next, we show that for each fixed $\theta \in [-r, 0]$, $\{f_\gamma(\theta) : \gamma \in \Gamma\}$ is precompact in S. Indeed, $\{T(t + \theta)\phi_\gamma(0) : \gamma \in \Gamma\}$ is precompact since $t + \theta > 0$ and $\|\phi_\gamma(0)\|_S$ is bounded (independent of γ). We will show that

$$L = \left\{ \int_0^{t+\theta} T(t + \theta - s)Fu_s(\phi_\gamma)ds : \gamma \in \Gamma \right\}$$

is totally bounded. Observe that for $0 < c < t + \theta$, we have

$$\left\| \int_{t+\theta-c}^{t+\theta} T(t + \theta - s)Fu_s(\phi_\gamma)ds \right\|_S \leq cMe^{\mu t} \tag{4.1.14}$$

for all $\gamma \in \Gamma$. By Lemma 4.1.1, if $0 < c < t + \theta$, then

$$L_c = \left\{ \int_0^{t+\theta-c} T(t + \theta - s)Fu_s(\phi_\gamma)ds : \gamma \in \Gamma \right\}$$

is precompact in S. This fact, together with (4.1.14), yields the precompactness of L. Now applying the well-known Arzela-Ascoli theorem to $\{f_\gamma; \gamma \in \Gamma\}$ concludes the proof. $\square$

For each scalar λ, define a linear operator $\Delta(\lambda) : \mathcal{D}(A_T) \to S$ by

$$\Delta(\lambda)x = A_T x - \lambda x + F(e^{\lambda \cdot} x), \qquad x \in \mathcal{D}(A_T),$$

where $e^{\lambda \cdot} x \in C_r$ is defined by

$$(e^{\lambda \cdot} x)(\theta) = e^{\lambda \theta} x, \qquad \theta \in [-r, 0].$$

Note that we still use C_r here to denote its complexification. In particular, we will say that λ satisfies the "characteristic equation" of (4.1.1) provided $\Delta(\lambda)x = 0$ for some $x \neq 0$. Suppose D is an arbitrary linear operator and let $\sigma(D)$, $P\sigma(D)$ denote the spectrum and point spectrum sets of D, respectively. Recall that $U(t)$, $t \geq 0$, is the strongly continuous semigroup of bounded linear operators on C_r defined by the solutions of (4.1.1).

Lemma 4.1.2 *For $t > r$, $\sigma(U(t))$ is a countable set and is compact with only one possible accumulation point, 0, and if $\mu \neq 0 \in \sigma(U(t))$, then $\mu \in P\sigma(U(t))$.*

Proof The lemma follows immediately from Proposition 2.4 and Theorem 6.26 in Kato [1], p. 185. $\square$

Lemma 4.1.3 *For $t > r$, $P\sigma(U(t)) = e^{tP\sigma(A_U)}$ plus possibly $\{0\}$. More specifically, if $\mu = \mu(t) \in P\sigma(U(t))$ for some $t > r$ and $\mu \neq 0$, then there exists $\lambda \in P\sigma(A_U)$ such that $e^{\lambda t} = \mu$. Furthermore, if $\{\lambda_n\}$ consists of all distinct points in $P\sigma(A_U)$ such that $e^{\lambda_n t} = \mu$, then for arbitrary k, the kernel $Ker(U(t) - \mu I)^k$ is the closed linear extension of the linearly independent sets $Ker(A_U - \lambda_n I)^k$, where n ranges over $e^{\lambda_n t} = \mu$.*

Proof See Lemma 22.1 and the exercise after it in Hale [1]. $\square$

Lemma 4.1.4 *Let $S(t)$, $t \geq 0$, be an arbitrary strongly continuous semigroup of bounded linear operators on S and suppose that for some $s > 0$ the spectral radius ρ of $S(s)$ is not zero and $\tau = (1/s)\ln \rho$. Then for all $\gamma > 0$ there exists a constant $M(\gamma) \geq 1$ such that*

$$\|S(t)x\|_S \leq M(\gamma)e^{(\tau+\gamma)t}\|x\|_S \quad \text{for all} \quad t \geq 0, \ x \in S.$$

Proof See Lemma 22.2 in Hale [1], p. 112. $\square$

Lemma 4.1.5   *There exists a real number ν such that $\mathrm{Re}\,\lambda \leq \nu$ for all $\lambda \in \sigma(A_U)$ and if γ is any real number there exist only a finite number of $\lambda \in P\sigma(A_U)$ such that $\gamma \leq \mathrm{Re}\,\lambda$.*

Proof   The existence of the constant ν follows immediately from Propositions 4.1.2 and 4.1.3. Indeed, one can choose $\nu = \max\{0, \|F\| + \mu\}$. Assume that $\{\lambda_k\}$ is an infinite sequence of distinct points in $P\sigma(A_U)$ such that $\mathrm{Re}\,\lambda_k > \gamma$ for all k, where γ is a given real number. By Lemma 4.1.3, $e^{\lambda_k t} \in P\sigma(U(t))$ for a fixed $t > r$. If $\{e^{\lambda_k t}\}$ is infinite, then $P\sigma(U(t))$ has an accumulation point different from zero, a fact that contradicts Lemma 4.1.2. If $\{e^{\lambda_k t}\}$ is finite, then

$$e^{\lambda_{n_k} t} = \mu = \text{ constant for some infinite subsequence } \{\lambda_{n_k}\}.$$

Then $Ker(U(t) - \mu I)$ is infinite dimensional, since it contains the linearly independent sets $Ker(A_U - \lambda_{n_k} I)$ by Lemma 4.1.3. But this contradicts Theorem 5.7.3 in Hille and Phillips [1], which claims that the set $Ker(U(t) - \mu I)$ is finite dimensional. Thus the assumption is false and the proof is complete. $\Box$

Now we are in a position to state the desired stability results.

Proposition 4.1.5   *Suppose β is some real number such that if λ satisfies the characteristic equation of (4.1.1), $\mathrm{Re}\,\lambda \leq \beta$. Then for each $\gamma > 0$, there exists a constant $M(\gamma) \geq 1$ such that for all $t \geq 0$,*

$$\|U(t)\phi\|_{C_r} \leq M(\gamma) e^{(\beta + \gamma)t} \|\phi\|_{C_r}. \tag{4.1.15}$$

Proof   Suppose $\nu \neq 0 \in \sigma(U(t))$ where $t > r$ is some fixed number. By Lemma 4.1.2, $\nu \in P\sigma(U(t))$. Also by virtue of Lemma 4.1.3, $\nu = e^{\lambda t}$ where $\lambda \in P\sigma(A_U)$. Then there exists

$$\phi \neq 0 \in \mathcal{D}(A_U), \qquad \phi' - \lambda\phi = 0. \tag{4.1.16}$$

But this is equivalent to

$$\phi(\beta) = e^{\lambda\theta}\phi(0), \qquad \phi(0) \neq 0, \qquad \phi'_-(0) = A_T\phi(0) + F\phi. \tag{4.1.17}$$

Then $\Delta(\lambda)\phi(0) = 0$, and $\mathrm{Re}\,\lambda \leq \beta$ by assumption. Thus the spectral radius of $U(t)$ is less than or equal to $e^{t\beta}$ and (4.1.15) follows immediately by applying Lemma 4.1.4. $\Box$

Corollary 4.1.4   *Let β be the smallest real number such that if λ satisfies the characteristic equation of (4.1.1), then $\mathrm{Re}\,\lambda \leq \beta$. If $\beta < 0$, then for all*

$\phi \in C_r$, $\|U(t)\phi\|_{C_r} \to 0$, as $t \to \infty$. If $\beta = 0$, then there exists $\phi \neq 0 \in C_r$ such that $\|U(t)\phi\|_{C_r} = \|\phi\|_{C_r}$ for all $t \geq 0$. If $\beta > 0$, then there exists $\phi \in C_r$ such that $\|U(t)\phi\|_{C_r} \to \infty$ as $t \to \infty$.

Proof   The existence of β is a consequence of Lemma 4.1.5. The claim for $\beta < 0$ is immediate from (4.1.15). If $\beta = 0$, let $x \neq 0 \in \mathcal{D}(A_T)$ such that $\Delta(\lambda)x = 0$ where $Re\,\lambda = 0$ (such a λ exists by Lemma 4.1.5). As in (4.1.16) and (4.1.17), $\phi(t) = e^{\lambda t}\phi(0)$, $\phi(0) = x$ solves $(A_U - \lambda I)\phi = 0$, $\phi \neq 0$. Thus $U(t)\phi = e^{\lambda t}\phi$ and $\|U(t)\phi\|_{C_r} = |e^{it \cdot Im\,\lambda}|\|\phi\|_{C_r} = \|\phi\|_{C_r}$. If $\beta > 0$, let $x \neq 0 \in \mathcal{D}(A_T)$ such that $\Delta(\lambda)x = 0$ and $Re\,\lambda > 0$. Again $\phi(t) = e^{\lambda t}\phi(0)$, $\phi(0) = x$ solves $(A_U - \lambda I)\phi = 0$, $\phi \neq 0$. Thus $U(t)\phi = e^{\lambda t}\phi$ and $\|U(t)\phi\|_{C_r} = e^{Re\,\lambda t}\|\phi\|_{C_r} \to \infty$ as $t \to \infty$.   ☐

Finally, let us apply Corollary 4.1.4 to an example whose exact stability region can be specified.

Example 4.1.2   We wish to determine the exact region of stability of the linear partial differential equation

$$\begin{cases} \frac{\partial u(x,t)}{\partial t} = \frac{\partial^2 u(x,t)}{\partial x^2} - au(x,t) - bu(x,t-r), & 0 \leq x \leq \pi, \quad t \geq 0, \\ u(0,t) = u(\pi,t) = 0, & t \geq 0, \\ u(x,t) = \phi(t)(x), & 0 \leq x \leq \pi, \quad -r \leq t \leq 0, \end{cases} \quad (4.1.18)$$

as a function of a, b and r, where the solutions are in the mild sense of (4.1.1) for $S = L^2(0,\pi)$, and $A_T : S \to S$ is defined by

$$A_T y(x) = \frac{d^2 y}{dx^2}(x),$$

$$\mathcal{D}(A_T) = \left\{ y(\cdot) \in S : y(x), \frac{dy}{dx}(x) \text{ are absolutely continuous, } \frac{d^2 y}{dx^2}(x) \in S, \right.$$

$$\left. \text{and } y(0) = y(\pi) = 0 \right\}.$$

Then, it is well known that A_T is closable on S and the closure of A_T generates an analytic compact semigroup $\{T(t)\}_{t \geq 0}$ on S with $\|T(t)\| \leq e^{-t}$, $t \geq 0$.

Let $F : C_r \to S$ be given by $F\phi = -a\phi(0) - b\phi(-r)$. The characteristic values of (4.1.18) are specified by the equation

$$\Delta(\lambda)f = [A_T - (\lambda + a + be^{-\lambda r})I]f = 0$$

for $f \in \mathcal{D}(A_T)\backslash\{0\}$. Since the eigenvalues of A_T are $-n^2$, $n = 1, 2, \cdots$, we have from Corollary 4.1.4 that the null solution of (4.1.18) is asymptotically stable if and only if all the roots of the equations

$$\lambda + a + be^{-\lambda r} = -n^2, \qquad n = 1, 2, \cdots,$$

have negative real parts. The exact region of stability of (4.1.18) is obtained as an immediate consequence of the following useful result which is due to Hayes [1] and whose proof is referred to, for instance, Hale and Lunel [1], p. 416.

Proposition 4.1.6   *All the roots of the equation $(z + \mu)e^z + \nu = 0$, where μ and ν are real numbers, have negative real parts if and only if*

$$\mu > -1, \quad \mu + \nu > 0, \quad \nu < \rho \sin \rho - \mu \cos \rho,$$

where $\rho = \pi/2$ if $\mu = 0$, or ρ is the root of $\rho = -\mu \tan \rho$ in $(0, \pi)$ if $\mu \neq 0$.

4.1.2   Stable Semigroups (Infinite Delays)

In all attempts to obtain an extension of stability results in the last section to functional equations with infinite delays, one meets some difficulties immediately. For instance, one will find the usual phase space C_r, ($r = \infty$ at the moment) does not work appropriately for the stability analysis. Some essential results for stability such as Proposition 4.1.4 do not remain true any more due to the invalidation of the compactness of the solution semigroups. Therefore, all the investigation becomes more delicate. Consider the following linear functional differential equation with infinite retarder on S

$$\frac{du(t)}{dt} = A_T u(t) + F u^t, \quad t \geq 0,$$

or its integrated form

$$u(t) = T(t)\phi(0) + \int_0^t T(t - s)Fu^s ds, \quad t \geq 0, \\ u^0(\theta) = \phi(\theta), \quad \theta \leq 0, \tag{4.1.19}$$

where ϕ is an element in some phase space C_g to be specified below, $\{T(t)\}_{t \geq 0}$ is a compact analytic semigroup on the Banach space S with the generator A_T. $F : C_g \to S$ is a bounded linear operator and for each $u(\cdot) : (-\infty, \infty) \to S$ and $t \geq 0$, u^t is a mapping from $(-\infty, 0]$ to S defined by $u^t(\theta) = u(t + \theta)$ for $\theta \leq 0$.

For our stability purposes, we assume

$$g(s) = e^{-\gamma s}, \quad s \leq 0,$$

for some constant $\gamma > 0$. Define the following Banach space

$$C_g = \left\{ \phi : (-\infty, 0] \to S; \ \phi \text{ is continuous and } \lim_{s \to -\infty} e^{\gamma s}\|\phi(s)\|_S = 0 \right\},$$

equipped with the norm

$$\|\phi\|_{C_g} = \sup_{-\infty \leq s \leq 0} e^{\gamma s}\|\phi(s)\|_S.$$

It may be proved by the standard "method of steps" that for any $\phi \in C_g$, there exists one and only one $u(\cdot) = u(\cdot)(\phi) : (-\infty, \infty) \to S$ such that $u^0(\cdot) = \phi(\cdot)$ and (4.1.19) is satisfied for all $t \geq 0$. Moreover, let $U(t) : C_g \to C_g$ be given by $U(t)\phi = u^t(\phi)$, $t \geq 0$. Then it can be proved that $\{U(t)\}_{t \geq 0}$ is a C_0-semigroup on C_g with some infinitesimal generator A_U.

In contrast to the finite delay case, in general, $U(t) : C_g \to C_g$ is no longer compact. This fact simply implies that the spectrum of $U(t) : C_g \to C_g$ is much more complicated than its counterpart in the finite delay case. In order to handle its stability, we need the notions of the essential spectrum and its radius.

Let $G : \mathcal{D}(G) \subset S$ be a closed operator with a closed domain. We denote by $ess(G)$ the set of *essential spectrum* of G (cf. Browder [1]), and the radius of $ess(G)$ is denoted by $r_e(G)$. It was proved in Nussbaum [1] that

$$r_e(G) = \inf \left\{ k \in \mathbf{R}^1; \, \alpha(G(N)) \leq k\alpha(N) \text{ for every bounded subset } N \text{ of } S \right\},$$

where $\alpha(\cdot)$ is the Kuratowski measure of noncompactness. Moreover, it is known that if λ_0 belongs to the spectrum of G but not to the essential spectrum, then λ_0 is in the point spectrum of G.

Theorem 4.1.1 *Assume that there exist positive constants M and μ such that*

$$\|T(t)\| \leq M \cdot e^{-\mu t}, \qquad t \geq 0.$$

Then the radius of the essential spectrum of the solution semigroup $\{U(t)\}_{t \geq 0}$ of (4.1.19) satisfies

$$r_e(U(t)) \leq M \cdot e^{-\min\{\mu, \gamma\} t}, \qquad t \geq 0. \tag{4.1.20}$$

Proof Define first of all an operator $S(t) : C_g \to C_g$ as follows: for each $\phi \in C_g$, $S(t)\phi$ is the solution of the initial value problem $u^0 = \phi$ and $u(t) = T(t)\phi(0)$ for $t \geq 0$. Therefore, for each $t \geq 0$ and $\theta \in (-\infty, -t]$ we have $[U(t) - S(t)](\phi)(\theta) = 0$. By using the Arzela-Ascoli theorem and carrying out a similar argument to Proposition 4.1.4, one can show that the mapping $\tau \in [0, t] \to \int_0^\tau T(\tau - s)F(U(s)\phi)ds \in S$ is compact. Therefore, we have the conclusion that $U(t) - S(t) : C_g \to C_g$ is compact for $t > 0$, and from (4.1.20), we get $r_e(U(t)) = r_e(S(t)) \leq \|S(t)\|$. On the other hand, for any $\phi \in C_g$ we have

$$\|S(t)\phi\|_{C_g}$$
$$= \sup_{s\leq 0} e^{\gamma s}\|(S(t)\phi)(s)\|_S$$
$$= \max\left\{ \sup_{-t\leq s\leq 0} e^{\gamma s}\|T(t+s)\phi(0)\|_S,\ \sup_{s\leq -t} e^{\gamma s}\|\phi(t+s)\|_S\right\}$$
$$\leq \max\left\{ \sup_{-t\leq s\leq 0} M\cdot e^{\gamma s}e^{-\mu(t+s)}\|\phi(0)\|_S,\ e^{-\gamma t}\sup_{s\leq -t} e^{\gamma(t+s)}\|\phi(t+s)\|_S\right\}$$
$$\leq M\cdot e^{-\min\{\mu,\gamma\}t}\|\phi\|_{C_g}.$$

This proof is complete.   □

From the results above, it is obvious that if the semigroup $\{T(t)\}_{t\geq 0}$ is stable, the point spectrum of $\{U(t)\}_{t\geq 0}$ will determine its asymptotic behavior as $t \to \infty$. Before proceeding to derive our main stability results, let us first study some properties of the operator A_U.

Theorem 4.1.2   *Let A_U denote the infinitesimal generator of the semigroup $\{U(t)\}_{t\geq 0}$ on C_g.*

(i). *If $A_U\phi = \lambda\phi$ with $\phi \neq 0$, then $Re\,\lambda \geq -\gamma$ and $U(t)\phi = e^{\lambda t}\phi$. Moreover, $\phi(\theta) = e^{\lambda\theta}\phi(0)$ with $\phi(0) \in \mathcal{D}(A_T)$ and $\phi(0)$ satisfies the characteristic equation*

$$(\lambda I - A_T)x - F(e^{\lambda\cdot}x) = 0; \qquad (4.1.21)$$

(ii). *If $Re\,\lambda \geq -\gamma$ and (4.1.21) has a nontrivial solution, then $\lambda \in P\sigma(A_U)$;*

(iii). *If $\mu \in P\sigma(U(t))$ and $\mu \neq 0$, then there exists a finite number $\lambda \in P\sigma(A_U)$ such that $e^{\lambda t} = \mu$.*

Proof   (i). Let $z(t) = U(t)\phi$. As $\phi \in \mathcal{D}(A_U)$ and $A_U\phi = \lambda\phi$, we know that

$$\frac{dz(t)}{dt} = A_U z(t) = U(t)A_U\phi = U(t)\lambda\phi = \lambda z(t).$$

Therefore, $z(t) = e^{\lambda t}\phi$ for $t \geq 0$. On the other hand, we know that

$$z(t)(\theta) = (U(t)\phi)(\theta) = (U(t+\theta)\phi)(0)$$

for $\theta < 0$. This implies that $e^{\lambda t}\phi(\theta) = e^{\lambda(t+\theta)}\phi(0)$ for $\theta \leq 0$ from which it follows that $\phi(\theta) = e^{\lambda\theta}\phi(0)$ for $\theta \leq 0$. Since

$$F(U(t)\phi) = F(e^{\lambda t}e^{\lambda\cdot}\phi(0)) = e^{\lambda t}F(e^{\lambda\cdot}\phi(0)),$$

is, as a function of $t \geq 0$, locally Hölder continuous, we know by the standing theory of partial differential equations that $(U(t)\phi)(0)$ is indeed a solution of

$du(t)/dt = A_T u(t) + F(u^t)$. This implies that $\phi(0) \in \mathcal{D}(A_T)$ and $\lambda e^{\lambda t}\phi(0) = A_T e^{\lambda t}\phi(0) + e^{\lambda t}F(e^{\lambda \cdot}\phi(0))$, completing the proof of (i).

The verification of (ii) is straightforward, where $\lambda \in P\sigma(A_U)$ with the corresponding eigenvector $e^{\lambda \cdot}\phi(0)$.

(iii). The existence of λ is an immediate consequence of Lemma 4.1.3. We need only to show the finiteness of such $\lambda \in P\sigma(A_T)$. Clearly, all solutions to $e^{\lambda t} = \mu$ have the form

$$\lambda_n = \frac{\ln \mu}{t} + \frac{i2\pi n}{t}, \qquad n = 0, \pm 1, \pm 2, \cdots.$$

As A_T generates an analytic semigroup, there exists an integer n_0 such that for $|n| \geq n_0$, λ_n belongs to the resolvent set of A_T. This means that the characteristic equation (4.1.21) with $\lambda = \lambda_n$, $|n| \geq n_0$, is equivalent to

$$x = (\lambda I - A_T)^{-1}F(e^{\lambda \cdot}x). \tag{4.1.22}$$

If $\lambda_n \in P\sigma(A_U)$, then (4.1.22) has a solution x_n such that

$$\|x_n\|_S = \sup_{\theta \leq 0} \|e^{\lambda_n \theta}\phi(0)\|_S = 1.$$

However,

$$\begin{aligned} \|x_n\|_S &= \|(\lambda_n I - A_T)^{-1}F(e^{\lambda_n \cdot}x_n)\|_S \\ &\leq \frac{M}{|\lambda_n|}\|F\|\|e^{\lambda_n \cdot}x_n\|_S = \frac{M}{|\lambda_n|}\|F\| \to 0, \end{aligned}$$

as $n \to \infty$, a contradiction. This completes the proof. $\qquad\qquad$ □

The following theorem and its corollary will give useful information to ensure stability behavior on the semigroup $\{U(t)\}_{t\geq 0}$.

Theorem 4.1.3 *For any $\varepsilon > 0$, the set*

$$L := \left\{\lambda \in \mathbf{C}; \; Re\,\lambda > -\min(\mu, \gamma) + \varepsilon\right\}$$

contains only a finite number of points of $P\sigma(A_U)$.

Proof The set L is a subset of the resolvent set of A_T and thus if $\lambda \in L$, then the characteristic equation (4.1.21) is equivalent to (4.1.22). Therefore, $\lambda_0 \in P\sigma(A_U) \cap L$ if and only if $1 \in P\sigma(F(\lambda))$, where $F(\lambda)x := (\lambda I - A)^{-1}F(e^{\lambda \cdot}x)$ for each $x \in S$. As $F(\lambda_0) : S \to S$ is compact, 1 is an isolated point of the spectrum of $F(\lambda_0)$. Clearly, the function $\lambda \to F(\lambda)$ is analytic in L. So either $L \subset P\sigma(A_U)$ or $P\sigma(A_U)$ is isolated in L. But the first claim is impossible by virtue of Theorem 4.1.2. So $P\sigma(A_U)$ is isolated in L. A similar argument to that of (iii) in Theorem 4.1.2 also shows that $P\sigma(A_U) \cap L$ is finite. □

An immediate consequence of the above theorem is the following asymptotic stability criterion:

Corollary 4.1.5   *Suppose the conditions in Theorem 4.1.1 hold. If $Re \lambda < 0$ for any solution λ such that the characteristic equation (4.1.21) has a nontrivial solution, then there exist constants $C > 0$ and $\mu > 0$ such that*

$$\|U(t)\| \leq C \cdot e^{-\mu t}, \quad t \geq 0.$$

Proof   By virtue of Theorem 4.1.3, $\Delta := \sup Re\,(P\sigma(A_U)) < 0$. Theorem 4.1.2 implies

$$\sup\left\{|\lambda| : \lambda \in P\sigma(U(t))\right\} = e^{\Delta t}.$$

By Theorem 4.1.1, we have $\mu_1 > 0$ and $C_1 > 0$ such that $r(U(t)) \leq C_1 \cdot e^{-\mu_1 t}$. This yields the conclusion by using Lemma 4.1.4. □

As an illustration of the preceding results, let us investigate the following linear partial Lotka-Volterra integrodifferential equation.

Example 4.1.3   Consider the equation

$$\frac{\partial}{\partial t}u(x,t) = \nu \frac{\partial^2}{\partial x^2}u(x,t) + \int_{-\infty}^{t} k(t-s)u(x,s)ds,$$

$$t \geq 0, \quad x \in (0, \pi), \tag{4.1.23}$$

$$\frac{\partial}{\partial x}u(x,t) = 0, \quad x = 0, \pi,$$

where ν is a positive constant. As usual, we let $S = C([0, \pi]; \mathbf{R}^1)$ and define $A_T y(x) = \nu \frac{d^2 y}{dx^2}(x)$ for $y(\cdot) \in \mathcal{D}(A_T) = \{y(\cdot) \in C^2([0, \pi]; \mathbf{R}^1); \frac{d^2 y}{dx^2}(0) = \frac{d^2 y}{dx^2}(\pi) = 0\}$. Then the closure of A_T generates an analytic compact semigroup in S. We also assume that $k : [0, \infty) \to \mathbf{R}^1$ is continuous and there exists $\gamma > 0$ such that

$$\int_0^\infty e^{\gamma t}|k(t)|dt < \infty.$$

Define $F : C_g \to S$ by

$$F(\phi)(x) = \int_{-\infty}^{0} k(-\theta)\phi(x, \theta)d\theta, \quad x \in [0, \pi].$$

Then

$$\|F(\phi)\|_S = \sup_{x \in [0,\pi]} |F(\phi)(x)| \leq \|\phi\|_{C_g} \int_0^\infty e^{\gamma t}|k(t)|dt$$

for arbitrary $\phi \in C_g$. Consequently, the null solution of (4.1.23) is asymptotically stable if $Re\ \lambda < 0$ for all $\lambda \in \mathbf{C}$ such that the equation

$$\left[\lambda - \int_{-\infty}^{0} k(-u)e^{\lambda u}du\right]v(x) = \nu \cdot \frac{d^2v}{dx^2}(x), \quad v(0) = v(\pi) = 0,$$

has a nontrivial solution $v \in C^2([0, \pi]; \mathbf{R}^1)$. As the eigenvalues of A_T are given by $-\nu j^2$, $j = 0, 1, 2, \cdots$, we conclude that the null solution of (4.1.23) is asymptotically stable if $Re\ \lambda < 0$ for all solutions λ of the equations

$$\lambda - \int_{-\infty}^{0} k(-\theta)e^{\lambda \theta}d\theta = -\nu j^2, \quad j = 0, 1, \cdots. \tag{4.1.24}$$

Note that the latter is equivalent to requiring that the null solution of each of the following scalar ordinary differential equations

$$\frac{dy}{dt}(t) = -\nu j^2 y(t) + \int_{-\infty}^{t} k(t-s)y(s)ds$$

is asymptotically stable for $j = 0, 1, \cdots$.

What remains is to find sufficient conditions to ensure that $Re\ \lambda < 0$ for each solution of (4.1.24). To this end, let us introduce the following results whose proofs are referred to Lenhart and Travis [1]. Consider the equation

$$\frac{du}{dt}(t) = A_T u(t) + au(t) + \sum_{i=1}^{L} b_i \int_0^{\infty} u(t-s)dm_i(s),$$

where A_T generates a C_0-semigroup on some Banach space, b_i, $1 \leq i \leq L$, are real constants, and $m_i : [0, \infty) \to \mathbf{R}^1$ are given measures such that

$$\int_0^{\infty} |dm_i(s)| = 1, \quad \int_0^{\infty} dm_i(s) \geq 0, \quad 1 \leq i \leq L.$$

We assume that the spectrum of A_T contains only real numbers $\{\theta_n\}$ with $\theta_1 < 0$ being the largest. The above discussion has demonstrated the importance of studying the roots of the associated characteristic equation

$$\lambda = a + \theta_n + \sum_{i=1}^{L} b_i \int_0^{\infty} e^{-\lambda s}dm_i(s), \quad n = 1, 2, \cdots. \tag{4.1.25}$$

Theorem 4.1.4 *All the roots of the characteristic equations (4.1.25) have negative real parts for all m_i, $1 \leq i \leq L$, if and only if*

$$\sum_{i=1}^{L} |b_i| \leq -(a + \theta_1) \quad and \quad \sum_{i=1}^{L} b_i < -(a + \theta_1).$$

4.2   Stability Equivalence and Reduction of Neutral Equations

Bearing the deterministic results of the last section in mind, we are now ready to study stochastic stability of a class of neutral linear stochastic functional differential equations (4.2.6) below. Before moving forward, we want to mention that it turns somewhat inconvenient to use some of the results derived in Section 4.1 such as Theorem 4.1.3 to ensure stability of equations with infinite retarder when A_T as defined there does not generate a compact analytic semigroup. This fact suggests that a different scheme should be carried out to deal with more general (stochastic) evolution equations. In this section, we shall first establish some equivalent relations among L^2-stability, uniformly asymptotic and exponential stabilities for a class of stochastic differential equations (4.2.3) with infinite delays. As a consequence, we can apply these relations by means of proper transformations to the investigation of neutral stochastic evolution equations (4.2.6) in which we are especially interested.

4.2.1   Stability of Retarded Stochastic Systems

We assume $\{h_j\}$, $0 < h_1 < \cdots < h_j \cdots$, is a sequence of real numbers which satisfies

$$\lim_{n \to \infty} h_n = \infty,$$

and

$$\lim_{n \to \infty} \frac{\ln n}{h_n} = 0. \tag{4.2.1}$$

Let A be a linear operator, generally unbounded, and the infinitesimal generator of a C_0-semigroup $T(t)$, $t \geq 0$, of bounded linear operators on the separate Hilbert space H. Suppose $\{F_j\} \in \mathcal{L}(H)$ and $\{G_j\} \in \mathcal{L}(H, \mathcal{L}(K, H))$ are two families of bounded linear operators, respectively. To introduce a proper phase space for future equations, let us assume that there is a sequence of nonnegative numbers $\{d_j\}$ such that

$$\|F_j\| \vee \|G_j\| \leq d_j$$

where $a \vee b = \max\{a, b\}$ for any a, $b \in \mathbf{R}^1$, and

$$\sum_{j=1}^{\infty} d_j e^{2\delta h_j} < \infty \tag{4.2.2}$$

for some positive number δ. Let $(\Omega, \mathcal{F}, P)$ be a probability space equipped with a filtration $\{\mathcal{F}_t\}_{t \geq 0}$ which satisfies the usual conditions. For our stability purpose, we intend to introduce the following space $\underline{X}$:

$$\underline{X} := L^2(\Omega, H) \oplus \underline{L}^2((-\infty, 0) \times \Omega; H),$$

where

$$\underline{L}^2((-\infty, 0) \times \Omega; H) = \Big\{ \psi : (-\infty, 0] \times \Omega \to H \text{ is } \mathcal{F} - \text{measurable and}$$

$$\sum_{j=1}^{\infty} d_j^2 e^{2\delta h_j} \int_{-h_j}^{0} E\|\psi(s)\|_H^2 ds < \infty \Big\}$$

with norm

$$\|\psi\|_{\underline{L}^2}^2 = \sum_{j=1}^{\infty} d_j^2 e^{2\delta h_j} \int_{-h_j}^{0} E\|\psi(s)\|_H^2 ds.$$

It is straightforward to check that $\underline{X}$ is a Hilbert space with the inner product

$$\langle \tilde{\psi}_1, \tilde{\psi}_2 \rangle_{\underline{X}} = E\langle \psi_1(0), \psi_2(0) \rangle_H + \sum_{j=1}^{\infty} d_j^2 e^{2\delta h_j} \int_{-h_j}^{0} E\langle \psi_1(s), \psi_2(s) \rangle_H ds$$

where $\tilde{\psi}_1 = (\psi_1(0), \psi_1)$ and $\tilde{\psi}_2 = (\psi_2(0), \psi_2) \in \underline{X}$, and the norm

$$\|\tilde{\psi}\|_{\underline{X}} = \sqrt{\langle \tilde{\psi}, \tilde{\psi} \rangle_{\underline{X}}},$$

where $\tilde{\psi} = (\psi(0), \psi) \in \underline{X}$.

Our major concern is to investigate stability of a wide class of neutral stochastic evolution equations (4.2.6) in Subsection 4.2.4. To this end, let us first study the following stochastic difference differential equation over H: for any $t \geq 0$, $\tilde{\psi} \in \underline{X}$,

$$
\begin{cases}
v(t, \tilde{\psi}) = T(t)\psi(0) + \int_0^t T(t-s) \sum_{j=1}^{\infty} F_j v(s - h_j, \tilde{\psi}) ds \\
\qquad + \int_0^t T(t-s) \sum_{j=1}^{\infty} G_j v(s - h_j, \tilde{\psi}) dW_s, \quad t \geq 0, \\
v(0, \tilde{\psi}) = \psi(0) \in L^2(\Omega, H), \; v(t, \tilde{\psi}) = \psi(t) \in \underline{L}^2((-\infty, 0) \times \Omega; H), \; t < 0,
\end{cases}
$$
$$(4.2.3)$$

where W_t, $t \geq 0$, is some given K-valued Q-Wiener process with finite trace class covariance operator Q with respect to $\{\mathcal{F}_t\}_{t \geq 0}$.

It can be shown that under the condition (4.2.2), the equation (4.2.3) does make sense in $\underline{X}$. In particular, by a standard argument we may establish the following existence and uniqueness of (4.2.3) whose proof is referred to Liu [7].

Theorem 4.2.1   *Let $\|F_j\|$, $\|G_j\|$, $j = 1, 2, \cdots$, be two families of bounded linear operators such that (4.2.2) holds. Then the equation (4.2.3) has a* *unique solution satisfying the property that there exist constants $C > 0$ and $\mu > 0$ such that for any real number $T \geq 0$, $\tilde{\psi} = (\psi(0), \psi) \in X$,*

$$\sup_{t \in [0,T]} E\|v(t, \tilde{\psi})\|_H^2 \leq C \cdot e^{\mu T} \|\tilde{\psi}\|_{\underline{X}}^2.$$

Definition 4.2.1   For $t \geq 0$, $\tilde{\psi} \in \underline{X}$ and the corresponding solution $v(t, \tilde{\psi})$ of (4.2.3), define $v^t(\tilde{\psi}) \in \underline{L^2}((-\infty, 0) \times \Omega; H)$ and $\tilde{v}^t(\tilde{\psi}) \in \underline{X}$, $t \geq 0$, as

$$v^t(\tilde{\psi}) = \{v(t + s, \tilde{\psi}) : s \in (-\infty, 0)\}, \quad \tilde{v}^t(\tilde{\psi}) = (v(t, \tilde{\psi}), v^t(\tilde{\psi})),$$

and map $\tilde{U}(t) : \underline{X} \to \underline{X}$, $t \geq 0$, as

$$\tilde{U}(t)\tilde{\psi} = \tilde{v}^t(\tilde{\psi}), \quad \tilde{\psi} \in \underline{X}.$$

Clearly, for any $t \geq 0$, $\tilde{U}(t)$ is a linear mapping. The following corollary which may be easily deduced from Theorem 4.2.1 shows that $\{\tilde{U}(t)\}$, $t \geq 0$, is actually a family of continuous linear operators from $\underline{X}$ into itself.

Corollary 4.2.1   *For $t \geq 0$ and $\tilde{\psi} \in \underline{X}$,*

$$\|\tilde{U}(t)\tilde{\psi}\|_{\underline{X}} \leq c \cdot e^{\nu t} \|\tilde{\psi}\|_{\underline{X}}$$

for some constants $c \geq 1$ and $\nu > 0$.

In addition to the properties shown in Corollary 4.2.1, we can actually prove that the bounded linear operator family $\tilde{U}(t)$, $t \geq 0$, is a strongly continuous semigroup (see Liu [7] for more details).

Theorem 4.2.2   *For the family $\tilde{U}(t) : \underline{X} \to \underline{X}$, $t \geq 0$, defined in Definition 4.2.1, we have that for arbitrary $\tilde{\psi} \in \underline{X}$,*

(i) $\tilde{U}(t)\tilde{U}(s)\tilde{\psi} = \tilde{U}(t + s)\tilde{\psi}$ for any $s, t \in [0, \infty)$;
(ii) there exist constants $C \geq 1$ and $\mu > 0$ such that

$$\|\tilde{U}(t)\| \leq C \cdot e^{\mu t}, \quad t \geq 0;$$

(iii) $\tilde{U}(t)$ is strongly continuous in the mean square sense, i.e., for any $\tilde{\psi} \in \underline{X}$,

$$\lim_{t \to 0^+} \|\tilde{U}(t)\tilde{\psi} - \tilde{\psi}\|_{\underline{X}}^2 = 0.$$

To establish stability of solutions for the equation (4.2.3) with initial data $\tilde{\psi} = (\psi(0), \psi) \in \underline{X}$, we first derive the following lemma which is a generalization of the usual uniform boundedness principle for continuous linear operators.

Let S be a real separable Banach space. A map $\rho(\cdot) : S \to \mathbf{R}^1$ is a *seminorm* if it satisfies the following conditions:

(i) $|\rho(x + y)| \leq |\rho(x)| + |\rho(y)|$ for any x, $y \in S$;

(ii) $|\alpha\rho(x)| = |\rho(\alpha x)|$ for any $\alpha \geq 0$, $x \in S$.

Lemma 4.2.1 *Let $\rho_n(\cdot)$, $n \geq 1$, be a family of continuous semi-norms on the Banach space S such that for each $x \in S$, $\sup_n \rho_n(x) < \infty$. Then there exists a positive constant $C < \infty$ such that for all $x \in S$,*

$$\sup_n \rho_n(x) \leq C\|x\|_S.$$

Proof The proof is a straightforward variant of the corresponding arguments of Lemma 13, p. 53 in Dunford and Schwartz [1]. □

Definition 4.2.2 *The null solution of (4.2.3) is said to be L^2-stable in mean if for every $\tilde{\psi} \in X$, we have*

$$\int_0^\infty E\|v(t, \tilde{\psi})\|_H^2 dt < \infty.$$

A direct application of Lemma 4.2.1 to Definition 4.2.2 produces the following statement which is equivalent to the above relation.

Lemma 4.2.2 *If the null solution of (4.2.3) is L^2-stable in mean, then there exists a constant $C > 0$ such that for each $\tilde{\psi} \in X$,*

$$\int_0^\infty E\|v(t, \tilde{\psi})\|_H^2 dt \leq C\|\tilde{\psi}\|_X^2 < \infty.$$

Proof For each $t \geq 0$, define on the Hilbert space X the seminorm

$$\rho_n(\tilde{\psi}) = \left(\int_0^n E\|v(t, \tilde{\psi})\|_H^2 dt \right)^{1/2}$$

for each $n \geq 1$. It is easy to see that for each fixed n, $\rho_n(\cdot)$ is a continuous seminorm by linear property of the solution (4.2.3). Thus, by virtue of Lemma 4.2.1,

$$\sup_n \rho_n(\tilde{\psi}) \leq C^{1/2} \cdot \|\tilde{\psi}\|_X$$

for some $C > 0$. □

Let $0 < r < \infty$. For the purpose of our stability analysis, we restrict the initial space of (4.2.3) to a closed Hilbert subspace of X

$$X_r := \left\{ \tilde{\psi} : \tilde{\psi} \in X \text{ and } \|\psi(t)\|_H = 0 \text{ a.s. on } (-\infty, -r) \right\}.$$

Theorem 4.2.3   *For arbitrary initial data $\tilde{\psi} \in \underline{X}_r$, let $v(t, \tilde{\psi})$, $t \in \mathbf{R}_+$, be the solution of the equation (4.2.3). Then the following three notions of stability are equivalent:*

- *(i). The null solution of (4.2.3) is L^2-stable in mean;*
- *(ii). The null solution of (4.2.3) is uniformly asymptotically stable in mean square in the sense of Definition 2.3.1;*
- *(iii). The null solution of (4.2.3) is exponentially stable in mean square.*

The proofs of the theorem are divided into several lemmas and propositions below.

Proposition 4.2.1   *For arbitrary $\tilde{\psi} \in \underline{X}_r$, assume that the null solution of (4.2.3) is L^2-stable in mean. Then it is also stable in mean square and the following relation*

$$\lim_{t \to \infty} E\|v(t, \tilde{\psi})\|_H^2 = 0$$

holds.

In order to prove this, we first show the following lemma.

Lemma 4.2.3   *Let δ' be an arbitrarily given real number, then any solution of (4.2.3) satisfies the equation*

$$
\begin{aligned}
v(t, \tilde{\psi}) = \ & e^{\delta' t} T(t)\psi(0) - \delta' \int_0^t e^{\delta'(t-s)} T(t-s) v(s, \tilde{\psi}) ds \\
& + \int_0^t e^{\delta'(t-s)} T(t-s) \sum_{j=1}^{\infty} F_j v(s - h_j, \tilde{\psi}) ds \qquad (4.2.4) \\
& + \int_0^t e^{\delta'(t-s)} T(t-s) \sum_{j=1}^{\infty} G_j v(s - h_j, \tilde{\psi}) dW_s
\end{aligned}
$$

for any $t \geq 0$ and $\tilde{\psi} \in \underline{X}_r$.

Proof   Indeed, substituting

$$
\begin{aligned}
v(t, \tilde{\psi}) = \ & T(t)\psi(0) + \int_0^t T(t-s) \sum_{j=1}^{\infty} F_j v(s - h_j, \tilde{\psi}) ds \\
& + \int_0^t T(t-s) \sum_{j=1}^{\infty} G_j v(s - h_j, \tilde{\psi}) dW_s
\end{aligned}
$$

into the right-hand side of (4.2.4) and using Fubini-type theorem for stochastic integrals, i.e., Proposition 1.3.4, we obtain

$$
e^{\delta' t} T(t)\psi(0) - \delta' \int_0^t e^{\delta'(t-s)} T(t-s) \Big[T(s)\psi(0) +
$$

$$
\int_0^s T(s-u) \sum_{j=1}^\infty F_j v(u-h_j, \tilde\psi) du + \int_0^s T(s-u) \sum_{j=1}^\infty G_j v(u-h_j, \tilde\psi) dW_u \Big] ds
$$

$$
+ \int_0^t e^{\delta'(t-s)} T(t-s) \sum_{j=1}^\infty F_j v(s-h_j, \tilde\psi) ds
$$

$$
+ \int_0^t e^{\delta'(t-s)} T(t-s) \sum_{j=1}^\infty G_j v(s-h_j, \tilde\psi) dW_s
$$

$$
= e^{\delta' t} T(t)\psi(0) + (1 - e^{\delta' t}) T(t)\psi(0)
$$

$$
- \delta' \int_0^t e^{\delta'(t-s)} \int_0^s T(t-u) \sum_{j=1}^\infty F_j v(u-h_j, \tilde\psi) du\, ds
$$

$$
- \delta' \int_0^t e^{\delta'(t-s)} \int_0^s T(t-u) \sum_{j=1}^\infty G_j v(u-h_j, \tilde\psi) dW_u\, ds
$$

$$
+ \int_0^t e^{\delta'(t-s)} T(t-s) \sum_{j=1}^\infty F_j v(s-h_j, \tilde\psi) ds
$$

$$
+ \int_0^t e^{\delta'(t-s)} T(t-s) \sum_{j=1}^\infty G_j v(s-h_j, \tilde\psi) dW_s
$$

$$
= T(t)\psi(0) + \int_0^t T(t-u) \sum_{j=1}^\infty F_j v(u-h_j, \tilde\psi) du
$$

$$
- \int_0^t e^{\delta'(t-u)} T(t-u) \sum_{j=1}^\infty F_j v(u-h_j, \tilde\psi) du
$$

$$
+ \int_0^t T(t-u) \sum_{j=1}^\infty G_j v(u-h_j, \tilde\psi) dW_u
$$

$$
- \int_0^t e^{\delta'(t-u)} T(t-u) \sum_{j=1}^\infty G_j v(u-h_j, \tilde\psi) dW_u
$$

$$
+ \int_0^t e^{\delta'(t-s)} T(t-s) \sum_{j=1}^\infty F_j v(s-h_j, \tilde\psi) ds
$$

$$
+ \int_0^t e^{\delta'(t-s)} T(t-s) \sum_{j=1}^\infty G_j v(s-h_j, \tilde\psi) dW_s
$$

$$= T(t)\psi(0) + \int_0^t T(t-u)\sum_{j=1}^{\infty} F_j v(u - h_j, \tilde{\psi})du$$

$$+ \int_0^t T(t-u)\sum_{j=1}^{\infty} G_j v(u - h_j, \tilde{\psi})dW_u = v(t, \tilde{\psi}).$$

This shows that any solution of (4.2.3) satisfies (4.2.4).   □

Proof of Proposition 4.2.1.   First assume that the system (4.2.3) is L^2-stable in mean, i.e., there exists a constant $M_1 > 0$ such that $\int_0^{\infty} E\|v(t, \tilde{\psi})\|_H^2 dt \le M_1\|\tilde{\psi}\|_{\underline{X}}^2$ by Lemma 4.2.2. Suppose $\|T(t)\| \le Me^{\mu t}$, $t \ge 0$, for some constants $M \ge 1$, $\mu > 0$, and let $\delta' = -2\mu$ in Lemma 4.2.3, then we have from (4.2.4) that

$$E\|v(t, \tilde{\psi})\|_H^2$$

$$\le 16\Big\{ M^2 e^{-4\mu t} E\|\psi(0)\|_H^2 + 4\mu^2 E\Big(\int_0^t Me^{-\mu(t-s)}\|v(s, \tilde{\psi})\|_H ds \Big)^2$$

$$+ E\Big(\int_0^t Me^{-\mu(t-s)} \sum_{j=1}^{\infty} \|F_j\| \cdot \|v(s - h_j, \tilde{\psi})\|_H ds \Big)^2$$

$$+ E\Big\| \int_0^t e^{-2\mu(t-s)}T(t-s)\sum_{j=1}^{\infty} G_j v(s - h_j, \tilde{\psi})dW_s \Big\|_H^2 \Big\}$$

$$\le C_1\Big\{ \|\tilde{\psi}\|_{\underline{X}}^2 + \int_0^{\infty} E\|v(s, \tilde{\psi})\|_H^2 ds$$

$$+ \Big(\sum_{j=1}^{\infty} \frac{1}{j^2} \Big) \sum_{j=1}^{\infty} \Big[j^2 d_j^2 \Big(\int_0^{\infty} E\|v(s, \tilde{\psi})\|_H^2 ds + \int_{-h_j}^0 E\|\psi(s)\|_H^2 ds \Big) \Big] \Big\}$$

which, by using the assumptions of Proposition 4.2.1, immediately implies

$$E\|v(t, \tilde{\psi})\|_H^2 \le C_2\Big\{ \|\tilde{\psi}\|_{\underline{X}}^2 + M_1\|\tilde{\psi}\|_{\underline{X}}^2$$

$$+ \sum_{j=1}^{\infty} \Big[d_j^2 e^{2\delta h_j}\Big(\|\tilde{\psi}\|_{\underline{X}}^2 + \int_{-h_j}^0 E\|\psi(s)\|_H^2 ds \Big) \Big] \Big\} \le C_3\|\tilde{\psi}\|_{\underline{X}}^2,$$

where C_1, C_2 and C_3 are some proper positive constants. Therefore, the null solution of (4.2.3) is stable in mean square. Now let $0 \le t_0 \le t$, we have

$$E\|v(t, \tilde{\psi})\|_H^2 \le 16\Big\{ E\Big\| e^{-2\mu(t-t_0)}T(t - t_0)v(t_0, \tilde{\psi}) \Big\|_H^2$$

$$+ E\Big\| 2\mu \int_{t_0}^t e^{-2\mu(t-s)}T(t-s)v(s, \tilde{\psi})ds \Big\|_H^2$$

$$+E\Big\|\int_{t_0}^{t}e^{-2\mu(t-s)}T(t-s)\sum_{j=1}^{\infty}F_j v(s-h_j,\tilde{\psi})ds\Big\|_H^2$$

$$+E\Big\|\int_{t_0}^{t}e^{-2\mu(t-s)}T(t-s)\sum_{j=1}^{\infty}G_j v(s-h_j,\tilde{\psi})dW_s\Big\|_H^2\Big\}$$

$$\le 16\Big\{C_3 M^2 e^{-\mu(t-t_0)}\|\tilde{\psi}\|_{\underline{X}}^2 + 2\mu M^2\Big(\int_{t_0}^{\infty}E\|v(s,\tilde{\psi})\|_H^2 ds\Big)$$

$$+M^2\Big(\int_{t_0}^{\infty}e^{-2\mu s}ds\Big)\Big(\sum_{j=1}^{\infty}\frac{1}{j^2}\Big)\sum_{j=1}^{\infty}j^2 d_j^2\int_{t_0-h_j}^{\infty}E\|v(s,\tilde{\psi})\|_H^2 ds$$

$$+M^2 trQ\Big(\sum_{j=1}^{\infty}\frac{1}{j^2}\Big)\sum_{j=1}^{\infty}j^2 d_j^2\int_{t_0-h_j}^{\infty}E\|v(s,\tilde{\psi})\|_H^2 ds\Big\}$$

$$\le C_4 e^{-\mu(t-t_0)}\|\tilde{\psi}\|_{\underline{X}}^2 + C_4\int_{t_0}^{\infty}E\|v(s,\tilde{\psi})\|_H^2 ds$$

$$+C_4\sum_{j=1}^{J}j^2 d_j^2\Big(\int_{t_0-h_j}^{\infty}E\|v(s,\tilde{\psi})\|_H^2 ds\Big)$$

$$+C_4\sum_{j=J+1}^{\infty}j^2 d_j^2\Big(\int_{t_0-h_j}^{\infty}E\|v(s,\tilde{\psi})\|_H^2 ds\Big)$$

where C_4 is some positive constant. Let $\varepsilon > 0$ be an arbitrarily given constant. By virtue of Theorem 4.2.1 and L^2-stability in mean of (4.2.3), the last term of the right hand side of the above inequality can be made less than $\varepsilon/3$ if J is sufficiently large. The L^2-stability in mean implies that the second and the third terms can be made less than $\varepsilon/3$ if t_0 is sufficiently large. Finally, the first term can be made less than $\varepsilon/3$ if $t - t_0$ is sufficiently large. Therefore, if t is large then $E\|v(t,\tilde{\psi})\|_H^2 < \varepsilon$. This concludes the proof. $\quad\Box$

Proposition 4.2.2   *Let $0 < r < \infty$ and the null solution of (4.2.3) be L^2-stable in mean. Then for arbitrarily given $\varepsilon > 0$, there exists $T(\varepsilon) > 0$ such that for all $\tilde{\psi} \in \underline{X}_r$,*

$$\|\tilde{U}(t)\tilde{\psi}\|_{\underline{X}}^2 \le \varepsilon\|\tilde{\psi}\|_{\underline{X}}^2 \quad if \quad t \ge T(\varepsilon).$$

In other words, the null solution of (4.2.3) is uniformly asymptotically stable in mean square.

Proof   Suppose (4.2.3) is L^2-stable in mean. For any $\tilde{\psi} \in \underline{X}_r$, we have by using Fubini's theorem that

$$\int_0^\infty \|v^t(\tilde{\psi})\|_{L^2}^2 dt = \sum_{j=1}^\infty d_j^2 e^{2\delta h_j} \int_0^\infty \int_{-h_j}^0 E\|v(t+u,\tilde{\psi})\|_H^2 du dt$$

$$= \sum_{j=1}^\infty d_j^2 e^{2\delta h_j} \int_{-h_j}^0 \int_{-r}^\infty E\|v(t,\tilde{\psi})\|_H^2 dt du$$

$$= \left(\sum_{j=1}^\infty h_j d_j^2 e^{2\delta h_j}\right)\left(\int_0^\infty E\|v(t,\tilde{\psi})\|_H^2 dt + \int_{-r}^0 E\|\psi(t)\|_H^2 dt\right).$$

It is easy to see that $\sum_{j=1}^\infty h_j d_j^2 e^{2\delta h_j} < \infty$ due to (4.2.2). We also have

$$\int_{-r}^0 E\|\psi(t)\|_H^2 dt \le \frac{1}{d_J^2} \cdot e^{-2\delta h_J} \|\tilde{\psi}\|_X^2$$

where h_J is the first number of $\{h_j\}$ such that $h_J \ge r$. Hence, there exists constant $M > 0$ such that for all $\tilde{\psi} \in X_r$,

$$\int_0^\infty \|\tilde{U}(t)\tilde{\psi}\|_X^2 dt = \int_0^\infty E\|v(t,\tilde{\psi})\|_H^2 dt + \int_0^\infty \|v^t(\tilde{\psi})\|_{L^2}^2 dt \le M\|\tilde{\psi}\|_X^2. \quad (4.2.5)$$

Without loss of generality, we suppose $0 < \varepsilon < 1$. Since $\|\tilde{U}(t)\tilde{\psi}\|_X^2$ is continuous, there exists, using (4.2.5), a first time $t_0(\varepsilon,\tilde{\psi}) > 0$ such that for any $\tilde{\psi} \in X_r$ with $\|\tilde{\psi}\|_X = 1$,

$$\|\tilde{U}(t_0(\varepsilon,\tilde{\psi}))\tilde{\psi}\|_X^2 = \varepsilon,$$

and for $t \in [0, t_0(\varepsilon,\tilde{\psi}))$,

$$\|\tilde{U}(t)\tilde{\psi}\|_X^2 > \varepsilon.$$

By virtue of (4.2.5), we obtain the inequality

$$\varepsilon t_0(\varepsilon,\tilde{\psi}) \le \int_0^{t_0(\varepsilon,\tilde{\psi})} \|\tilde{U}(t)\tilde{\psi}\|_X^2 dt \le M,$$

for any $\tilde{\psi} \in X_r$ with $\|\tilde{\psi}\|_X = 1$, which immediately yields

$$t_0(\varepsilon,\tilde{\psi}) \le \frac{M}{\varepsilon} =: T(\varepsilon).$$

Hence, as a result of Proposition 4.2.1, we have if $t \ge T(\varepsilon)$, then for any $\tilde{\psi} \in X_r$ with $\|\tilde{\psi}\|_X = 1$,

$$\|\tilde{U}(t)\tilde{\psi}\|_X^2 = \|\tilde{U}(t - t_0(\varepsilon,\tilde{\psi}))\tilde{U}(t_0(\varepsilon,\tilde{\psi}))\tilde{\psi}\|_X^2 \le M\|\tilde{U}(t_0(\varepsilon,\tilde{\psi}))\tilde{\psi}\|_X^2 = M\varepsilon,$$

which proves the proposition. □

Observe that the implication (iii) $\Longrightarrow$ (i) in Theorem 4.2.3 is straightforward. Therefore, to conclude our proofs, it suffices to show the null solution of (4.2.3) is exponentially stable in mean square if it is uniformly asymptotically stable in mean square. But this is also immediate by carrying out a similar argument to Theorem 2.3.1. Therefore, the proof of Theorem 4.2.3 is complete.

The following result gives a very useful criterion for the system (4.2.3) to be L^2-stable in mean by using a Lyapunov function method.

Theorem 4.2.4   *The system (4.2.3) is L^2-stable in mean if there exists a positive self-adjoint operator $Q \in \mathcal{L}(H \oplus H)$ such that for any $\tilde{\psi} \in \underline{X}_r$,*

$$\frac{d}{dt}\langle Q\tilde{v}^t(\tilde{\psi}), \tilde{v}^t(\tilde{\psi})\rangle_{\underline{X}} \leq -E\|v(t, \tilde{\psi})\|_H^2.$$

Proof   Assume that there exists a positive self-adjoint operator $Q \in \mathcal{L}(H \oplus H)$ satisfying the above inequality. Then

$$\int_0^\infty E\|v(t, \tilde{\psi})\|_H^2 dt \leq \langle Q\tilde{\psi}, \tilde{\psi}\rangle_{\underline{X}} \leq \|Q\|\|\tilde{\psi}\|_{\underline{X}}^2,$$

where $\tilde{v}^0(\tilde{\psi}) = \tilde{\psi} \in \underline{X}_r$. $\square$

4.2.2   Stability of Neutral Stochastic Systems

A remarkable consequence of studying Equation (4.2.3) is that we can apply these results derived in Subsection 4.2.1, for instance, Theorem 4.2.3, to a wide class of linear neutral stochastic evolution equations of retarded type to establish their stability properties. Define

$$\underline{C}_r = \left\{ \phi \in C([-r, 0]; L^2(\Omega, H)) : \phi \in \mathcal{F}, \max_{s \in [-r,0]} E\|\phi(s)\|_H^2 < \infty \right\}$$

with norm

$$\|\phi\|_{\underline{C}_r}^2 = \max_{s \in [-r,0]} E\|\phi(s)\|_H^2 \quad \text{for any} \quad \phi \in \underline{C}_r.$$

Consider the following abstract linear stochastic neutral functional differential equation:

$$u(t, \phi) - \sum_{j=1}^m D_j u(t - r_j, \phi) = T(t)\left[\phi(0) - \sum_{j=1}^m D_j \phi(-r_j)\right]$$

$$+ \int_0^t T(t - s) \sum_{j=1}^m A_j u(s - r_j, \phi) ds$$

$$+ \int_0^t T(t - s) \sum_{j=1}^m B_j u(s - r_j, \phi) dW_s, \quad t \geq 0,$$

$$u_0(\cdot) = \phi(\cdot) \in \underline{C}_r,$$

(4.2.6)

where $\{T(t)\}_{t \geq 0}$ is some given C_0-semigroup defined on the Hilbert space H with inner product $\langle \cdot, \cdot \rangle_H$ and $0 < r_1 < r_2 < \cdots < r_m = r$ are constants. Further A_j and D_j are bounded linear operators from H into H, the B_j are bounded linear operators from H into $\mathcal{L}(K, H)$, $j = 1, \cdots, m$, and $W_t, t \geq 0$, is a given standard Q-Wiener process on the Hilbert space K with finite trace class covariance operator Q.

By employing a sequence of steps as in the proof of Theorem 4.2.1, it is possible to establish that for each $\phi \in \underline{C}_r$, there exists a unique $u(\cdot, \phi) \in C([-r, \infty); L^2(\Omega; H))$ satisfying (4.2.6) and $\{U(t)\}_{t \geq 0}$ defined by $U(t)\phi = u_t(\phi)$ is a strongly continuous semigroup from $\underline{C}_r$ into $\underline{C}_r$ in the sense of Theorem 4.2.2.

To apply the stability results obtained in Subsection 4.2.1 to Equation (4.2.6), we first study some properties of initial data between the equations (4.2.3) and (4.2.6). For arbitrary $\phi \in \underline{C}_r$, $r > 0$, define the transform

$$\psi(s) = \phi(s) - \sum_{j=1}^{m} D_j \phi(s - r_j), \quad s \in [-r, 0],$$

$$\psi(s) = 0, \quad s < -r.$$

(4.2.7)

Then, we have the following property.

Lemma 4.2.4   *Let $\mathcal{R}$ be the transform defined by (4.2.7), and denote it by $\mathcal{R}\phi = \tilde{\psi} = (\psi(0), \psi)$. Then $\mathcal{R}$ is a bounded linear operator from $\underline{C}_r$ into $\underline{X}_r$, $r > 0$.*

Proof   The linearity is clear. To prove boundedness, note that $\|\psi(s)\|_H = 0$ almost surely if $s < -r$ and

$$\sup_{-r \leq s \leq 0} \|\tilde{\psi}(s)\|_{\underline{X}}^2 \leq 2\|\phi\|_{\underline{C}_r}^2 \left[1 + \sum_{j=1}^{m} \|D_j\|^2 \right].$$

Hence,

$$\|\tilde{\psi}\|_{\underline{X}}^2 = E\|\psi(0)\|_H^2 + \sum_{j=1}^{\infty} d_j^2 e^{2\delta h_j} \int_{-h_j}^{0} E\|\psi(s)\|_H^2 ds$$

$$\leq E\|\psi(0)\|_H^2 + \sum_{j=1}^{\infty} d_j^2 e^{2\delta h_j} \int_{-r}^{0} E\|\psi(s)\|_H^2 ds$$

$$\leq 2\|\phi\|_{\underline{C}_r}^2 \left[1 + \sum_{j=1}^{m} \|D_j\|^2 \right] \left(1 + r \sum_{j=1}^{\infty} d_j^2 e^{2\delta h_j} \right) < \infty.$$

Thus, there exists a constant $M > 0$ such that

$$\|\tilde{\psi}\|_{\underline{X}} = \|\mathcal{R}\phi\|_{\underline{X}} \le M \|\phi\|_{\underline{C}_r}.$$

$\square$

By virtue of Lemma 4.2.4, we may define a Hilbert subspace $\underline{M}_r = \overline{\mathcal{R}\underline{C}_r} \subset \underline{X}_r$ of $\underline{X}$ which is the closure of $\mathcal{R}\underline{C}_r$ in the Hilbert space $\underline{X}$. Then $\underline{M}_r$ will be taken as the initial data space for the equation (4.2.6) in the remainder of this section.

For each solution $u(\cdot, \phi)$ of (4.2.6), we may extend its domain to $(-\infty, \infty)$ by letting

$$u(t, \phi) = 0, \quad t \le -r.$$

Making a change of variables

$$v(t) = u(t) - \sum_{j=1}^{m} D_j u(t - r_j), \quad t \in \mathbf{R}^1, \tag{4.2.8}$$

we get

$$u(t) = v(t) + \sum_{k=1}^{\infty} \tilde{D}_k v(t - \tilde{r}_k),$$

where for each $k \ge 1$, $\tilde{D}_k : H \to H$ is a bounded linear operator, $0 < \tilde{r}_1 < \tilde{r}_2 < \cdots$ and each $\tilde{r}_k$ is of the form $n_1 r_1 + n_2 r_2 + \cdots + n_m r_m$ for some nonnegative integers $n_1, n_2, \cdots, n_m$ such that the above equality makes sense.

Under the above change of variables, $v(t)$ satisfies the following initial value problem of retarded stochastic evolution equation with infinite delay

$$
\begin{aligned}
v(t, \tilde{\psi}) =\ & T(t)\psi(0) \\
& + \int_0^t T(t - s) \sum_{j=1}^{m} A_j \left[v(s - r_j, \tilde{\psi}) + \sum_{k=1}^{\infty} \tilde{D}_k v(s - \tilde{r}_k - r_j, \tilde{\psi}) \right] ds \\
& + \int_0^t T(t - s) \sum_{j=1}^{m} B_j \left[v(s - r_j, \tilde{\psi}) + \sum_{k=1}^{\infty} \tilde{D}_k v(s - \tilde{r}_k - r_j, \tilde{\psi}) \right] dW_s, \\
v(t, \tilde{\psi}) =\ & \psi(t), \ t \le 0, \quad \text{where } \tilde{\psi} = (\psi(0), \psi) \in \underline{M}_r.
\end{aligned}
$$

$$\tag{4.2.9}$$

To make our arguments compatible with the framework established in the previous subsection, let's assume that there exists a sequence of nonnegative numbers $\{\tilde{d}_j\}$, $j = 1, 2, \cdots$, such that

$$\|\tilde{D}_j\| \le \tilde{d}_j, \quad \sum_{j=1}^{\infty} \tilde{d}_j e^{2\delta \tilde{r}_j} < \infty, \tag{4.2.10}$$

for some number $\delta > 0$. Suppose $\{F_j\} : H \to H$ and $\{G_j\} : H \to \mathcal{L}(K, H)$ are two given sequences of bounded linear operators and $\{h_j\}$ is a given increasing sequence of positive numbers such that

$$\sum_{j=1}^{m} A_j \left[v(s - r_j, \tilde{\psi}) + \sum_{k=1}^{\infty} \tilde{D}_k v(s - \tilde{r}_k - r_j, \tilde{\psi}) \right] = \sum_{j=1}^{\infty} F_j v(s - h_j, \tilde{\psi}),$$

$$\sum_{j=1}^{m} B_j \left[v(s - r_j, \tilde{\psi}) + \sum_{k=1}^{\infty} \tilde{D}_k v(s - \tilde{r}_k - r_j, \tilde{\psi}) \right] = \sum_{j=1}^{\infty} G_j v(s - h_j, \tilde{\psi}).$$

Then (4.2.9) can be rewritten in the form of (4.2.3) with initial value space $\underline{M}_r$:

$$v(t, \tilde{\psi}) = T(t)\psi(0) + \int_0^t T(t - s) \sum_{j=1}^{\infty} F_j v(s - h_j, \tilde{\psi}) ds$$

$$+ \int_0^t T(t - s) \sum_{j=1}^{\infty} G_j v(s - h_j, \tilde{\psi}) dW_s, \qquad (4.2.11)$$

$$v(t, \tilde{\psi}) = \psi(t), \ t \le 0, \ \text{where} \ \tilde{\psi} = (\psi(0), \psi) \in \underline{M}_r.$$

The following proposition shows that the formulation of $\{F_j\}$, $\{G_j\}$ and $\{h_j\}$, $j = 1, 2 \cdots$, is justified in defining the equation (4.2.11).

Proposition 4.2.3   *Suppose the relation (4.2.10) holds and $\{h_j\}$ is defined as in (4.2.11). Let $d_j = \|F_j\| \vee \|G_j\|$, $j = 1, 2, \cdots$, where $a \vee b = \max\{a, b\}$ for any $a, b \ge 0$, then $\sum_{j=1}^{\infty} d_j e^{2\delta h_j} < \infty$ for some positive constant δ.*

Proof   Note that

$$F_j = A_{j_1} \tilde{D}_{\alpha_{j_1}} + \cdots + A_{j_n} \tilde{D}_{\alpha_{j_n}}, \quad 1 \le \alpha_{j_i} < \infty, \ 1 \le j_i \le m, \ 1 \le i \le n \le m,$$

$$G_j = B_{j_1} \tilde{D}_{\alpha_{j_1}} + \cdots + B_{j_n} \tilde{D}_{\alpha_{j_n}}, \quad 1 \le \alpha_{j_i} < \infty, \ 1 \le j_i \le m, \ 1 \le i \le n \le m,$$

$$\tilde{r}_{j_1} + r_{j_1} = \tilde{r}_{j_2} + r_{j_2} = \cdots = \tilde{r}_{j_n} + r_{j_n} = h_j.$$

Therefore,

$$\|F_j\| \vee \|G_j\| \le \max_{1 \le i \le m} (\|A_i\| \vee \|B_i\|) \left[\tilde{d}_{\alpha_{j_1}} + \cdots + \tilde{d}_{\alpha_{j_n}} \right],$$

and

$$\sum_{j=1}^{\infty} (\|F_j\| \vee \|G_j\|) \cdot e^{2\delta h_j} \le \max_{1 \le i \le m} (\|A_i\| \vee \|B_i\|) \sum_{j=1}^{\infty} \left[\tilde{d}_{\alpha_{j_1}} + \cdots + \tilde{d}_{\alpha_{j_n}} \right] e^{2\delta h_j}$$

$$\le (m e^{2\delta r}) \max_{1 \le i \le m} (\|A_i\| \vee \|B_i\|) \sum_{j=1}^{\infty} \tilde{d}_j e^{2\delta \tilde{r}_j} < \infty.$$

This completes the proof. ☐

To apply the stability results in Subsection 4.2.1 such as Theorem 4.2.3 to (4.2.11), we need to prove the validity of (4.2.1) for the sequence $\{h_j\}$ defined in (4.2.11). In fact, we have the following result whose proof is referred to the appendix.

Proposition 4.2.4 *Let $\{h_n\}$ be the sequence defined in the equation (4.2.11); then*

$$\lim_{n\to\infty} \frac{\ln n}{h_n} = 0.$$

Next, let us study an example to illustrate the procedure from (4.2.6) to (4.2.11).

Example 4.2.1 Consider the following stochastic neutral difference differential equation

$$
\begin{aligned}
d\big[u(x,t) - Du(x,t-r)\big] = {} & \frac{\partial^2}{\partial x^2}\big[u(x,t) - Du(x,t-r)\big]dt \\
& + \sum_{j=1}^{m} A_j u(x,t-r_j)dt \\
& + \sum_{j=1}^{m} B_j u(x,t-r_j)dB_t, \quad t \geq 0, \\
u(0,t) = u(\pi,t) = {} & 0, \quad t \geq 0,
\end{aligned}
\tag{4.2.12}
$$

where $x \in [0,\pi]$, $t \geq 0$, $0 < r_1 < r_2 < \cdots < r_m = r$, D, A_j and B_j, $j = 1, 2, \cdots, m$, are real numbers with $|D| < 1$, and B_t, $t \geq 0$, is some one-dimensional standard Brownian motion. Let

$$v(x,t) = u(x,t) - Du(x,t-r), \quad x \in [0,\pi], \quad t \in \mathbf{R}_+.$$

Then

$$u(x,t) = v(x,t) + \sum_{j=1}^{\infty} D^j v(x,t-jr)$$

and the first equation of (4.2.12) can be rewritten as

$$
\begin{aligned}
dv(x,t) = {} & \frac{\partial^2}{\partial x^2}v(x,t)dt + \sum_{j=1}^{m} A_j\Big[v(x,t-r_j) + \sum_{k=1}^{\infty} D^k v(x,t-kr-r_j)\Big]dt \\
& + \sum_{j=1}^{m} B_j\Big[v(x,t-r_j) + \sum_{k=1}^{\infty} D^k v(x,t-kr-r_j)\Big]dB_t.
\end{aligned}
$$

In that case we have

$$\|\tilde{D}_j\| = |D^j|, \qquad \tilde{r}_j = jr,$$

so

$$\sum_{j=1}^{\infty} \tilde{d}_j e^{2\delta \tilde{r}_j} = \sum_{j=1}^{\infty} |D^j|(e^{2\delta r})^j = \sum_{j=1}^{\infty} \left(|D|e^{2\delta r} \right)^j < \infty$$

if $\delta > 0$ is sufficiently small so that $|D|e^{2\delta r} < 1$.

Now we are in a position to obtain the main stability results of (4.2.6) in the section. We first establish an equivalent result of stability between (4.2.6) and (4.2.11).

Theorem 4.2.5   *The system (4.2.6) is L^2-stable in mean if and only if the system (4.2.11) is L^2-stable in mean.*

Proof   Firstly, assume the system (4.2.11) is L^2-stable in mean. Let $\phi \in \underline{C}_r$ and for $\tilde{\psi} = \mathcal{R}\phi \in \underline{M}_r$,

$$u(t, \phi) = v(t, \tilde{\psi}) + \sum_{j=1}^{\infty} \tilde{D}_j v(t - \tilde{r}_j, \tilde{\psi})$$

where

$$\|\tilde{D}_j\| = \tilde{d}_j, \qquad \sum_{j=1}^{\infty} \tilde{d}_j e^{2\delta \tilde{r}_j} < \infty, \qquad \delta > 0.$$

Therefore, for any $t \geq 0$,

$$\begin{aligned}
E\|u(t,\phi)\|_H^2 = \ & E\|v(t,\tilde{\psi})\|_H^2 + 2\sum_{j=1}^{\infty} E\langle v(t,\tilde{\psi}), \tilde{D}_j v(t - \tilde{r}_j, \tilde{\psi})\rangle_H \\
& + 2 \sum_{1=j<k}^{\infty} E\langle \tilde{D}_j v(t - \tilde{r}_j, \tilde{\psi}), \tilde{D}_k v(t - \tilde{r}_k, \tilde{\psi})\rangle_H \qquad (4.2.13) \\
& + \sum_{j=1}^{\infty} E\|\tilde{D}_j v(t - \tilde{r}_j, \tilde{\psi})\|_H^2.
\end{aligned}$$

Integrating both sides of (4.2.13), using Lemma 4.2.4 and the assumptions of

Theorem 4.2.5, we obtain that for some positive constants M, C_1, C_2,

$$\int_0^\infty E\|u(t,\phi)\|_H^2 dt$$

$$\leq 2\sum_{j=1}^\infty \int_0^\infty \left(E\|v(t,\tilde\psi)\|_H^2 E\|\tilde D_j v(t-\tilde r_j,\tilde\psi)\|_H^2\right)^{\frac{1}{2}} dt$$

$$+2\sum_{j<k}^\infty \int_0^\infty \left(E\|\tilde D_j v(t-\tilde r_j)\|_H^2 E\|\tilde D_k v(t-\tilde r_k)\|_H^2\right)^{\frac{1}{2}} dt$$

$$+C_1\|\tilde\psi\|_{\underline X}^2 + \sum_{j=1}^\infty \tilde d_j^2 \int_{-\tilde r_j}^\infty E\|v(t,\tilde\psi)\|_H^2 dt$$

$$\leq C_1\|\tilde\psi\|_{\underline X}^2 + 2\sum_{j=1}^\infty \tilde d_j M^{1/2}\|\tilde\psi\|_{\underline X}\left(M\|\tilde\psi\|_{\underline X}^2 + \int_{-\tilde r_j}^0 E\|\psi(t)\|_H^2 dt\right)^{\frac{1}{2}} \quad (4.2.13)$$

$$+2\sum_{1=j<k}^\infty \tilde d_j \tilde d_k \left(M\|\tilde\psi\|_{\underline X}^2 + \int_{-\tilde r_j}^0 E\|\psi(t)\|_H^2 dt\right)^{\frac{1}{2}}$$

$$\cdot\left(M\|\tilde\psi\|_{\underline X}^2 + \int_{-\tilde r_k}^0 E\|\psi(t)\|_H^2 dt\right)^{\frac{1}{2}}$$

$$+\sum_{j=1}^\infty \tilde d_j^2\left(M\|\tilde\psi\|_{\underline X}^2 + \int_{-\tilde r_j}^0 E\|\psi(t)\|_H^2 dt\right)$$

$$\leq C_2\left\{\|\tilde\psi\|_{\underline X}^2 + \sum_{j=1}^\infty \tilde d_j\left(\int_{-\tilde r_j}^0 E\|\psi(t)\|_H^2 dt\right)^{\frac{1}{2}}\right\}^2 < \infty,$$

which is precisely the definition of L^2-stability in mean of the solution of (4.2.6).

We now assume that the system (4.2.6) is L^2-stable in mean. For arbitrarily given $\tilde\psi \in R\underline C_r$, let $\tilde\psi = R\phi$ for some $\phi \in \underline C_r$. Then, from (4.2.8) we have for any $t \geq 0$,

$$E\|v(t,\tilde\psi)\|_H^2 = E\|u(t,\phi)\|_H^2 - 2\sum_{j=1}^m E\langle u(t,\phi), D_j u(t-r_j,\phi)\rangle_H$$

$$+2\sum_{1=j<k}^m E\langle D_j u(t-r_j,\phi), D_k u(t-r_k,\phi)\rangle_H \quad (4.2.14)$$

$$+\sum_{j=1}^m E\|D_j u(t-r_j,\phi)\|_H^2.$$

Integrating both sides of (4.2.14) over $[0,\infty)$, we obtain that for some constant $C > 0$,

$$\int_0^\infty E\|v(t,\tilde\psi)\|_H^2 dt \le \int_0^\infty E\|u(t,\phi)\|_H^2 dt + 2\Big(\int_0^\infty E\|u(t,\phi)\|_H^2 dt\Big)^{\frac{1}{2}}$$

$$\cdot \sum_{j=1}^m \|D_j\| \Big(\int_0^\infty E\|u(t,\phi)\|_H^2 dt + \int_{-r_j}^0 E\|\phi(t)\|_H^2 dt\Big)^{\frac{1}{2}}$$

$$+2 \sum_{1=j<k}^m \|D_j\| \cdot \|D_k\| \Big(\int_0^\infty E\|u(t,\phi)\|_H^2 dt + \int_{-r}^0 E\|\phi(t)\|_H^2 dt\Big)$$

$$+\sum_{j=1}^m \|D_j\|^2 \Big(\int_0^\infty E\|u(t,\phi)\|_H^2 dt + \int_{-r_j}^0 E\|\phi(t)\|_H^2 dt\Big)$$

$$\le C\|\phi\|_{\underline{C}_r}^2 + 2C^{1/2}\|\phi\|_{\underline{C}_r} \sum_{j=1}^m \|D_j\| \Big(C\|\phi\|_{\underline{C}_r}^2 + \int_{-r}^0 E\|\phi(t)\|_H^2 dt\Big)^{1/2}$$

$$+\Big\{2\sum_{1=j<k}^m \|D_j\| \cdot \|D_k\| + \sum_{j=1}^m \|D_j\|^2\Big\}\Big(C\|\phi\|_{\underline{C}_r}^2 + \int_{-r}^0 E\|\phi(t)\|_H^2 dt\Big) < \infty.$$

Let

$$\rho_n(\tilde\psi) = \Big(\int_0^n E\|v(t,\tilde\psi)\|_H^2 dt\Big)^{1/2} < \infty, \quad n \ge 0, \quad \tilde\psi \in \mathcal{R}\underline{C}_r.$$

Since $\mathcal{R}\underline{C}_r$ is dense in $\underline{M}_r$, it is easy to see that $\rho_n(\cdot)$ may be extended to a family of continuous semi-norms on $\underline{M}_r$. Then by Lemma 4.2.1, there exists a constant $C > 0$ such that for any $\tilde\psi \in \underline{M}_r$,

$$\rho(\tilde\psi) = \Big(\int_0^\infty E\|v(t,\tilde\psi)\|_H^2 dt\Big)^{1/2} \le C^{1/2}\|\tilde\psi\|_{\underline{X}}.$$

That is, the null solution of (4.2.11) is L^2-stable in mean.   □

Lastly, we shall show the following counterpart of Theorem 4.2.3, which states stability equivalent relations for the equation (4.2.6).

Theorem 4.2.6   *Under the relation (4.2.10), the following three notions of stability are equivalent:*

(i). The null solution of (4.2.6) is L^2-stable in mean;
(ii). The null solution of (4.2.6) is uniformly asymptotically stable in mean square;
(iii). The null solution of (4.2.6) is exponentially stable in mean square.

Proof   All the proofs are quite similar to those in Theorem 4.2.3 except for the implication (i) $\Longrightarrow$ (ii). Assume the null solution of (4.2.6) is L^2-stable in mean, then by Theorem 4.2.5, the null solution of (4.2.11) is L^2-stable in mean. We can conclude by using Theorem 4.2.3 that for arbitrarily given $\varepsilon > 0$, there exists number $T(\varepsilon) \geq 0$ such that

$$\|\tilde{U}(t)\tilde{\psi}\|_{\underline{X}}^2 \leq \varepsilon\|\tilde{\psi}\|_{\underline{X}}^2 \tag{4.2.15}$$

for any $\tilde{\psi} \in M_r$ if $t \geq T(\varepsilon)$. However, for any given $\phi \in \underline{C}_r$, we have $\tilde{\psi} = \mathcal{R}\phi$ where $\tilde{\psi} \in M_r$. Moreover, by Lemma 4.2.4 we know that there exists a constant $C > 0$ such that

$$\|\tilde{\psi}\|_{\underline{X}}^2 \leq C\|\phi\|_{\underline{C}_r}^2. \tag{4.2.16}$$

Also for $\tilde{\psi} = \mathcal{R}\phi$,

$$E\|u(t,\phi)\|_H^2 = E\|v(t,\tilde{\psi}) + \sum_{j=1}^{\infty} \tilde{D}_j v(t-\tilde{r}_j,\tilde{\psi})\|_H^2$$

$$\leq 2E\|v(t,\tilde{\psi})\|_H^2 + 2E\left(\sum_{j=1}^{\infty} \tilde{d}_j\|v(t-\tilde{r}_j,\tilde{\psi})\|_H\right)^2$$

$$\leq 2E\|v(t,\tilde{\psi})\|_H^2 + 2\left(\sum_{j=1}^{\infty} \tilde{d}_j\right)\left[\sum_{j=1}^{J} \tilde{d}_j E\|v(t-\tilde{r}_j,\tilde{\psi})\|_H^2\right.$$

$$\left. + \sum_{j=J+1}^{\infty} \tilde{d}_j E\|v(t-\tilde{r}_j,\tilde{\psi})\|_H^2\right].$$

Let $\sum_{j=1}^{\infty} \tilde{d}_j = M < \infty$. Since, in view of Theorem 4.2.3, $\tilde{U}(t)$, $t \geq 0$, is uniformly asymptotically stable in mean square, it follows that for arbitrarily given $\varepsilon > 0$, we can find J such that

$$\sum_{j=J+1}^{\infty} \tilde{d}_j E\|v(t-\tilde{r}_j,\tilde{\psi})\|_H^2 \leq \frac{\varepsilon}{2MC}\|\phi\|_{\underline{C}_r}^2.$$

Then, by (4.2.15) and (4.2.16) we can find $T(\varepsilon) \geq 0$ such that

$$E\|v(t,\tilde{\psi})\|_H^2 \leq \frac{\varepsilon}{4}\|\phi\|_{\underline{C}_r}^2,$$

and

$$\sum_{j=1}^{J} \tilde{d}_j E\|v(t-\tilde{r}_j,\tilde{\psi})\|_H^2 \leq \frac{\varepsilon}{4M}\|\phi\|_{\underline{C}_r}^2$$

if $t \geq T(\varepsilon)$. Hence, we see that for the given $\varepsilon > 0$ above, there exists $T(\varepsilon) \geq 0$ such that for $\sigma \in [-r,0]$, $E\|u(t+\sigma,\phi)\|_H^2 \leq \varepsilon\|\phi\|_{\underline{C}_r}^2$ whenever $t \geq T(\varepsilon)$, i.e., $U(t)$, $t \geq 0$, is uniformly asymptotically stable in mean square. $\quad\square$

Example 4.2.2   We now consider the following stochastic neutral functional differential equation as an illustrative example to close this section.

$$d\left[X(x,t) - \frac{X(x,t-1)}{2}\right] = \frac{\partial^2}{\partial x^2}\left[X(x,t) - \frac{X(x,t-1)}{2}\right]dt + \frac{X(x,t-1)}{2}dB_t,$$
$$0 \leq x \leq 1, \quad t \geq 0,$$

$$(4.2.17)$$

subject to the boundary condition

$$X(0,t) = X(1,t) = 0, \quad t \geq 0,$$

and the initial condition

$$X(x,t) = \phi(x,t), \quad 0 \leq x \leq 1, \quad -1 \leq t \leq 0,$$

where $\phi \in C([0,1] \times [-1,0]; \mathbf{R}^1)$ and B_t, $t \geq 0$, is a standard one dimensional Brownian motion.

Let

$$y(x,t) = X(x,t) - \frac{1}{2}X(x,t-1),$$

then

$$X(x,t) = y(x,t) + \frac{1}{2}y(x,t-1) + \frac{1}{2^2}y(x,t-2) + \cdots.$$

So, (4.2.17) can be reduced to a retarded equation with infinite delay

$$dy(x,t) = \frac{\partial^2}{\partial x^2}y(x,t)dt + \sum_{j=1}^{\infty}\frac{1}{2^j}y(x,t-j)dB_t,$$
$$0 \leq x \leq 1, \quad t \geq 0,$$

$$(4.2.18)$$

subject to the boundary condition

$$y(0,t) = y(1,t) = 0, \quad 0 \leq x \leq 1, \quad t \geq 0,$$

and the initial condition

$$y(x,t) = \psi(x,t), \quad 0 \leq x \leq 1, \quad t \leq 0.$$

Assume $\psi : [0,1] \times (-\infty, 0] \to \mathbf{R}^1$ has the following Fourier series expansion

$$\psi(x,t) = \sum_{n=1}^{\infty}\psi_n(t)\sin(n\pi x);$$

we seek solutions of (4.2.17) of the form

$$y(x,t) = \sum_{n=1}^{\infty}y_n(t)\sin(n\pi x).$$

Then

$$dy_n(t) = -n^2\pi^2 y_n(t) + \sum_{j=1}^{\infty} \frac{1}{2^j} y_n(t-j)dB_t.$$

We can regard the totality of (4.2.17) as a stochastic difference differential equation in the sequence space $H = \{(l_n) : \sum_{j=1}^{\infty} l_n^2 < \infty\}$ of the form

$$dv(t,\tilde{\psi}) = Av(t,\tilde{\psi}) + \sum_{j=1}^{\infty} D_j v(t-j,\tilde{\psi})dB_t,$$

where A is the unbounded operator which takes the j-th coordinate of H onto itself multiplied by $-(j\pi)^2$, and D_j is the bounded operator which maps each element of H into its multiple of $\frac{1}{2^j}$. In this case, $h_j = j$, and if we let $\delta = \ln 2/4$, then

$$\sum_{j=1}^{\infty} \|D_j\| e^{2\delta h_j} = \sum_{j=1}^{\infty} \frac{1}{2^j} (\sqrt{2})^j = \sum_{j=1}^{\infty} \left(\sqrt{\frac{1}{2}}\right)^j < \infty.$$

Let

$$\Lambda(\tilde{v}^t(\tilde{\psi})) = \frac{\|v(t,\tilde{\psi})\|_H^2}{2} - \frac{1}{2} \sum_{n=1}^{\infty} \sum_{j=1}^{\infty} \int_0^t \frac{1}{2^{2j}} v_n^2(s-j,\tilde{\psi})ds,$$

where

$$v(t,\tilde{\psi}) = (v_1(t,\tilde{\psi}), v_2(t,\tilde{\psi}), \cdots, v_n(t,\tilde{\psi}), \cdots) \in H.$$

Then along a dense set of trajectories $\tilde{v}^t(\tilde{\psi})$, we have

$$\frac{d(E\Lambda(\tilde{v}^t(\tilde{\psi})))}{dt} = \sum_{n=1}^{\infty} \left[-n^2\pi^2 Ev_n^2(t,\tilde{\psi}) \right] + \frac{1}{2} \sum_{n=1}^{\infty} \sum_{j=1}^{\infty} \frac{Ev_n^2(t-j,\tilde{\psi})}{2^{2j}}$$

$$-\frac{1}{2} \sum_{n=1}^{\infty} \sum_{j=1}^{\infty} \frac{Ev_n^2(t-j,\tilde{\psi})}{2^{2j}}$$

$$\leq -\sum_{n=1}^{\infty} Ev_n^2(t,\tilde{\psi}) = -E\|v(t,\tilde{\psi})\|_H^2.$$

Thus, for a dense set of points $\{\psi\}$ in H,

$$0 \leq E\Lambda(\tilde{v}^t(\tilde{\psi})) \leq E\Lambda(\tilde{\psi}) - \int_0^{\infty} E\|v(t,\tilde{\psi})\|_H^2 dt.$$

However,

$$\Lambda(\tilde{\psi}) = \frac{\|\psi(0)\|_H^2}{2}$$

is a continuous function on $\underline{X}$ and, in fact, $\Lambda(\tilde{\psi}) \leq 1/2 \cdot \|\tilde{\psi}\|_X$ for all $\tilde{\psi} \in \underline{X}$. Therefore,

$$\int_0^{\infty} E\|v(t,\tilde{\psi})\|_H^2 dt \leq \Lambda(\tilde{\psi}) \leq 1/2 \cdot \|\tilde{\psi}\|_X^2$$

for all $\tilde{\psi} \in \underline{X}_1$. This proves that the system (4.2.17) is L^2-stable in mean.

4.3 Decay Criteria of Stochastic Delay Differential Equations

In this section, we shall investigate decay of strong solutions for a class of nonlinear stochastic delay differential equations. Typically, we are concerned with the following system on V^*:

$$
\begin{cases}
u(t, \phi) = \phi(0) + \int_0^t A(s, u(s, \phi), u(s - \tau(s), \phi))ds \\
\qquad\qquad + \int_0^t B(s, u(s, \phi), u(s - \tau(s), \phi))dW_s, \quad \forall t \geq 0, \\
u(t, \phi) = \phi(t) \in V, \qquad t \in [-r, 0],
\end{cases}
\tag{4.3.1}
$$

where $\phi(t, \omega) : [-r, 0] \times \Omega \to V$, $r \geq 0$, is some given initial datum such that $\phi(t)$ is $\mathcal{F}_0$-measurable for any $t \in [-r, 0]$ and $\sup_{-r \leq s \leq 0} E\|\phi(s)\|_V^2 < \infty$. $\tau : [0, \infty) \to [0, r]$ is a certain differentiable function satisfying $d\tau(t)/dt \leq 0$ for all $t \geq 0$. $A(t, \cdot, \cdot) : V \times V \to V^*$ and $B(t, \cdot, \cdot) : V \times V \to \mathcal{L}(K, H)$ are two families of measurable nonlinear operators satisfying that $t \in [0, \infty) \to A(t, x, y) \in V^*$, $t \in [0, \infty) \to B(t, x, y) \in \mathcal{L}(K, H)$ are Lebesgue measurable for any $x, y \in V$.

For the purpose of existence and uniqueness of strong solutions, the following assumptions similar to those in Section 1.3 are imposed on (4.3.1): for any $T \geq 0$, there exist constants $\alpha > 0$, $p > 1$ and $\theta, \lambda, \gamma \in \mathbf{R}^1$ such that

(a) (Coercivity).

$$
2\langle x, A(t, x, y)\rangle_{V,V^*} + \|B(t, x, y)\|_{\mathcal{L}_2^0}^2
$$

$$
\leq -\alpha\|x\|_V^p + \lambda\|x\|_H^2 + \theta\|y\|_H^2 + \gamma, \quad \forall x, y \in V, \quad 0 \leq t \leq T,
\tag{4.3.2}
$$

where $\|\cdot\|_{\mathcal{L}_2^0}$ denotes the Hilbert-Schmidt norm

$$
\|B(t, x, y)\|_{\mathcal{L}_2^0}^2 = tr(B(t, x, y)QB(t, x, y)^*);
$$

(b) (Growth). There exists a constant $C > 0$ such that

$$
\|A(t, x, y)\|_{V^*} \leq C(1 + \|x\|_V^{p-1} + \|y\|_V^{p-1}), \quad \forall x, y \in V, \quad 0 \leq t \leq T;
\tag{4.3.3}
$$

(c) (Monotonicity). For arbitrary $x_1, x_2, y_1, y_2 \in V$, and $0 \leq t \leq T$,

$$
2\langle x_1 - x_2, A(t, x_1, y_1) - A(t, x_2, y_2)\rangle_{V,V^*} + \|B(t, x_1, y_1) - B(t, x_2, y_2)\|_{\mathcal{L}_2^0}^2
$$

$$
\leq \lambda\Big(\|x_1 - x_2\|_H^2 + \|y_1 - y_2\|_H^2\Big).
$$

(d) (Continuity). The map $(\xi, \eta) \in \mathbf{R}^1 \to \langle w, A(t, x + \xi u, y + \eta v)\rangle_{V,V^*} \in \mathbf{R}^1$ is continuous for all $x, y, u, v, w \in V$ and $0 \leq t \leq T$;

(e) (Lipschitz). There exists a constant $L > 0$ such that

$$\|B(t, u, v) - B(t, \tilde{u}, \tilde{v})\|_{\mathcal{L}_2^0} \leq L\Big(\|u - \tilde{u}\|_V + \|v - \tilde{v}\|_V\Big), \tag{4.3.4}$$
$$\forall u, \tilde{u}, v, \tilde{v} \in V, \ 0 \leq t \leq T.$$

Let $M^p(a, b; V)$ denote the space of all V-valued processes $(X_t)_{t \in [a,b]}$, $-\infty < a \leq b < \infty$, which are measurable from $[a, b] \times \Omega$ into V and satisfy

$$\int_a^b E\|X_t\|_V^p \, dt < \infty.$$

It may be proved (see Appendix) that for each given initial datum

$$\phi \in M^p(-r, 0; V) \cap L^2(\Omega; C(-r, 0; H)),$$

there exists a unique process

$$u(t, \phi) \in M^p(-r, T; V) \cap L^2(\Omega \times [-r, T]; H)$$

(strong solution) satisfying Equation (4.3.1) and $u(\cdot, \phi) \in C(0, T; H)$ almost surely where $C(a, b; H)$ denotes the space of all continuous functions from $[a, b]$ into H. If T is replaced by ∞, $u(t, \phi)$, $-r \leq t < \infty$, is called a global strong solution of (4.3.1). Unless otherwise specified, we always suppose that there exists a unique global strong solution of (4.3.1), which satisfies the above conditions (a)–(e). In particular, we will show below that a proper version of the coercive condition (a) plays a role of decay criterion.

4.3.1 Nonlinear Coercive Conditions for Decay

For the purpose of decay, we formulate the following coercive condition on the equation (4.3.1):

(H10) There exist constants $\alpha > 0$, $\theta \in \mathbf{R}_+$, $\lambda \in \mathbf{R}^1$, and a nonnegative continuous function $\gamma(t)$, $t \in \mathbf{R}_+$, and $\mu > 0$ such that

$$2\langle x, A(t, x, y)\rangle_{V,V^*} + \|B(t, x, y)\|_{\mathcal{L}_2^0}^2$$
$$\leq -\alpha\|x\|_V^p + \lambda\|x\|_H^2 + \theta\|y\|_H^2 + \gamma(t), \quad x, y \in V,$$

where $p \geq 2$ and $\gamma(t)e^{\mu t}$ is integrable on $[0, \infty)$.

Also recall that there exists a positive constant $\beta > 0$ such that

$$\|x\|_H \leq \beta\|x\|_V, \quad \forall x \in V. \tag{4.3.5}$$

Then we are in a position to state our stability results in the following form.

Theorem 4.3.1 *Suppose the condition (H10) holds and assume $u(t, \phi)$, $t \geq 0$, is a global strong solution of the equation (4.3.1) with initial datum*

$\phi \in M^2(-r, 0; V) \cap L^2(\Omega; C(-r, 0; H))$, *then there exist constants* $\tau > 0$, $C = C(\phi) > 0$ *such that*

$$E\|u(t, \phi)\|_H^2 \leq C(\phi) \cdot e^{-\tau t}, \qquad \forall t \geq 0, \tag{4.3.6}$$

if either one of the following hypotheses holds

(i) $\lambda < 0$, $-\lambda > \theta$, $(\forall p \geq 2)$.
(ii) $\nu > \theta$ *where* $\nu = \alpha/\beta^2 - \lambda$, *(in the particular case* $p = 2$).

Proof   We only show the case (ii). Case (i) can be similarly proved. Firstly, from (4.3.5) and (H10) it is easy to deduce that for any $t \geq 0$,

$$2\langle x, A(t, x, y)\rangle_{V, V^*} + \|B(t, x, y)\|_{\mathcal{L}_2^0}^2$$
$$\leq -\nu\|x\|_H^2 + \theta\|y\|_H^2 + \gamma(t), \quad x, y \in V. \tag{4.3.7}$$

Since $\nu > \theta$, it is possible to find a suitable number $\varepsilon \in (0, \nu)$ such that

$$\theta e^{\varepsilon r} \leq \nu - \varepsilon. \tag{4.3.8}$$

Applying Itô's formula to the strong solution $u(t, \phi)$, $t \geq 0$, yields that

$$e^{\varepsilon t}\|u(t)\|_H^2 - \|\phi(0)\|_H^2$$
$$= \varepsilon \int_0^t e^{\varepsilon s}\|u(s)\|_H^2 ds + 2\int_0^t e^{\varepsilon s}\langle u(s), A(s, u(s), u(s - \tau(s)))\rangle_{V, V^*} ds$$
$$+ 2\int_0^t e^{\varepsilon s}\langle u(s), B(s, u(s), u(s - \tau(s)))dW_s\rangle_H$$
$$+ \int_0^t e^{\varepsilon s} tr(B(s, u(s), u(s - \tau(s)))QB(s, u(s), u(s - \tau(s)))^*) ds.$$

Now, define an increasing sequence of stopping times

$$\sigma_n = \begin{cases} \inf\left\{t > 0 : \left|\int_0^t e^{\varepsilon s}\langle u(s), B(s, u(s), u(s - \tau(s)))dW_s\rangle_H\right| > n\right\}, \\ \infty \qquad \text{if the set is empty.} \end{cases}$$

Clearly, $\sigma_n \uparrow \infty$, as $n \to \infty$, and

$$\int_0^t e^{\varepsilon s}\langle u(s), B(s, u(s), u(s - \tau(s)))dW_s\rangle_H, \qquad t \in \mathbf{R}_+,$$

is a continuous localmartingale, so it follows that for any $n \geq 1$,

$$E\left(\int_0^{t \wedge \sigma_n} e^{\varepsilon s}\langle u(s), B(s, u(s), u(s - \tau(s)))dW_s\rangle_H\right) = 0, \qquad t \in \mathbf{R}_+.$$

Therefore, it deduces from (4.3.7) that for all $t \geq 0$,

$$Ee^{\varepsilon(t \wedge \sigma_n)} \|u(t \wedge \sigma_n)\|_H^2$$

$$\leq E\|\phi(0)\|_H^2 - (\nu - \varepsilon)E \int_0^{t \wedge \sigma_n} e^{\varepsilon s} \|u(s)\|_H^2 ds + \theta E \int_0^{t \wedge \sigma_n} e^{\varepsilon s} \|u(s - \tau(s))\|_H^2 ds$$

$$+ E \int_0^{t \wedge \sigma_n} \gamma(s) e^{[\varepsilon + \mu - (\mu \wedge \varepsilon)]s} ds$$

$$\leq E\|\phi(0)\|_H^2 - (\nu - \varepsilon)E \int_0^{t \wedge \sigma_n} e^{\varepsilon s} \|u(s)\|_H^2 ds$$

$$+ \theta e^{\varepsilon r} E \int_0^{t \wedge \sigma_n} e^{\varepsilon(s - \tau(s))} \|u(s - \tau(s))\|_H^2 ds + C_1 \cdot e^{[\varepsilon - (\mu \wedge \varepsilon)]t}$$

where $C_1 = \int_0^\infty \gamma(s) e^{\mu s} ds < \infty$. Using (4.3.8) and the fact that $\tau'(t) \leq 0$, $t \geq 0$, we deduce that there exists a constant $C_2 > 0$ such that

$$Ee^{\varepsilon(t \wedge \sigma_n)} \|u(t)\|_H^2 \leq C_2 - (\nu - \varepsilon)E \int_0^{t \wedge \sigma_n} e^{\varepsilon s} \|u(s)\|_H^2 ds$$

$$+ \theta e^{\varepsilon r} E \int_0^{t \wedge \sigma_n} e^{\varepsilon s} \|u(s)\|_H^2 ds + C_1 e^{[\varepsilon - (\mu \wedge \varepsilon)]t}$$

$$\leq C_2 + C_1 e^{[\varepsilon - (\mu \wedge \varepsilon)]t}.$$

Letting n tend to infinity in the above shows that there exists a constant $C_3 = C_3(r, \phi) > 0$ such that

$$E\|u(t, \phi)\|_H^2 \leq C_3 \cdot e^{-(\mu \wedge \varepsilon)t}.$$

In other words, the strong solution (4.3.1) is exponentially decayable in mean square and the proof is now complete. $\square$

If we consider the non-delay formulation, i.e., $\theta = 0$, Theorem 3.2.1 is an immediate deduction from Theorem 4.3.1. In particular, we may formulate the following fractional power type of coercivity condition to get a similar decay criterion to Corollary 3.2.1:

(H11) There exist constants $\alpha > 0$, $\lambda \in \mathbf{R}^1$, $0 \leq \sigma \leq 1$ and nonnegative continuous functions $\gamma(t)$, $\zeta(t)$, $t \in \mathbf{R}_+$, and $\mu > 0$ such that

$$2\langle x, A(t, x, y)\rangle_{V,V^*} + \|B(t, x, y)\|_{\mathcal{L}_2^0}^2$$
$$\leq -\alpha\|x\|_V^p + \lambda\|x\|_H^2 + \zeta(t)\|y\|_H^{2\sigma} + \gamma(t), \quad x, y \in V, \tag{4.3.9}$$

where $p \geq 2$ and $\zeta(t)e^{\mu t}$, $\gamma(t)e^{\mu t}$, $t \geq 0$, are both integrable on $[0, \infty)$.

Corollary 4.3.1 *Suppose that (H11) holds and $u(t, \phi)$, $t \geq 0$, is a global strong solution to the equation (4.3.1) with initial datum $\phi \in M^2(-r, 0; V) \cap$*

$L^2(\Omega; C(-r, 0; H))$; *then there exist constants* $\tau > 0$, $C = C(\phi) > 0$ *such that*

$$E\|u(t, \phi)\|_H^2 \leq C(\phi) \cdot e^{-\tau t}, \qquad \forall t \geq 0, \qquad (4.3.10)$$

if either one of the following hypotheses holds:

(i). $\lambda < 0$, $(\forall p \geq 2)$.
(ii). $\nu > 0$ *where* $\nu = \alpha/\beta^2 - \lambda$, *(particularly, for $p = 2$).*

By carrying out a similar argument to that of Theorem 3.2.2, we can actually obtain almost sure pathwise decay under the conditions (H10) and (H11) above. Precisely, we have the following:

Theorem 4.3.2   *Suppose that the condition (H10) or (H11) holds. Then there exist a subset $\Omega_0 \subset \Omega$ with $P(\Omega_0) = 0$ and a random variable $T(\omega) \geq 0$ such that for each $\omega \in \Omega\backslash\Omega_0$, the strong solution $u(t, \phi)$ of (4.3.1) satisfies*

$$\|u(t, \phi)\|_H \leq M(\phi) \cdot e^{-\lambda t}, \qquad \forall t \geq T(\omega),$$

for some positive constants $M = M(\phi) > 0$ and $\lambda > 0$.

Example 4.3.1   Consider the following semilinear stochastic partial differential equation:

$$dy(t, x) = \frac{\partial^2}{\partial x^2}y(t, x)dt + e^{-t/2}y(t - r, x)^{\frac{1}{3}}dt + \sqrt{\mu}\frac{y(t, x)}{1 + |y(t - r, x)|}dB_t,$$

$$t \geq 0, \qquad x \in (0, 1),$$

$$y(t, x) = \phi(t, x), \quad 0 \leq x \leq 1, \quad t \in [-r, 0]; \quad y(t, 0) = y(t, 1) = 0, \quad t \geq 0,$$

$$(4.3.11)$$

where $\phi \in C^2([0, 1] \times [-r, 0]; \mathbf{R}^1)$ and $\mu > 0$, $r > 0$ are two positive numbers. B_t, $t \geq 0$, is a real standard Brownian motion. We can set this problem in our formulation by taking $H = L^2[0, 1]$, $V = H_0^1([0, 1])$, $K = \mathbf{R}^1$, $A(t, u, v) = \frac{d^2}{dx^2}u(x) + e^{-t/2}v(x)^{\frac{1}{3}}$ and $B(t, u, v) = \sqrt{\mu} \cdot u(x)/(1 + |v(x)|)$, $u, v \in V$.

It is easy to deduce that for sufficiently small $\delta > 0$ and $u, v \in V$,

$$2\langle u, A(t, u, v)\rangle_{V,V^*} + \|B(t, u, v)\|_{\mathcal{L}_2^0}^2$$
$$\leq -2\pi^2\|u\|_H^2 + (\delta + \mu)\|u\|_H^2 + 1/\delta \cdot e^{-t}\|v\|_H^{2/3}. \qquad (4.3.12)$$

Therefore, whenever $2\pi^2 > \delta + \mu > 0$, or equivalently, $2\pi^2 > \mu > 0$ (note that $\delta > 0$ is an arbitrary positive number), we easily deduce from Corollary 4.3.1 and Theorem 4.3.2 that for an arbitrary delay interval $[-r, 0]$, $r > 0$, the strong solution is exponentially decayable in the mean square and also in the almost sure senses.

4.3.2 Linear Stability Conditions

The criterion (H10) in Subsection 4.3.1 is quite useful to ensure exponential stability of the nonlinear system (4.3.1). If one is concerned with linear stochastic delay differential equations, it is possible to derive less restrictive conditions than (H10). Similarly to those in Subsection 2.2.2, a generalization of Theorem 2.2.5 with time delays can be formulated.

Consider the following linear stochastic delay differential equation on V^*

$$\begin{cases} u(t,\phi) = \phi(0) + \int_0^t Au(s,\phi)ds + \int_0^t Bu(s-\tau(s),\phi)dW_s, & t \geq 0, \\ u(t,\phi) = \phi(t) \in V, & t \in [-r,0], \end{cases}$$

(4.3.13)

where $\phi(t) : [-r,0] \times \Omega \to V$, $r \geq 0$, is some given initial datum such that $\phi(t)$ is $\mathcal{F}_0$-measurable for any $t \in [-r,0]$ and $\sup_{-r \leq s \leq 0} E\|\phi(s)\|_V^2 < \infty$. $\tau : [0,\infty) \to [0,r]$ is a continuously differentiable function (of delay) satisfying $0 \leq \tau(t) \leq t+r$, $d\tau(t)/dt \leq 0$ for all $t \geq 0$. $A \in \mathcal{L}(V,V^*)$ and $B \in \mathcal{L}(V,\mathcal{L}(K,H))$ satisfy the condition that for some constants $\alpha > 0$, $\lambda \in \mathbf{R}^1$,

$$2\langle x, Ax\rangle_{V,V^*} + \langle x, \Delta(I)x\rangle_{V,V^*} \leq -\alpha\|x\|_V^2 + \lambda\|x\|_H^2, \quad \forall x \in V, \quad (4.3.14)$$

where for arbitrary $P \in \mathcal{L}(H,H)$, $\Delta(P) \in \mathcal{L}(V,V^*)$ is defined by

$$\langle y, \Delta(P)x\rangle_{V,V^*} = tr[B(x)^*PB(y)Q], \quad x, y \in V.$$

It is known that under the condition (4.3.14), there exists a unique strong solution of (4.3.13). Moreover, A generates a strongly continuous semigroup $T(t)$, $t \geq 0$, and the strong solution is also a mild solution. On the other hand, note that under (4.3.14), $T(t)$ maps H into V and there is a constant $C > 0$ such that

$$\int_0^\infty e^{-2\lambda t}\|T(t)x\|_H^2 dt \leq C\|x\|_V^2, \quad \forall x \in V$$

(see Lions [1], Chap. IV, Theorem 1.1). Then the operators $BT(t)$ and $T(t)B$ defined by

$$(BT(t))(x)(y) = B(T(t)x)(y), \quad \forall x \in V, \quad y \in K,$$

$$(T(t)B)(x)(y) = T(t)(B(x)(y)), \quad \forall x \in V, \quad y \in K,$$

belong to $\mathcal{L}(V,\mathcal{L}(K,H))$. We say that B and $T(t)$ *commute* if

$$(BT(t))(x) = (T(t)B)(x), \quad \forall x \in V.$$

Theorem 4.3.3 *Suppose that the following relations hold:*

(H12) $\qquad \exists C \geq 1, \gamma > 0: \ \|T(t)\| \leq C \cdot e^{-\gamma t}, \quad t \geq 0,$

(H13)
$$\left\| \int_0^\infty T^*(t)\Delta(I)T(t)dt \right\| < 1.$$

Assume also that B commutes with T(t). Then there exist positive constants μ, M such that the strong solution of (4.3.13) satisfies

$$E\|u(t,\phi)\|_H^2 \le M \cdot \|\phi\|_1^2 e^{-\mu t}, \quad t \ge 0,$$

where $\|\phi\|_1^2 = \max\{E\|\phi(0)\|_H^2, \int_{-r}^0 E\|\phi(t)\|_H^2 dt\}.$

Proof   Since $u(t,\phi)$ is the strong solution of (4.3.13), it is also the mild solution and satisfies

$$u(t,\phi) = T(t)\phi(0) + \int_0^t T(t-s)Bu(s-\tau(s),\phi)dW_s, \quad t \ge 0,$$

which immediately implies

$$\|u(t,\phi)\|_H^2 = \|T(t)\phi(0)\|_H^2 + \left\| \int_0^t T(t-s)Bu(s-\tau(s),\phi)dW_s \right\|_H^2$$
$$+ 2\left\langle T(t)\phi(0), \int_0^t T(t-s)Bu(s-\tau(s),\phi)dW_s \right\rangle_H, \quad t \ge 0.$$

Hence, it is easy to deduce that for any $t \ge 0$,

$$E\|u(t,\phi)\|_H^2 = \|T(t)\phi(0)\|_H^2 + E\left\| \int_0^t T(t-s)Bu(s-\tau(s),\phi)dW_s \right\|_H^2,$$

since $T(t)\phi(0)$ is $\mathcal{F}_0$-measurable, and consequently

$$E\left\langle T(t)\phi(0), \int_0^t T(t-s)Bu(s-\tau(s),\phi)dW_s \right\rangle_H = 0.$$

By standard properties of stochastic integrals and since B commutes with $T(t)$, we have

$$E\left\| \int_0^t T(t-s)Bu(s-\tau(s),\phi)dW_s \right\|_H^2$$
$$= \int_0^t E\left[tr\left((T(t-s)Bu(s-\tau(s),\phi))^*T(t-s)Bu(s-\tau(s),\phi)Q \right) \right] ds$$
$$= \int_0^t E\langle T^*(t-s)\Delta(I)T(t-s)u(s-\tau(s),\phi), u(s-\tau(s),\phi) \rangle_H ds.$$

Let $\lambda > 0$ be a constant to be determined later. From the last equation,

$$\int_0^\infty e^{\lambda t}E\|u(t,\phi)\|_H^2 dt = \int_0^\infty e^{\lambda t}E\|T(t)\phi(0)\|_H^2 dt$$
$$+ \int_0^\infty e^{\lambda t}\int_0^t E\langle T^*(t-s)\Delta(I)T(t-s)u(s-\tau(s),\phi), u(s-\tau(s),\phi) \rangle_H dsdt.$$

$$(4.3.15)$$

Evaluating the first term on the right hand side of (4.3.15), we obtain by (H12)

$$\int_0^\infty e^{\lambda t} E\|T(t)\phi(0)\|_H^2 dt \le \frac{C^2}{2\gamma-\lambda}\|\phi\|_1^2, \qquad (4.3.16)$$

if λ is such a number $0 < \lambda < 2\gamma$. Also, Fubini's theorem and the change of variables $u = s - \tau(s)$ yield

$$\int_0^\infty e^{\lambda t} \int_0^t E\langle T^*(t-s)\Delta(I)T(t-s)u(s-\tau(s),\phi), u(s-\tau(s),\phi)\rangle_H \, dsdt$$

$$= \int_0^\infty e^{\lambda s} \int_0^\infty e^{\lambda t} E\langle T^*(t)\Delta(I)T(t)u(s-\tau(s),\phi), u(s-\tau(s),\phi)\rangle_H \, dtds$$

$$\le \left\|\int_0^\infty e^{\lambda t} T^*(t)\Delta(I)T(t)dt\right\| \int_0^\infty e^{\lambda s} E\|u(s-\tau(s),\phi)\|_H^2 ds$$

$$\le f(\lambda)e^{\lambda r} \int_{-r}^\infty e^{\lambda s} E\|u(s,\phi)\|_H^2 ds$$

$$\le f(\lambda)e^{\lambda r}\|\phi\|_1^2 + f(\lambda)e^{\lambda r} \int_0^\infty e^{\lambda s} E\|u(s,\phi)\|_H^2 ds, \qquad (4.3.17)$$

where

$$f(\lambda) = \left\|\int_0^\infty e^{\lambda t} T^*(t)\Delta(I)T(t)dt\right\|.$$

Using (4.3.15)–(4.3.17), we deduce that

$$\int_0^\infty e^{\lambda t} \int_0^t E\langle T^*(t-s)\Delta(I)T(t-s)u(s-\tau(s),\phi), u(s-\tau(s),\phi)\rangle_H \, dsdt$$

$$\le f(\lambda)e^{\lambda r}\left(1 + \frac{C^2}{2\gamma-\lambda}\right)\|\phi\|_1^2 + f(\lambda)e^{\lambda r}$$

$$\cdot \int_0^\infty e^{\lambda t} \int_0^t E\langle T^*(t-s)\Delta(I)T(t-s)u(s-\tau(s),\phi), u(s-\tau(s),\phi)\rangle_H \, dsdt.$$

The continuity of functions defined by integrals depending on parameters and (H13) yield

$$\lim_{\lambda\to 0+} f(\lambda)e^{\lambda r} < 1.$$

Then, we can choose λ properly such that $0 < \lambda < 2\gamma$, $f(\lambda)e^{\lambda r} < 1$. Consequently, there exists a constant $C_1 > 0$, depending only on λ, such that

$$\int_0^\infty e^{\lambda t} \int_0^t E\langle T^*(t-s)\Delta(I)T(t-s)u(s-\tau(s),\phi), \ u(s-\tau(s),\phi)\rangle_H dsdt$$

$$\le C_1\|\phi\|_1^2. \qquad (4.3.18)$$

From (4.3.16) and (4.3.18) it follows that for each $\lambda > 0$ small enough, there exists a positive constant $M_1 = M_1(\lambda)$ such that

$$\int_0^\infty e^{\lambda t} E\|u(t, \phi)\|_H^2 dt \le M_1 \|\phi\|_1^2. \tag{4.3.19}$$

Then, applying Itô's formula to the process $e^{\lambda t}\|u(t, \phi)\|_H^2$, $t \ge 0$, and carrying out a similar argument to Theorem 3.2.1, yields the required results.   □

Remark   Under the same conditions as in Theorem 4.3.3, we may also obtain almost sure pathwise exponential decay of (4.3.13). If B and $T(t)$ do not commute, it is possible to formulate less general conditions than those in Theorem 4.3.3 to ensure the desired exponential decay. The reader is referred to Caraballo [1] for more details in this respect.

4.4   Razumikhin Type Stability Theorems

In this section, we will investigate stability of mild solutions for semilinear stochastic functional differential equations, based on the ideas of constructing Lyapunov functions rather than functionals, in the spirit of Razumikhin in finite dimensions.

Consider the following stochastic retarded evolution equation in the Hilbert space H:

$$\begin{cases} u(t) = T(t)\phi(0) + \int_0^t T(t-s)F(u_s)ds + \int_0^t T(t-s)G(u_s)dW_s, \\ \qquad\qquad\qquad\qquad\qquad\qquad\qquad\qquad\qquad t \ge 0, \quad (4.4.1) \\ u_0(\cdot) = \phi(\cdot) \in C_r = C([-r, 0]; H), \end{cases}$$

where $T(t)$, $t \ge 0$, is some C_0-semigroup of bounded linear operators on H with unbounded infinitesimal generator A, and $F : C([-r, 0]; H) \to H$ and $G : C([-r, 0]; H) \to \mathcal{L}(K, H)$ are two measurable nonlinear mappings satisfying $F(0) = 0$ and $G(0) = 0$. To ensure the existence and uniqueness of solutions for (4.4.1), we impose the following Lipschitz condition on the coefficients F and G: for some $k > 0$,

$$\|F(x) - F(y)\|_H + \|G(x) - G(y)\| \le k\|x - y\|_{C_r}, \tag{4.4.2}$$

where $x, y \in C_r$. Since Itô's formula is only applicable to strong solutions, we introduce the following approximating system of (4.4.1) as we did in Section 3.3 for non delay stochastic systems:

$$\begin{cases} du(t) = Au(t)dt + R(n)F(u_t)dt + R(n)G(u_t)dW_t, \quad t \ge 0, \\ u(t) = R(n)\phi(t) \in \mathcal{D}(A), \quad t \in [-r, 0], \end{cases} \tag{4.4.3}$$

where $n_0 \leq n \in \rho(A)$ for some natural number n_0, the resolvent set of A and $R(n) = nR(n, A)$. It is possible to derive the following approximation result by a similar argument to Proposition 1.3.6.

Proposition 4.4.1   *Under the hypothesis (4.4.2), the equation (4.4.3) has for each $n \geq n_0$ a unique strong solution $u^n(t) \in \mathcal{D}(A)$ lying in $L^p(\Omega; C(0, \infty; H))$, $p \geq 2$. Moreover, for any $T \geq 0$, $u^n(t) \to u(t)$ of (4.4.1) almost surely as $n \to \infty$ uniformly with respect to $[0, T]$.*

In infinite dimensional spaces, for the purpose of deriving stability results, the construction of appropriate Lyapunov functionals rather than functions is a natural generalization of the Lyapunov direct method in finite dimensional spaces. We next derive a stability result of (4.4.1) to show that the situation in treating time delay stochastic systems by using this approach could become complicated. Let $\Lambda(\cdot) : C([-r, 0]; H) \to \mathbf{R}_+$ be an arbitrary continuous nonlinear functional and $u(t, \phi)$ be the solution of (4.4.1) with initial datum $\phi \in C([-r, 0]; H)$.

Proposition 4.4.2   *Suppose $v(\cdot)$, $l(\cdot) : \mathbf{R}_+ \to \mathbf{R}_+$ are two continuous nondecreasing functions, $v(t)$ and $l(t)$ are positive for $t > 0$, $v(t)$ is convex with $v(0) = l(0) = 0$. If there is a continuous nonlinear functional $\Lambda(\cdot) : C([-r, 0]; H) \to \mathbf{R}_+$ such that*

$$\begin{cases} v(\|\phi(0)\|_H^2) \leq \Lambda(\phi) \leq l(\|\phi\|_{C_r}^2), & \forall \phi \in C([-r, 0]; H), \\ E\Lambda(u_t(\phi)) \leq E\Lambda(u_s(\phi)), & \forall t \geq s \geq 0. \end{cases} \tag{4.4.4}$$

Then the null solution of (4.4.1) is stable in mean square.

Proof   For any $\varepsilon > 0$, there is a $0 < \delta = \delta(\varepsilon) < \varepsilon$ such that $l(\delta) < v(\varepsilon)$. If $\|\phi\|_{C_r}^2 \leq \delta$, then the inequalities on $\Lambda(\cdot)$ above imply

$$v(E\|u(t, \phi)\|_H^2) \leq Ev(\|u(t, \phi)\|_H^2) \leq E\Lambda(u_t(\phi)) \leq \Lambda(\phi) \leq l(\|\phi\|_{C_r}^2)$$
$$\leq l(\delta) < v(\varepsilon), \quad t \geq 0,$$

by the properties of the mean. Therefore, $E\|u(t, \phi)\|_H^2 < \varepsilon$ for any $t \geq 0$.   ☐

In spite of the formal simplicity of the above result, it is hard to apply Proposition 4.4.2 directly to practical problems even though H is finite dimensional. The reason is twofold. On the one hand, instead of the usual Lyapunov functions in finite dimensional spaces, a Lyapunov functional as above must be constructed properly, a case which is usually not easy to handle. On the other hand, the condition (4.4.4) is actually difficult to check because of the inclusion of the solution itself which is not known explicitly in most situations. One of the most effective ways to deal with these problems is to study stability of retarded systems of the type (4.4.1) by using a method introduced by Razumikhin.

Let $C^2(H; \mathbf{R}_+)$ denote the family of all nonnegative functions $\Lambda(x)$ on H which are twice continuously Fréchet differentiable. If $\Lambda \in C^2(H; \mathbf{R}_+)$, define an operator $\mathbf{L}\Lambda$ for any $\phi \in C_r$ with $\phi(0) \in \mathcal{D}(A)$, by

$$(\mathbf{L}\Lambda)(\phi) = \langle \Lambda'_x(\phi(0)), A\phi(0) + F(\phi)\rangle_H + \frac{1}{2} tr\left[\Lambda''_{xx}(\phi(0))G(\phi)QG^*(\phi)\right], \quad (4.4.5)$$

where Q is the covariance operator of the Wiener process W. Then we can assert the following results.

Theorem 4.4.1   *Let $p \geq 2$. Assume that there exists a continuous nonlinear function $\Lambda \in C^2(H; \mathbf{R}_+)$ and constants $c_i > 0$, $i = 1, \cdots, 4$, $\lambda > 0$ such that for any $x \in H$,*

$$c_1\|x\|_H^p \leq \Lambda(x) \leq c_2\|x\|_H^p, \quad \|\Lambda'_x(x)\|_H \leq c_3\|x\|_H^{p-1}, \quad (4.4.6)$$

$$\|\Lambda''_{xx}(x)\| \leq c_4\|x\|_H^{p-2}, \quad E(\mathbf{L}\Lambda)(\psi) \leq -\lambda E\Lambda(\psi(0)), \quad (4.4.7)$$

provided $\psi \in C(\Omega \times [-r, 0]; H)$ with $\psi(0) \in \mathcal{D}(A)$ satisfies

$$E\Lambda(\psi(\theta)) < qE\Lambda(\psi(0)) \quad \text{for all} \ -r \leq \theta \leq 0$$

for some $q > 1$. Then the null solution of (4.4.1) is exponentially stable in p-th moment. Moreover, for arbitrary $\phi \in C(\Omega \times [-r, 0]; H)$,

$$E\|u(t, \phi)\|_H^p \leq \frac{c_2}{c_1} E\|\phi\|_{C_r}^p e^{-\gamma t}, \quad \forall t \geq 0,$$

where $\gamma = \min\left\{\lambda, \frac{\log(q)}{r}\right\}$.

Proof   Fix the initial data $\phi \in C(\Omega \times [-r, 0]; H)$ and write $u(t, \phi) = u(t)$ simply. Let $\varepsilon \in (0, \gamma)$ be one arbitrary number and set $\bar{\gamma} = \gamma - \varepsilon$. Define

$$U(t) = \max_{-r \leq \theta \leq 0}\left[e^{\bar{\gamma}(t+\theta)}E\Lambda(u(t+\theta))\right], \quad t \geq 0.$$

Obviously, $U(t)$ is well defined and continuous. We claim that

$$D^+U(t) = \limsup_{h \to 0^+} \frac{U(t+h) - U(t)}{h} \leq 0 \quad \text{for} \ t \geq 0. \quad (4.4.8)$$

To show this, for each fixed $t_0 \geq 0$, define

$$\bar{\theta} = \max\left\{\theta \in [-r, 0] : U(t_0) = e^{\bar{\gamma}(t_0+\theta)}E\Lambda(u(t_0 + \theta))\right\}.$$

Obviously, $\bar{\theta}$ is well defined, $\bar{\theta} \in [-r, 0]$ and

$$U(t_0) = e^{\bar{\gamma}(t_0 + \bar{\theta})} E\Lambda(u(t_0 + \bar{\theta})).$$

If $\bar{\theta} < 0$, one has

$$e^{\bar{\gamma}(t_0 + \theta)} E\Lambda(u(t_0 + \theta)) < e^{\bar{\gamma}(t_0 + \bar{\theta})} E\Lambda(u(t_0 + \bar{\theta})) \quad \text{for all} \quad \bar{\theta} < \theta \leq 0.$$

It is therefore easy to observe that for any $h > 0$ small enough,

$$e^{\bar{\gamma}(t_0 + h)} E\Lambda(u(t_0 + h)) \leq e^{\bar{\gamma}(t_0 + \bar{\theta})} E\Lambda(u(t_0 + \bar{\theta})).$$

Hence

$$U(t_0 + h) \leq U(t_0) \quad \text{and} \quad D^+ U(t_0) \leq 0.$$

If $\bar{\theta} = 0$, then

$$e^{\bar{\gamma}(t_0 + \theta)} E\Lambda(u(t_0 + \theta)) \leq e^{\bar{\gamma} t_0} E\Lambda(u(t_0)) \quad \text{for any} \quad \theta \in [-r, 0].$$

Therefore,

$$\begin{aligned} E\Lambda(u(t_0 + \theta)) &\leq e^{-\bar{\gamma}\theta} E\Lambda(u(t_0)) \\ &\leq e^{\bar{\gamma} r} E\Lambda(u(t_0)) \quad \text{for any} \quad \theta \in [-r, 0]. \end{aligned} \quad (4.4.9)$$

In the case of $E\Lambda(u(t_0)) = 0$, (4.4.6) and (4.4.9) imply that $u(t_0 + \theta) = 0$ for all $\theta \in [-r, 0]$ almost surely. Recall that $F(0) = 0$ and $G(0) = 0$, it then follows that $u(t_0 + h) = 0$ almost surely for all $h > 0$, hence $U(t_0 + h) = 0$ and $D^+ U(t_0) = 0$. On the other hand, in the case of $E\Lambda(u(t_0)) > 0$, (4.4.9) implies

$$\begin{aligned} E\Lambda(u(t_0 + \theta)) &\leq e^{\bar{\gamma} r} E\Lambda(u(t_0)) \\ &< q E\Lambda(u(t_0)) \quad \text{for any} \quad \theta \in [-r, 0] \end{aligned}$$

as $e^{\bar{\gamma} r} < q$. Let $\nu = q - e^{\bar{\gamma} r} > 0$, it then follows from the continuity of $E\Lambda(u(t))$ and (4.4.6) that for some $h > 0$ small enough,

$$E\Lambda(u(t_0 + \theta)) \leq \left(e^{\bar{\gamma} r} + \frac{\nu}{2} \right) E\Lambda(u(t_0)) \quad \text{for any} \quad \theta \in [0, h].$$

Now we need to introduce the strong solution sequence $\{u^n(t)\}$ of (4.4.3) such that by Proposition 4.4.1, $u^n(t) \to u(t)$ in $C(0, T; H)$, $\forall T \geq 0$, uniformly with respect to t as $n \to \infty$ almost surely. Consequently, for some constant $\delta \in \left(0, \frac{\nu}{4 + 2\nu} E\Lambda(u(t_0)) \right)$, there are a sufficiently small constant $h > 0$ and number $N > 0$ large enough such that for $n \geq N$, it follows that for any $s \in [t_0, t_0 + h]$,

$$\begin{aligned} E\Lambda(u(s)) &> E\Lambda(u(t_0)) - \delta > 0, \\ E\Lambda(u(s + \theta)) &< E\Lambda(u(t_0 + \theta)) + \delta, \quad \theta \in [-r, 0], \end{aligned}$$

$$e^{\bar{\gamma}r}E\Lambda(u(t_0)) < e^{\bar{\gamma}r}E\Lambda(u(s)) + \delta,$$
$$E\Lambda(u^n(s)) > E\Lambda(u(s)) - \delta > 0,$$
$$e^{\bar{\gamma}r}E\Lambda(u(s)) < e^{\bar{\gamma}r}E\Lambda(u^n(s)) + \delta,$$
$$E\Lambda(u^n(s+\theta)) < E\Lambda(u(s+\theta)) + \delta, \quad \theta \in [-r, 0].$$

These immediately imply

$$
\begin{aligned}
E\Lambda(u^n(s+\theta)) &< e^{\bar{\gamma}r}E\Lambda(u^n(s)) + 4\delta \\
&< e^{\bar{\gamma}r}E\Lambda(u^n(s)) + \nu(E\Lambda(u(s)) - \delta) \\
&< e^{\bar{\gamma}r}E\Lambda(u^n(s)) + \nu E\Lambda(u^n(s)) \\
&= qE\Lambda(u^n(s)) \quad \text{for any } \theta \in [-r, 0].
\end{aligned}
\tag{4.4.10}
$$

In addition to (4.4.7), (4.4.10) implies that

$$E(\mathbf{L}\Lambda)(u^n_s) \le -\lambda E\Lambda(u^n(s)), \quad \forall s \in [t_0, t_0 + h]. \tag{4.4.11}$$

Applying Itô's formula to the function $e^{\bar{\gamma}t}\Lambda(u)$ along the strong solutions $u^n(t)$ of (4.4.2), we can derive, by using (4.4.11), that for any $\bar{h} \in [0, h]$,

$$
\begin{aligned}
e^{\bar{\gamma}(t_0+\bar{h})} &E\Lambda(u^n(t_0+\bar{h})) \\
&\le e^{\bar{\gamma}t_0}E\Lambda(u^n(t_0)) + (\bar{\gamma} - \lambda)\int_{t_0}^{t_0+\bar{h}} e^{\bar{\gamma}s}E\Lambda(u^n(s))ds \\
&\quad + \int_{t_0}^{t_0+\bar{h}} e^{\bar{\gamma}s}E\langle\Lambda'_x(u^n(s)), (R(n) - I)F(u^n_s)\rangle_H ds \\
&\quad + \frac{1}{2}\int_{t_0}^{t_0+\bar{h}} Ee^{\bar{\gamma}s}tr[\Lambda''_{xx}(u^n(s))R(n)G(u^n_s)Q(R(n)G(u^n_s))^*]ds \\
&\quad - \frac{1}{2}\int_{t_0}^{t_0+\bar{h}} Ee^{\bar{\gamma}s}tr[\Lambda''_{xx}(u^n(s))G(u^n_s)QG(u^n_s)^*]ds,
\end{aligned}
$$

which, letting $n \to \infty$, immediately yields

$$
\begin{aligned}
e^{\bar{\gamma}(t_0+\bar{h})}E\Lambda(u(t_0+\bar{h})) &\le e^{\bar{\gamma}t_0}E\Lambda(u(t_0)) + \int_{t_0}^{t_0+\bar{h}}(\bar{\gamma} - \lambda)e^{\bar{\gamma}s}E\Lambda(u(s))ds \\
&\le e^{\bar{\gamma}t_0}E\Lambda(u(t_0)).
\end{aligned}
$$

$$\tag{4.4.12}$$

Then it must be the case that

$$e^{\bar{\gamma}s}E\Lambda(u(s)) \le e^{\bar{\gamma}t_0}E\Lambda(u(t_0)), \quad \forall s \in [t_0, t_0 + h].$$

So it must hold that $EU(t_0 + h) = EU(t_0)$ for any $h > 0$ sufficiently small, and hence $D^+U(t_0) = 0$. Since t_0 is arbitrary, the inequality (4.4.8) is shown to hold for any $t \ge 0$. It now follows immediately from (4.4.8) that

$$U(t) \le U(0), \quad \forall t \ge 0.$$

Also, (4.4.6) implies

$$e^{\bar{\gamma}t}E\Lambda(u(t)) \le U(t) \le U(0) \le c_2 E\|\phi\|_{C_r}^p, \quad \forall t \ge 0.$$

Note that ε is arbitrary, it thus follows that

$$E\Lambda(u(t)) \le c_2 E\|\phi\|_{C_r}^p e^{-\gamma t}, \quad \forall t \ge 0,$$

which, by virtue of (4.4.6), immediately yields

$$E\|u(t)\|_H^p \le \frac{c_2}{c_1} E\|\phi\|_{C_r}^p e^{-\gamma t}, \quad \forall t \ge 0.$$

$\square$

Under the same conditions as in Theorem 4.4.1, one can deduce the almost sure exponential stability of (4.4.1) by carrying out a similar argument to Theorem 3.2.2.

Next we will apply Theorem 4.4.1 to a class of stochastic delay differential equations of the following form:

$$u(t) = T(t)\phi(0) + \int_0^t T(t-s)f(u(s-\tau_1(s)))ds$$

$$+ \int_0^t T(t-s)g(u(s-\tau_2(s)))dW_s, \quad t \ge 0, \qquad (4.4.13)$$

$$u_0(\cdot) = \phi(\cdot) \in C(\Omega \times [-r,0]; H),$$

where $\tau_i(\cdot): \mathbf{R}_+ \to [0,r]$, $i = 1, 2$, are both continuous, and

$$f(\cdot): H \to H \quad \text{and} \quad g(\cdot): H \to \mathcal{L}(K,H),$$

$f(0) = 0$, $g(0) = 0$, are two nonlinear measurable mappings satisfying the usual Lipschitz continuous conditions.

Corollary 4.4.1 *Let $p \ge 2$ and assume that there exists a nonlinear function $\Lambda \in C^2(H; \mathbf{R}_+)$, constants $c_i > 0$, $i = 1, \cdots, 4$, $\lambda > 0$, $\lambda_1 > 0$ and $\lambda_2 > 0$ such that for any $x \in H$,*

$$c_1\|x\|_H^p \le \Lambda(x) \le c_2\|x\|_H^p, \quad \|\Lambda_x'(x)\|_H \le c_3\|x\|_H^{p-1}, \qquad (4.4.14)$$

$$\|\Lambda_{xx}''(x)\| \le c_4\|x\|_H^{p-2}, \qquad (4.4.15)$$

and

$$\langle \Lambda_x'(\phi(0)), A\phi(0) + f(\phi(-\tau_1(0)))\rangle_H + \frac{1}{2}tr\left[\Lambda_{xx}''g(\phi(-\tau_2(0)))Qg^*(\phi(-\tau_2(0)))\right]$$

$$\le -\lambda\Lambda(\phi(0)) + \lambda_1\Lambda(\phi(-\tau_1(0))) + \lambda_2\Lambda(\phi(-\tau_2(0))),$$

$$(4.4.16)$$

for any $\phi \in C_r$ with $\phi(0) \in \mathcal{D}(A)$. If $\lambda > \lambda_1 + \lambda_2$, then the null solution of (4.4.13) is exponentially stable in p-th moment. Its p-th moment Lyapunov exponent should not be greater than $(q\sum_{i=1}^{2}\lambda_i - \lambda)$, where $q \in (1, \lambda/\sum_{i=1}^{2}\lambda_i)$ is the unique root of $\lambda - q\sum_{i=1}^{2}\lambda_i = \log(q)/r$. Furthermore, the null solution is also exponentially stable in the almost sure sense.

Proof   Define for $\phi \in C([-r, 0]; H)$,

$$F(\phi) = f(\phi(-\tau_1(0))) \quad \text{and} \quad G(\phi) = g(\phi(-\tau_2(0))).$$

Then (4.4.13) can be handled in the framework of (4.4.1). Moreover, in that case the operator $\mathbf{L}\Lambda$ defined in (4.4.5) becomes

$$(\mathbf{L}\Lambda)(\phi) = \langle \Lambda'_x(\phi(0)), A\phi(0) + f(\phi(-\tau_1(0)))\rangle_H$$
$$+ \frac{1}{2}tr\left[\Lambda''_{xx}(\phi(0))g(\phi(-\tau_2(0)))Qg^*(\phi(-\tau_2(0)))\right]$$

for any $\phi \in C_r$ with $\phi(0) \in \mathcal{D}(A)$. If $\psi \in C(\Omega \times [-r, 0]; H)$ with $\psi(0) \in \mathcal{D}(A)$ satisfies

$$E\Lambda(\psi(\theta)) < qE\Lambda(\psi(0)) \quad \text{for all } \theta \in [-r, 0]$$

for some $q > 1$, then by Condition (4.4.16),

$$E(\mathbf{L}\Lambda)(\psi) \leq -\lambda E\Lambda(\psi(0)) + \lambda_1 E\Lambda(\psi(-\tau_1(0))) + \lambda_2 E\Lambda(\psi(-\tau_2(0)))$$
$$\leq \left(q\sum_{i=1}^{2}\lambda_i - \lambda\right)E\Lambda(\psi(0)).$$

So, by virtue of Theorem 4.4.1, the null solution of (4.4.13) is exponentially stable in the p-th moment and also in the almost sure senses. Its p-th moment Lyapunov exponent should not be greater than $(q\sum_{i=1}^{2}\lambda_i - \lambda)$. The proof is now complete.   □

Last, let us discuss an example to close this section.

Example 4.4.1   Consider the semilinear stochastic partial differential equation with finite time lags r_1, r_2 $(r > r_1, r_2 > 0)$,

$$dZ(t, x) = \mu\frac{\partial^2}{\partial x^2}Z(t, x)dt + \left[\nu\int_{-r_1}^{0}Z(t+\theta, x)h(\theta)d\theta\right]dt$$
$$+\alpha(Z(t, x))Z(t - r_2, x)dB_t, \quad (\mu > 0, \nu > 0),$$
$$Z(t, 0) = Z(t, \pi) = 0, \quad t \geq 0,$$
$$Z(s, x) = \phi(s, x), \quad s \in [-r, 0], \ x \in [0, \pi],$$
$$\phi(s, \cdot) \in H = L^2(0, \pi), \ \phi(\cdot, x) \in C([-r, 0]; \mathbf{R}^1),$$

$$(4.4.17)$$

where B_t is a standard one dimensional Brownian motion. Here $\alpha(\cdot) : \mathbf{R}^1 \to \mathbf{R}^1$, $h(\cdot) : [-r_1, 0] \to \mathbf{R}^1$ are two bounded Lipschitz continuous functions such that $|\alpha(x)| \leq L$, $|h(\theta)| \leq M$, $L > 0$, $M > 0$, for any $x \in \mathbf{R}^1$, $\theta \in [-r_1, 0]$.

Let $A = \mu \partial^2 / \partial x^2$ with the domain

$$\mathcal{D}(A) = \left\{ u \in L^2(0, \pi) : \frac{\partial u}{\partial x}, \frac{\partial^2 u}{\partial x^2} \in L^2(0, \pi), \, u(0) = u(\pi) = 0 \right\},$$

so it is easy to deduce

$$\langle Au, u \rangle_H \leq -\mu \|u\|_H^2, \quad u \in \mathcal{D}(A).$$

On the other hand, suppose

$$E\|\psi(\theta)\|_H^2 \leq qE\|\psi(0)\|_H^2, \quad q > 1, \quad \forall \theta \in [-r, 0],$$

for $\psi \in C(\Omega \times [-r, 0]; H)$ with $\psi(0) \in \mathcal{D}(A)$. It is clear that

$$E\left\langle \psi(0), \nu \int_{-r_1}^0 \psi(\theta) h(\theta) d\theta \right\rangle_H \leq q^{1/2}(\nu r_1 M) E\|\psi(0)\|_H^2$$

and

$$E\|\alpha(\psi(0))\psi(-r_2)\|_H^2 \leq qL^2 E\|\psi(0)\|_H^2.$$

Then letting $\Lambda(u) = \|u\|_H^2$, we have by a direct computation that

$$E(\mathbf{L}\Lambda)(\psi) \leq \left(-2\mu + 2q^{1/2}(\nu r_1 M) + qL^2 \right) E\|\psi(0)\|_H^2.$$

By virtue of Theorem 4.4.1 and letting $q \to 1$, one gets that if $2\mu > 2(\nu r_1 M) + L^2$, then for arbitrary $0 \leq r_2 \leq r$, the null solution of (4.4.17) is exponentially stable in the mean square and almost sure senses.

4.5   Notes and Comments

All the main results in Section 4.1 are due to Travis and Webb [1] and Milota [1]. The discussion of characteristic equations with infinite delay is based on the work of Lenhart and Travis [1]. In deterministic cases, the reduction of a neutral equation to a retarded equation with infiinite delay and the notion of the L^2-stability was developed by Datko [3], [4]. The stochastic version presented in Section 4.2 is due to Liu [7] which is in the spirit of Datko's work. The established result that L^2-stability implies asymptotic stability allows one to apply Lyapunov functional method to obtain stability criteria. Staffans [1] has shown that an ordinary neutral functional differential equation with a linear stable difference operator can be reduced to a retarded equation with infinite delay. It is hoped that this reduction can be extended to a class of stochastic partial neutral functional differential equations and applied to the study of their stability.

The main results in Section 4.3 are taken from Caraballo, Liu and Truman [2] and Caraballo [1]. There exists some work on stability for mild solutions of nonlinear stochastic evolution equations with time delays. For instance, Jahanipur [1] considered stability of a class of stochastic delay evolution equations with monotone nonlinearity.

In finite dimensional spaces, the Razumikhin type arguments take advantages of the structure of $\mathbf{R}^n$ while the Lyapunov functional type ones need handle difficulties caused by the infinite dimensional nature of C_r. This is clearly illustrated by an example in Levin and Nohel [1]. The Razumikhin ideas were firstly extended to obtain a version for stochastic systems by Taniguchi [2] and subsequently by Mao [3], [4] in finite dimensional spaces. Theorem 4.4.1 and Corollary 4.4.1 are discussed and established in Liu and Shi [1] in the spirit of Mao [3] to deduce stability (in the sense of mild solutions) for more general stochastic functional differential equations in infinite dimensions.

Chapter 5

Some Related Topics of Stability and Applications

In this chapter, we shall present some selected topics in connection with stability properties of stochastic differential equations in infinite dimensions. The choice of the material reflects our own personal reference and the treatment here is somewhat sketchy. Some chosen material touches upon specific stochastic systems which could be regarded as a potential starting point for research in this area. Some other material reveals interesting and important relationships between the main topic, stability, of this book and topics from other branches of science or technology such as stochastic control. In particular, in Section 5.1 we shall begin with an exposition of stability for a class of stochastic parabolic equations with singular noise, e.g., distributed one. The problem of optimal feedback control and its relation with stablity of stochastic linear systems are studied in Section 5.2. Section 5.3 is wholly devoted to the investigation of feedback stabilization for a class of nonlinear stochastic models. In Sections 5.4 and 5.5, the stability results of the previous chapters are applied to some important models in chemical dynamics, fluid dynamics and mathematical population biology.

5.1 Parabolic Equations with Boundary and Pointwise Noise

In order to motivate our theory, suppose $\mathcal{O}$ is a bounded open domain in $\mathbf{R}^n$ with smooth boundary $\partial\mathcal{O}$. Let

$$-A(x, \partial) = \sum_{|\alpha|\leq 2} a_\alpha(x)\partial^\alpha, \qquad x \in \mathcal{O},$$

be a uniformly strongly elliptic operator of order two with smooth coefficients $a_\alpha(x)$.

Consider a class of nonhomogeneous partial differential equations of mixed type with boundary noise:

$$\frac{\partial y(t,x)}{\partial t} = -A(x,\partial)y(t,x), \quad x \in \mathcal{O}, \quad t \geq 0,$$

$$y(0,x) = y_0(x), \quad x \in \mathcal{O}, \qquad\qquad (5.1.1)$$

$$\frac{\partial y(t,x)}{\partial n} + a(x)y(t,x) = g(x)f(y(t,x))\dot{B}_t, \quad x \in \partial\mathcal{O}, \quad t \geq 0,$$

where $y_0(x) \in L^2(\mathcal{O})$, $g(x) \in L^2(\partial\mathcal{O})$, $\partial/\partial n$ is the outward normal derivative, $a(x)$ is a real positive function defined on $\partial\mathcal{O}$, $f(\cdot)$ is a certain real function defined on $\mathbf{R}^1$ and $\dot{B}_t$ is some given white noise.

We take $H = L^2(\mathcal{O})$ and define $\mathcal{D}(A)$ as the closure in $H^2(\mathcal{O})$ of the subspace of $C^2(\overline{\mathcal{O}})$ which consists of functions ϕ satisfying the boundary condition $\partial\phi/\partial n + a\phi = 0$. Let $-A$ be the restriction of $-A(x,\partial)$ on $\mathcal{D}(A)$. Then it is possible to show that $-A$ generates an analytic semigroup $T(t)$ of bounded operators (see, for instance, Yosida [1]). Let M be the mapping: $L^2(\partial\mathcal{O}) \to L^2(\mathcal{O})$ defined by $y(\cdot) = Mg(\cdot)$, where $y(x)$ is the solution of

$$A(x,\partial)y(x) = 0, \quad x \in \mathcal{O},$$

$$\frac{\partial y(x)}{\partial n} + a(x)y(x) = g(x), \quad x \in \partial\mathcal{O}.$$

It can be deduced that (5.1.1) is described by the abstract stochastic evolution equations on H

$$Y_t = T(t)y_0 + \int_0^t AT(t-s)Mgf(Y_s)dB_s,$$

$$= T(t)y_0 + \int_0^t A^\theta T(t-s)bf(Y_s)dB_s, \quad t \geq 0, \qquad (5.1.2)$$

where $\theta = 1/4 + \varepsilon$, $\varepsilon > 0$ and $b = A^{1-\theta}Mg \in H$. The new feature of this formulation is that the generator A appears in the integral term, and it will turn out that we can apply equations of (5.1.2) type to systems with not only boundary but point noise.

In this section, we are mainly concerned about the stability of a class of stochastic evolution equations. The results of this discussion can be applied to some models such as (5.1.1). Precisely, let $T(t)$ be a strongly continuous analytic semigroup of negative type on a real separable Hilbert space H and $-A$ be its infinitesimal generator. Then fractional power A^γ, $0 < \gamma < 1$, is well defined and that for each $0 < T < \infty$ there exists a constant $M = M(T) > 0$ such that

$$\|A^\gamma T(t)\| \leq M/t^\gamma \quad \text{for any} \ \ 0 < \gamma < 1 \ \ \text{and} \ \ t \in (0,T]. \qquad (5.1.3)$$

Consider the following stochastic evolution equation

$$Y_t = T(t)y_0 + \int_0^t A^\theta T(t-s)bf(Y_s)dB_s, \quad t \geq 0,$$

$$Y_0 = y_0 \in H, \qquad\qquad (5.1.4)$$

where $0 \le \theta < 1/2$, $b \in H$, $f(y)$ is a real Lipschitz continuous function on $\mathcal{D}(A^{\eta})$ for some $0 \le \eta < 1$, $f(0) = 0$, and B_t is a real standard Brownian motion defined on some probability space $(\Omega, \mathcal{F}, \{\mathcal{F}_t\}_{t \ge 0}, P)$. The existence, uniqueness and regularity of (5.1.4) have been established in Ichikawa [7]. To study stability of (5.1.4), following the ideas in Section 2.2 we consider the operator integral equation

$$
\begin{aligned}
\langle P(t)y, y \rangle_H &= \langle FT(T-t)y, T(T-t)y \rangle_H + \int_t^T \Big[\langle MT(r-t)y, T(r-t)y \rangle_H \\
&\quad + f^2(T(r-t)y) \langle (A^*)^{\theta} P(r) A^{\theta} b, b \rangle_H \Big] dr, \\
P(T) &= F, \quad y \in H, \ 0 \le t \le T, \ T \ge 0,
\end{aligned}
$$

(5.1.5)

where A^* is the dual operator of A, and the associated equation

$$
\begin{aligned}
&\langle Q(t)z, z \rangle_H \\
&= \langle FT(T-t)A^{\theta}z, T(T-t)A^{\theta}z \rangle_H + \int_t^T \Big[\langle MT(r-t)A^{\theta}z, T(r-t)A^{\theta}z \rangle_H \\
&\quad + \langle Q(r)b, b \rangle_H f^2(T(r-t)A^{\theta}z) \Big] dr, \\
&Q(T) = (A^*)^{\theta} F A^{\theta}, \quad z \in H, \ 0 \le t \le T, \ T \ge 0,
\end{aligned}
$$

(5.1.6)

In the remaining of this section, we will always suppose $0 \le \beta = \theta + \eta < 1/2$. In particular, we have the following existence and uniqueness results of (5.1.5) and (5.1.6).

Proposition 5.1.1 *There exists a unique solution to the equation (5.1.5) $P(t) \in \mathcal{L}(\mathcal{D}(A^{-\theta}), \mathcal{D}((A^*)^{\theta}))$ or (5.1.6) $Q(t) \in \mathcal{L}(H)$ which is nonnegative and strongly continuous on $[0, T)$ with*

$$
\|P(t)\|_{\mathcal{L}(\mathcal{D}(A^{-\theta}), \mathcal{D}((A^*)^{\theta}))} \le \frac{C}{(T-t)^{2\theta}}, \qquad \|Q(t)\| \le \frac{C}{(T-t)^{2\theta}}
$$

for some $C = C(T) > 0$, where $\| \cdot \|_{\mathcal{L}(\mathcal{D}(A^{-\theta}), \mathcal{D}((A^*)^{\theta}))}$ is the operator norm in $\mathcal{L}(\mathcal{D}(A^{-\theta}), \mathcal{D}((A^*)^{\theta}))$. Moreover, if $F = 0$, both $P(t)$ and $Q(t)$ are strongly continuous on $[0, T]$. The relation between $P(t)$ and $Q(t)$ is given by

$$
P(t) = (A^*)^{-\theta} Q(t) A^{-\theta}, \qquad t \in [0, T],
$$

$$
Q(t) = (A^*)^{\theta} P(t) A^{\theta}, \qquad t \in [0, T].
$$

Proof It suffices to show, say, the uniqueness and existence of $Q(t)$. The other claims could be similarly proved. To show uniqueness of $Q(t)$, we consider

$$
\langle R(t)z, z \rangle_H = \int_t^T \langle R(r)b, b \rangle_H f^2(T(r-t)A^{\theta}z) dr.
$$

This yields

$$\|R(t)\| \leq C \int_t^T \frac{\|R(r)\|}{(r-t)^{2\beta}} dr \quad \text{for some} \quad C > 0.$$

Thus by using Gronwall's inequality, we obtain $\|R(t)\| = 0$ almost surely with respect to t. To establish a solution of (5.1.6), we define

$$\langle Q_0(t)z, z \rangle_H = \langle FT(T-t)A^\theta z, T(T-t)A^\theta z \rangle_H$$
$$+ \int_t^T \langle MT(r-t)A^\theta z, T(r-t)A^\theta z \rangle_H dr$$

and iterate a sequence of nonnegative operators

$$\langle Q_n(t)z, z \rangle_H = \langle Q_0(t)z, z \rangle_H + \int_t^T \langle Q_{n-1}(r)b, b \rangle_H f^2(T(r-t)A^\theta z) dr.$$

Then $Q_n(t)$ is well defined and continuous on $[0, T)$. Moreover for each $t \in [0, T)$, $Q_n(t)$ is monotonically increasing in n. But we have an estimate

$$\|Q_n(t)\| \leq \frac{\bar{C}}{(T-t)^{2\theta}} + \bar{C} \int_t^T \frac{\|Q_{n-1}(r)\|}{(r-t)^{2\beta}} dr$$

for some number $\bar{C} > 0$. Thus we obtain

$$\|Q_n(t)\| \leq \frac{C}{(T-t)^{2\theta}}$$

for some number $C = C(T) > 0$ independent of n. Hence, for each t there exists a strong limit $Q(t)$ of $Q_n(t)$ and $Q(t)$ has the desired properties. □

To examine the stability of the null solution of (5.1.4), it is useful to consider more general initial value y_0, e.g., $y_0 \in \mathcal{D}(A^{-\theta})$. In fact, we have the following:

Proposition 5.1.2   *For each $0 \leq t \leq T$ and $y_0 \in Y$,*

$$E\langle P(t)Y_t(y_0), Y_t(y_0) \rangle_H = E\langle FY_T(y_0), Y_T(y_0) \rangle_H$$
$$+ \int_t^T E\langle MY_r(y_0), Y_r(y_0) \rangle_H dr.$$

Moreover, for each $0 \leq t \leq T$ and $y_0 \in \mathcal{D}(A^{-\theta})$, $0 \leq \theta + \eta < 1/2$,

$$(y_0, P(0)y_0)_{\mathcal{D}((A^*)^\theta)} = E\langle FY_T(y_0), Y_T(y_0) \rangle_H + \int_0^T E\langle MY_r(y_0), Y_r(y_0) \rangle_H dr,$$

where $(\cdot, \cdot)_{\mathcal{D}((A^*)^\theta)}$ denotes the duality between $\mathcal{D}(A^{-\theta})$ and $\mathcal{D}((A^*)^\theta)$.

Proof   The proof is similar to that of Proposition 2.2.2.   ☐

Now we can derive our stability results for mild solutions of (5.1.4).

Theorem 5.1.1   *The three statements below are equivalent:*

(i) The null solution of (5.1.4) is L^2-stable in mean, i.e.,

$$\int_0^\infty E\|Y_t(y_0)\|_H^2 dt < \infty \quad \text{for any } y_0 \in \mathcal{D}(A^{-\theta}). \tag{5.1.7}$$

(ii) There exists a solution $0 \leq P \in \mathcal{L}(\mathcal{D}(A^{-\theta}), \mathcal{D}((A^*)^\theta)) \cap \mathcal{L}(H)$ to the Lyapunov equation

$$2\langle Py, -Ay\rangle_H + \langle (A^*)^\theta P A^\theta b, b\rangle_H f^2(y) = -\|y\|_H^2, \quad y \in \mathcal{D}(A). \tag{5.1.8}$$

(iii) For each $0 < T < \infty$,

$$E\|Y_t(y_0)\|_H^2 \leq \begin{cases} Ct^{-2\theta}\|y_0\|_{-\theta}^2, & 0 < t \leq T, \\ Ce^{-\alpha t}\|y_0\|_{-\theta}^2, & t > T, \end{cases} \text{ for some } \alpha > 0, \ C = C(T) > 0, \tag{5.1.9}$$

where $\|\cdot\|_{-\theta}$ is the norm of $\mathcal{D}(A^{-\theta})$.

Proof   We first assume (i) holds, then

$$\int_0^\infty E\|Y_t(A^\theta z)\|_H^2 dt < \infty \quad \text{for any } z \in H.$$

Thus from Proposition 5.1.2 with $F = 0$, $M = I$, we have

$$\langle (A^*)^\theta P_T(0) A^\theta z, z\rangle_H = \int_0^T E\|Y_t(A^\theta z)\|_H^2 dt,$$

where $P_T(t)$ is the solution of (5.1.5). Thus

$$(A^*)^\theta P_T(0) A^\theta \uparrow Q \geq 0, \quad \text{as } T \to \infty, \quad \text{in } \mathcal{L}(H).$$

Let $P = (A^*)^{-\theta} Q A^{-\theta}$, then $P_T(0) \to P$ in a strong sense, as $T \to \infty$, in $\mathcal{L}(H) \cap \mathcal{L}(\mathcal{D}(A^{-\theta}), \mathcal{D}((A^*)^\theta))$. Since $P_T(t) = P_{T-t}(0)$, we conclude that P satisfies (5.1.8). Hence, (i) implies (ii). Conversely, suppose P is a solution of (5.1.8). Then by Proposition 5.1.2, we have

$$(y_0, Py_0)_{\mathcal{D}((A^*)^\theta)} = E\langle PY_T(y_0), Y_T(y_0)\rangle_H + \int_0^T E\|Y_t(y_0)\|_H^2 dt.$$

Thus (i) is true. Now we assume (iii) holds, then (i) follows immediately. Finally, we assume (i) holds and show (iii). To this end, using Theorem 3.1.1 we obtain

$$E\|Y_t(y_0)\|_H^2 \leq C_1 \cdot e^{-\alpha t}\|y_0\|_H^2, \quad y_0 \in H,$$

for some $\alpha > 0$, $C_1 > 0$. Then for each $0 < T < \infty$, we have

$$E\|Y_t(y_0)\|_H^2 \leq C_1 \cdot e^{-\alpha(t-T)} E\|Y_T(y_0)\|_H^2 \quad \text{for any } t \geq T.$$

On the other hand, we have from Proposition 5.1.2 that

$$E\|Y_t(y_0)\|_H^2 \leq C_2 t^{-2\theta}\|y_0\|_{-\theta}^2, \quad t \in (0, T]$$

for some $C_2 > 0$. Combining these two estimates, we obtain (5.1.9).   □

Remark   The condition (5.1.7) may be replaced by

$$\int_0^\infty E\|Y_t(y_0)\|_H^2 dt \leq C\|y_0\|_{-\theta}^2 \quad \text{for some } C > 0.$$

If $y_0 \in H$, then (i) or (ii) implies

$$E\|Y_t(y_0)\|_H^2 \leq M \cdot e^{-\alpha t}\|y_0\|_H^2 \quad \text{for some } M \geq 1 \text{ and } \alpha > 0.$$

Next we give sufficient conditions for stability. Note that, in the present situation, the Lyapunov equation (5.1.8) is equivalent to

$$(y_0, Py_0)_{\mathcal{D}((A^*)^\theta)} = \int_0^\infty \left[\|T(t)y_0\|_H^2 + \langle (A^*)^\theta PA^\theta b, b \rangle_H f^2(T(t)y_0)\right] dt,$$
$$\tag{5.1.10}$$

where $y_0 \in \mathcal{D}(A^{-\theta})$. We may equally consider as before

$$\langle Qy, y \rangle_H = \int_0^\infty \left[\|T(t)A^\theta y\|_H^2 + \langle Qb, b \rangle_H f^2(T(t)A^\theta y)\right] dt, \quad y \in H. \quad (5.1.11)$$

As an immediate consequence, we obtain:

Proposition 5.1.3   *If the inequality*

$$\|b\|_H^2 \int_0^\infty |f(T(t)A^\theta y)|^2 dt \leq \rho\|y\|_H^2, \quad y \in H, \tag{5.1.12}$$

for some $0 \leq \rho < 1$ holds, then there exists a unique nonnegative solution to (5.1.10) and hence to (5.1.11).

Corollary 5.1.1   *Suppose $T(t)y = \sum_{n=1}^\infty e^{-\lambda_n t}\psi_n \langle \psi_n, y \rangle_H$ for some orthonormal basis $\{\psi_n\}$ of H and $\lambda_n > 0$. If $f(y) = \langle f, y \rangle_H$ for some $f \in H$, then (5.1.12) is written as*

$$\|b\|_H^2 \sum_{n=1}^\infty f_n^2/2\lambda_n^{1-2\theta} \leq \rho < 1, \quad f_n = \langle f, \psi_n \rangle_H. \tag{5.1.13}$$

Finally, we examine two examples to illustrate our theory derived above.

Example 5.1.1 Firstly, consider the example (5.1.1) at the beginning of this section. It is known that (5.1.1) may be formulated in terms of the semigroup model (5.1.2). We may take $\beta = 1/2 - \varepsilon$ for any small $\varepsilon > 0$. If we assume $f(y) = \langle f, y \rangle_H$ for some $f \in H$, then there exists a unique $\{\mathcal{F}_t\}_{t \geq 0}$-adapted solution of (5.1.2) in the space

$$L^2((0,T) \times \Omega; \mathcal{D}(A^{1/4 - \varepsilon})) \cap C((0,T]; L^2(\Omega; \mathcal{D}(A^{1/4 - \varepsilon}))) \cap C([0,T]; L^{4 - \varepsilon}(\Omega; H)),$$

more suggestively,

$$L^2((0,T) \times \Omega; \mathcal{D}(A^{1/4 -})) \cap C((0,T]; L^2(\Omega; \mathcal{D}(A^{1/4 -}))) \cap C([0,T]; L^{4 -}(\Omega; H)).$$

If we take $\mathcal{O} = (0,1)$ and $A = -d^2/dx^2 + 1$, $\mathcal{D}(A) = \{y \in H^2(0,1) : y'(0) = y'(1) = 0\}$, then (5.1.1) may represent

$$\begin{cases} \partial y(t,x)/\partial t = \partial^2 y(t,x)/\partial x^2 - y(t,x), & x \in (0,1), \quad t \geq 0, \\ y(0,x) = y_0(x), & x \in (0,1), \\ \partial y(t,0)/\partial x = -k\langle f, y(t,0)\rangle_H \dot{B}_t, \quad \partial y(t,1)/\partial x = 0, & t \geq 0. \end{cases} \tag{5.1.14}$$

It is easy to see that the solution of

$$y''(x) - y(x) = 0, \quad y'(0) = a, \quad y'(1) = 0,$$

is given by

$$m(x) = a\cosh(1-x)/\sinh 1.$$

Note that

$$T(t)y = e^{-t}\langle y, 1\rangle_H + 2\sum_{n=1}^{\infty} e^{-(1 + n^2\pi^2)t} \cos n\pi x \langle \cos n\pi x, y\rangle_H$$

and

$$m_0 = \int_0^1 m(x)dx = a$$

$$m_n = \int_0^1 \sqrt{2}\cos n\pi x \cdot m(x)dx = \sqrt{2}a/(1 + n^2\pi^2).$$

If we assume $\|f\|_H = 1$, then (5.1.13) with $\theta = 1/2$ yields

$$k^2\left[\frac{1}{2} + \sum_{n=1}^{\infty}\left(\frac{n\pi}{1 + n^2\pi^2}\right)^2\right] < 1.$$

This is satisfied if

$$k^2\left[\frac{1}{2} + \frac{1}{\pi^2}\sum_{n=1}^{\infty}\frac{1}{n^2}\right] < 1.$$

Thus if $k^2 < 3/2$, the mild solution of (5.1.14) is exponentially stable in mean square.

Example 5.1.2   Consider the stochastic parabolic equation

$$\begin{cases} dy(t,x) = [\partial^2 y(t,x)/\partial x^2]dt + k\delta(x - \xi)\langle f, y\rangle_H dB_t, & 0 < x, \ \xi < 1, \\ y(t,0) = y(t,1) = 0, & y(0,x) = y_0(x), \end{cases}$$

(5.1.15)

where $\delta(\cdot)$ is the usual delta function at zero. In this case we take $H = L^2(0,1)$ and $A = -d^2/dx^2$, $\mathcal{D}(A) = H_0^1(0,1) \cap H^2(0,1)$. The semigroup model for (5.1.15) is

$$Y_t = T(t)y_0 + \int_0^t AT(t - s)gk\langle f, Y_s\rangle_H dB_s,$$

(5.1.16)

where g is the Green's function

$$g(x,\xi) = \begin{cases} (1 - \xi)x, \ 0 \leq x < \xi, \\ (1 - x)\xi, \ \xi \leq x \leq 1. \end{cases}$$

Since

$$\int_0^1 \sqrt{2}\sin n\pi x \cdot g(x,\xi)dx = \sqrt{2}\sin n\pi\xi/(n\pi)^2,$$

we have $g \in \mathcal{D}(A^{3/4-})$. Hence, (5.1.16) is written as

$$Y_t = T(t)y_0 + \int_0^t A^\theta T(t - s)bk\langle f, Y_s\rangle_H dB_s,$$

where $\theta = 1/4 + \varepsilon$, $\varepsilon > 0$ and $b = A^{1-\theta}g$. Thus, all the conditions in Ichikawa [7] to ensure the existence and uniqueness of mild solutions are satisfied. To obtain a sufficient condition for stability we assume $\|f\|_H = 1$ and apply (5.1.13) with $\theta = 1/2$ to get

$$\left(\frac{k}{\pi}\right)^2 \sum_{n=1}^\infty \frac{1}{n^2} < 1.$$

Hence, if $k^2 < 6$, the system is exponentially stable in mean square.

5.2   Stochastic Stability and Quadratic Control

Let X_t^u, $t \geq 0$, be a family of properly defined processes in H with $X_0^u = x_0$. The parameter u associated with each member is termed a control. Generally, each u may be identified with a specific member of a given family of functions of X_t^u which takes values of the form $u = K(t, X_t^u)$ depending on t and X_t^u. The object of control theory is to select the control so that the corresponding process possesses some desired properties. For instance, one object may be to transfer an initial state to some target set of H, for instance, $\{0\}$, that is, choose a value of u so that $X_t^u \to 0$ as t approaches some time, possibly infinity. This may be developed further by associating to each control a cost $J(u; x_0)$. A problem of optimal control is to select u so that $X_t^u \to 0$, and meanwhile minimizes the cost $J(u; x_0)$ with respect to other controls of a specified class of the so-called "admissible comparison controls".

For many years, much effort has been devoted to the study of optimal control and system theory for deterministic partial differential equations and functional differential equations, by Butkovskii [1], Curtain and Pritchard [1], Lions [2], Manitius [1] and Wang [2] among others. These equations are also known as distributed parameter systems and can be described by several mathematical models. In this section, we shall focus our attention on systems as above but subject to some random environmental effects. More precisely, we shall formulate stochastic distributed parameter systems as infinite dimensional stochastic differential equations. This formulation is helpful not only in considering the major concepts of controllability, observability and stability in system theory, but in posing various problems in optimal control and filtering. We take this approach based on the theory of semigroups and stochastic evolution equations. In particular, we shall employ dynamic programming methods which enable us to determine the optimal feedback control and study its relationship with the stability property of linear stochastic systems. The design of controls to ensure exponential stability of mild solutions of a class of semilinear stochastic evolution equations is discussed in Section 5.3.

5.2.1 Optimal Control on a Finite Interval

Let $(\Omega, \mathcal{F}, \{\mathcal{F}_t\}_{t \geq 0}, P)$ be a complete probability space and H, K_i, $i = 1, 2, 3$, real separable Hilbert spaces. Consider the controlled stochastic differential equation on H

$$\begin{cases} dX_t = (AX_t + Bu(t))dt + G(X_t)dW_t^1 + FdW_t^2 \\ \qquad + C(u(t))dW_t^3, \quad 0 \leq t \leq T, \\ X_0 = x_0 \in H, \end{cases} \tag{5.2.1}$$

where $0 \leq T < \infty$, A is the infinitesimal generator of a strongly continuous semigroup $T(t)$, $t \geq 0$, on H, $u(t)$ is a control with values in a real separable Hilbert space U, $B \in \mathcal{L}(U, H)$, $G \in \mathcal{L}(H, \mathcal{L}(K_1, H))$, $F \in \mathcal{L}(K_2, H)$, $C \in \mathcal{L}(U, \mathcal{L}(K_3, H))$, W_t^i are independent K_i-valued Wiener processes with covariance operators Q_i, $trQ_i < \infty$, $i = 1, 2, 3$. For each $u(t)$, it is easy to show as in Theorem 1.3.4 that there exists a unique mild solution of (5.2.1) in $C(0, T; L^2(\Omega, \mathcal{F}, P; H))$.

We can define the *quadratic cost functional* as follows:

$$J(u; x_0) = E\langle LX_T, X_T\rangle_H + \int_0^T E\Big\{\langle MX_t, X_t\rangle_H + \langle Nu(t), u(t)\rangle_U\Big\}dt \quad (5.2.2)$$

where $0 \le L$, $M \in \mathcal{L}(H)$ and $0 < N \in \mathcal{L}(U)$ with inverse $N^{-1} \in \mathcal{L}(U)$. Our control problem is then to minimize (5.2.2) over $L^2([0, T] \times \Omega; U)$. In finite dimensions optimal control of this type is well known and is given in terms of the solution of a Riccati equation. This is one of the best examples for which dynamic programming gives a complete solution. We shall show that this is the case in infinite dimensions as well. A *feedback control* is defined as a map $K(t, x) : [0, T] \times H \to U$ such that the integral equation given by (5.2.1) with $u(t) = K(t, X_t)$ has a unique continuous solution X_t, $0 \le t \le T$. If $u(t)$ of this form is adapted to $\{\mathcal{F}_t^W\}$, the σ-field generated by $W_t^i - W_s^i$, $0 \le s \le t \le T$, $i = 1$, 2, 3 with $\int_0^T E\|u(t)\|_U^2 dt < \infty$, we say that $u(t)$ is *admissible*. Note that the feedback control $u = K(t, x)$ is admissible if $K(t, x) : [0, T] \times H \to U$ is measurable in t and satisfies

$$\begin{aligned}
\|K(t, x)\|_U &\le c(1 + \|x\|_H), & c &> 0, & \forall x &\in H, \\
\|K(t, x) - K(t, y)\|_U &\le c\|x - y\|_H, & c &> 0, & \forall x, y &\in H.
\end{aligned} \quad (5.2.3)$$

Indeed, the equation (5.2.1) with $u(t) = K(t, X_t)$ has a unique mild solution X_t, $0 \le t \le T$, and $u(t) = K(t, X_t)$ is admissible.

Now define

$$\begin{aligned}
\langle \Gamma(S)u, v\rangle_U &= tr(C^*(v)SC(u)Q_3), & S &\in \mathcal{L}(U), & u, v &\in U, \\
\langle \Delta(S)u, v\rangle_U &= tr(G^*(v)SG(u)Q_1), & S &\in \mathcal{L}(U), & u, v &\in U,
\end{aligned} \quad (5.2.4)$$

and

$$\begin{aligned}
(\mathbf{L}_u\Lambda)(x) = \langle \Lambda'(x), Ax + Bu\rangle_H + \frac{1}{2}\Big\{\langle \Gamma(\Lambda''(x))u, u\rangle_U + \langle \Delta(\Lambda''(x))x, x\rangle_H \\
+ tr(F^*\Lambda''(x)FQ_2)\Big\}, \qquad x \in \mathcal{D}(A), \quad u \in U,
\end{aligned}$$

for any twice Fréchet differentiable function $\Lambda(x)$ on H. The following theorem gives sufficient conditions for optimality.

Theorem 5.2.1   *Suppose there exists a feedback control $\bar{u} = \bar{K}(t, x)$ and a real function $\Lambda(t, x)$ on $[0, T] \times H$ with the properties:*

(a) *$\Lambda(t, x)$ is continuous on $[0, T] \times H$,*

(b) *$\Lambda(t, x)$ is twice Fréchet differentiable in x, moreover, $\Lambda_x'(t, x)x_1$ and $\langle \Lambda_{xx}''(t, x)x_1, x_2\rangle_H$ are continuous for all x_1, $x_2 \in H$,*

(c) *$\Lambda(t, x)$ is differentiable in t for any $x \in \mathcal{D}(A)$ and $\Lambda_t'(t, x)$ is continuous on $[0, T] \times \mathcal{D}(A)$,*

(d) *$|\Lambda(t, x)| + \|x\|_H\|\Lambda_x'(t, x)\|_H + \|x\|_H\|\Lambda_{xx}''(t, x)\| \le c(1 + \|x\|_H^2)$, $c > 0$, $x \in H$,*

(e) $\Lambda(T, x) = \langle Lx, x \rangle_H$, $x \in H$,

(f)

$$
\begin{aligned}
0 &= \Lambda'_t(t, x) + (\mathbf{L}_{\bar{u}}\Lambda)(t, x) + \langle Mx, x \rangle_H + \langle N\bar{u}, \bar{u} \rangle_U \\
&\leq \Lambda'_t(t, x) + (\mathbf{L}_u\Lambda)(t, x) + \langle Mx, x \rangle_H + \langle Nu, u \rangle_U
\end{aligned}
$$

for any $x \in \mathcal{D}(A)$ *and* $u \in U$,

(g) $\bar{K}(t, x)$ *is continuous and satisfies the condition (5.2.3).*

Then the feedback control $\bar{u} = \bar{K}(t, x)$ *is optimal and the minimal cost is* $J(\bar{u}; x_0) = \Lambda(0, x_0)$.

Proof Let $\bar{X}_t$, $0 \leq t \leq T$, be the mild solution of (5.2.1) with $\bar{u} = \bar{K}(t, x)$. Introducing an approximation of the form (2.2.7) to (5.2.1) and applying Itô's formula to $\Lambda(t, x)$, we can show as in Proposition 2.2.2

$$
E\Lambda(T, \bar{X}_T) - \Lambda(0, x_0) = -\int_0^T E\left\{ \langle M\bar{X}_t, \bar{X}_t \rangle_H + \langle N\bar{u}(t), \bar{u}(t) \rangle_U \right\} dt.
$$

Hence,

$$
\Lambda(0, x_0) = E\langle L\bar{X}_T, \bar{X}_T \rangle_H + \int_0^T E\left\{ \langle M\bar{X}_t, \bar{X}_t \rangle_H + \langle N\bar{u}(t), \bar{u}(t) \rangle_U \right\} dt = J(\bar{u}; x_0).
$$

Repeating the same procedure for the solution X_t of (5.2.1) with arbitrary admissible control $u(t)$, we obtain

$$
\Lambda(0, x_0) \leq E\langle LX_T, X_T \rangle_H + \int_0^T E\left\{ \langle MX_t, X_t \rangle_H + \langle Nu(t), u(t) \rangle_U \right\} dt = J(u; x_0).
$$

Here the inequality follows from (f).

To solve the quadratic cost problem (5.2.1) and (5.2.2), we need to find a function $\Lambda(t, x)$ and a control $\bar{u} = \bar{K}(t, x)$ satisfying the conditions of Theorem 5.2.1. We seek a function $\Lambda(t, x)$ of the form

$$
\Lambda(t, x) = \langle Q(t)x, x \rangle_H + q(t), \qquad Q(t) \in \mathcal{L}(H).
$$

Then (f) yields the following Riccati equation:

$$
\begin{aligned}
\frac{d}{dt}\langle Q(t)x, x \rangle_H &+ 2\langle Ax, Q(t)x \rangle_H + \langle Mx, x \rangle_H + \langle \Delta(Q(t))x, x \rangle_H \\
&- \langle Q(t)B[N + \Gamma(Q(t))]^{-1}B^*Q(t)x, x \rangle_H = 0, \quad x \in \mathcal{D}(A), \\
&\quad Q(T) = L,
\end{aligned}
\tag{5.2.5}
$$

$$
q(t) = \int_t^T tr(F^*Q(r)FQ_2)dr
\tag{5.2.6}
$$

and the feedback control:

$$\bar{u} = -[N + \Gamma(Q(t))]^{-1}B^*Q(t)x. \tag{5.2.7}$$

So if we can establish a solution of the Riccati equation (5.2.5) shown below, the proof is complete. ⬜

Theorem 5.2.2   *The Riccati equation (5.2.5) has a unique solution in the class of self-adjoint nonnegative strongly continuous $\mathcal{L}(H)$-valued functions. The control law (5.2.7) is optimal and the minimum cost is*

$$J(\bar{u}; x_0) = \langle Q(0)x_0, x_0 \rangle_H + \int_0^T tr(F^*Q(t)FQ_2)dt.$$

Proof   In a manner parallel to finite dimensional cases, take the sequence of linear differential equations:

$$\frac{d}{dt}\langle Q_0(t)x, x \rangle_H + 2\langle Ax, Q_0(t)x \rangle_H + \langle [M + \Delta(Q_0(t))]x, x \rangle_H = 0, \quad x \in \mathcal{D}(A),$$
$$Q_0(T) = L;$$
$$\frac{d}{dt}\langle Q_n(t)x, x \rangle_H + 2\langle [A - BK_{n-1}(t)]x, Q_n(t)x \rangle_H + \langle [M + \Delta(Q_n(t))]x, x \rangle_H$$
$$+\langle K_{n-1}^*(t)[N + \Gamma(Q_n(t))]K_{n-1}(t)x, x \rangle_H = 0, \quad x \in \mathcal{D}(A),$$
$$Q_n(T) = L;$$

$$\tag{5.2.8}$$

$$K_n(t) = [N + \Gamma(Q_n(t))]^{-1}B^*Q_n(t).$$

By a similar argument to Proposition 2.2.1, we can show that these equations are equivalent to the following integral equations

$$Q_0(t)x = \int_0^T T(r - t)[M + \Delta(Q_0(r))]T(r - t)xdr$$
$$+T^*(T - t)LT(T - t)x, \quad x \in H,$$
$$Q_n(t)x = \int_0^T U_n^*(r, t)\{M + \Delta(Q_n(r)) \tag{5.2.9}$$
$$+K_{n-1}^*(r)[N + \Gamma(Q_{n-1}(r))]K_{n-1}(r)\}U_n(r, t)xdr$$
$$+U_n^*(T, t)LU_n(T, t)x, \quad x \in H,$$

where $U_n(t, s)$ is the perturbation of $T(t)$ by $-BK_{n-1}(t)$. Moreover, they have a unique solution. As in Proposition 2.2.2 we can show

$$J(u_n; x_0) = \langle Q_n(0)x_0, x_0 \rangle_H + \int_0^T tr(F^*Q_n(t)FQ_2)dt$$

where u_n is the control law

$$u_n = -K_n(t)x, \qquad n = 1, 2, \cdots, \quad u_0 = 0.$$

Next we shall show $Q_{n-1}(t) \geq Q_n(t) \geq 0$, $n = 1, 2, \cdots$. Set $R_n(t) = Q_{n-1}(t) - Q_n(t)$, then it satisfies

$$\frac{d}{dt}\langle R_n(t)x, x\rangle_H + 2\langle [A - BK_{n-1}(t)]x, R_n(t)x\rangle_H + \langle \Delta(R_n(t))x, x\rangle_H$$
$$+\langle K_{n-1}^*(t)[N + \Gamma(Q_{n-1}(t))]K_{n-1}(t)x, x\rangle_H = 0, \quad x \in \mathcal{D}(A), \quad (5.2.10)$$

$$R_n(T) = 0;$$

$$R_n(t)x = \int_0^T U_n^*(r,t)\{\Delta(R_n(r))$$
$$+K_{n-1}^*(r)[N + \Gamma(Q_{n-1}(r))]K_{n-1}(r)\}U_n(r-t)x dr, \qquad (5.2.11)$$

$$x \in \mathcal{D}(A).$$

But (5.2.11) has a unique solution $R_n(t) \geq 0$, thus necessarily $Q_{n-1}(T) \geq Q_n(t)$. Since $Q_n(t)$, $n = 1, 2, \cdots$, is the sequence of monotonically decreasing nonnegative operators, there exists a limit $Q(t)$. Passing to the limit $n \to \infty$ in (5.2.9) and then differentiating it, we can show that $Q(t)$ satisfies (5.2.5). Letting $n \to \infty$ in (5.2.9) yields

$$Q(t)x = \int_0^T U^*(r,t)\{M + \Delta(Q(r)) + K^*(r)[N + \Gamma(Q(r))]K(r)\}U(r,t)x dr$$
$$+U^*(T,t)LU(T,t)x,$$
$$K(t) = [N + \Gamma(Q(t))]^{-1}B^*Q(t),$$

$$(5.2.12)$$

where $U(t,s)$ is the perturbation of $T(t)$ by $-BK(t)$ and we have used the strong convergence of $U_n(t,s)$. The uniqueness of a solution of (5.2.5) (and hence (5.2.12)) and the rest of the theorem follows from Theorem 5.2.1. $\quad\square$

5.2.2 Optimal Control on an Infinite Interval

In this subsection, we first take $F = 0$ in (5.2.1) and consider the resulting equation

$$\begin{cases} dX_t = (AX_t + Bu(t))dt + G(X_t)dW_t^1 + C(u(t))dW_t^3, & 0 \leq t < \infty, \\ X_0 = x_0 \in H, \end{cases}$$

$$(5.2.13)$$

and the quadratic cost functional

$$J(u; x_0) = \int_0^\infty E\Big\{\langle MX_t, X_t\rangle_H + \langle u(t), u(t)\rangle_U\Big\}dt. \qquad (5.2.14)$$

For admissible controls we take the class of feedback controls $u = K(t, x)$ such that

(i) $K(t,x): [0, \infty) \times H \to U$ is measurable and for some $c > 0$,

$$\|K(t,x)\|_U \le c(1 + \|x\|_H), \quad \|K(t,x) - K(t,y)\|_U \le c\|x - y\|_H,$$

for any x, $y \in H$, and

(ii) $E\|X_t\|_H^2 \to 0$ as $t \to \infty$, where X_t, $t \ge 0$, is the mild solution of (5.2.13) with $u = K(t,x)$.

Definition 5.2.1   The system (5.2.13) (or (A, B, C, G)) is *stabilizable* if there exists $K \in \mathcal{L}(H,U)$ such that the feedback law $u = -Kx$ yields a L^2-stable mild solution X_t in mean, i.e.,

$$\int_0^\infty E\|X_t(x_0)\|_H^2 dt < \infty.$$

In this case, we also say that $(A - BK, C, G)$ is stable.

If (A, B, C, G) is stabilizable, then the control problem (5.2.13), (5.2.14) is meaningful one. In particular, similarly to Theorem 2.2.1, we may obtain the following results:

Lemma 5.2.1 (A, B, C, G) *is stabilizable if and only if there exists* $K \in \mathcal{L}(H,U)$ *and* $0 \le P \in \mathcal{L}(H)$ *such that*

$$2\langle (A - BK)x, Px \rangle_H + \langle [K^*\Gamma(P)K + \Delta(P)]x, x \rangle_H = -\langle x, x \rangle_H, \quad x \in \mathcal{D}(A),$$

where $\Gamma(P)$ and $\Delta(P)$ are defined by (5.2.4).

Lemma 5.2.2   *If $(A - BK, C, G)$ is stable, then there exists $0 \le Q \in \mathcal{L}(H)$ such that*

$$\begin{aligned}
2\langle (A - BK)x, \ Qx \rangle_H &+ \langle (M + K^*NK)x, x \rangle_H \\
&+ \langle [\Delta(Q) + K^*\Gamma(Q)K]x, x \rangle_H = 0, \quad x \in \mathcal{D}(A).
\end{aligned} \quad (5.2.15)$$

Proof   Let $Q_T(t)$, $0 \le t \le T$, be the unique solution of

$$\begin{aligned}
\tfrac{d}{dt}\langle Q(t)x, \ x \rangle_H &+ \langle (A - BK)x, Q(t)x \rangle_H + \langle (M + K^*NK)x, x \rangle_H \\
&+ \langle [\Delta(Q(t)) + K^*\Gamma(Q(t))K]x, x \rangle_H = 0, \quad x \in \mathcal{D}(A), \quad (5.2.16)
\end{aligned}$$
$$Q(T) = 0.$$

Similarly to Proposition 2.2.2, we then have

$$\langle Q_T(0)x_0, x_0 \rangle_H = \int_0^T E\Big\{ \langle MX_t, X_t \rangle_H + \langle NKX_t, KX_t \rangle_H \Big\} dt$$

where X_t is the solution of (5.2.13) with $u = -Kx$. Since $(A - BK, C, G)$ is stable, $Q_T(0)$ is uniformly bounded in T. But $Q_T(0)$ is monotonically increasing and non-negative. So there exists a limit $Q \geq 0$ and it satisfies (5.2.15). □

Theorem 5.2.3 *Suppose that there exists an admissible control $\bar{u} = -\bar{K}(x)$ and a real-valued function $\Lambda(x)$ on H such that*

(a) *$\Lambda(x)$ is twice Fréchet differentiable and $\Lambda(x)$, $\Lambda'(x)$, $\Lambda''(x)$ are continuous,*

(b) *$|\Lambda(x)| + \|x\|_H \|\Lambda'(x)\|_H + \|x\|_H^2 \|\Lambda''(x)\| \leq c\|x\|_H^2$, $x \in H$, $c > 0$,*

(c)

$$
\begin{aligned}
0 &= (\mathbf{L}_{\bar{u}}\Lambda)(x) + \langle Mx, x\rangle_H + \langle N\bar{u}, \bar{u}\rangle_U \\
&\leq (\mathbf{L}_u\Lambda)(x) + \langle Mx, x\rangle_H + \langle Nu, u\rangle_U
\end{aligned}
$$

for any $x \in \mathcal{D}(A)$ and $u \in U$,

(d) *$\|\bar{K}(x) - \bar{K}(y)\|_U \leq c\|x - y\|_H$, $x, y \in H$, $c > 0$.*

Then $\bar{u} = -\bar{K}(x)$ is optimal and $J(\bar{u}; x_0) = \Lambda(x_0)$.

Proof As in Theorem 5.2.1 one can show

$$
E\Lambda(\bar{X}_t) - \Lambda(x_0) = -\int_0^t E\Big\{\langle M\bar{X}_r, \bar{X}_r\rangle_H + \langle N\bar{u}(r), \bar{u}(r)\rangle_U\Big\}dr
$$

where $\bar{X}_t$ is the mild solution of (5.2.13) with $u = \bar{u}$. Note that $|\Lambda(x)| \leq c\|x\|_H^2$ and $E\|\bar{X}_t\|_H^2 \to 0$ as $t \to \infty$. So

$$
\Lambda(x_0) = \int_0^\infty E\Big\{\langle M\bar{X}_t, \bar{X}_t\rangle_H + \langle N\bar{u}(t), \bar{u}(t)\rangle_U\Big\}dt = J(\bar{u}; x_0).
$$

In a similar way, for any admissible control u we obtain

$$
\Lambda(x_0) \leq \int_0^\infty E\Big\{\langle MX_t, X_t\rangle_H + \langle NX_t, X_t\rangle_H\Big\}dt = J(u; x_0).
$$

Now we seek a function $\Lambda(x)$ of the form

$$
\Lambda(x) = \langle Qx, x\rangle_H, \qquad 0 \leq Q \in \mathcal{L}(H).
$$

Then (c) above yields an algebraic Riccati equation

$$
2\langle Ax, Qx\rangle_H + \langle\{M + \Delta(Q) - QB[N + \Gamma(Q)]^{-1}B^*Q\}x, x\rangle_H = 0, \quad x \in \mathcal{D}(A), \tag{5.2.17}
$$

and the control law

$$
u = -Kx, \quad K = [N + \Gamma(Q)]^{-1}B^*Q. \tag{5.2.18}
$$

Hence, to conclude our proof, we need only establish conditions which guarantee the existence of a solution to (5.2.17) and the admissibility of $\bar{u} = -Kx$.

Claim: If there exist operators $K \in \mathcal{L}(H, U)$, $0 \le Q \in \mathcal{L}(H)$ satisfying (5.2.15), then the Riccati equation (5.2.17) has a solution. Indeed, let $Q_T(t)$, $Q'_T(t)$ be the solutions of (5.2.14) and of (5.2.5) with $L = 0$, respectively. Then

$$0 \le Q'_T(t) \le Q_T(t) \le Q.$$

Since $Q'_T(t)$ is monotonically increasing in T, there exists a limit of $Q'_T(t)$ as $T \to \infty$ which is independent of t and satisfies (5.2.17). ⬚

Now we are in a position to summarize one of the main results in this subsection as follows.

Theorem 5.2.4   *Suppose that there exists some $K \in \mathcal{L}(H, U)$ such that $(A - BK, C, G)$ is stable. Then there is a unique solution $Q \ge 0$ for the Riccati equation (5.2.17). Moreover, the optimal control for (5.2.13) is the feedback law (5.2.18) and $J(\bar{u}; x_0) = \langle Qx_0, x_0 \rangle_H$.*

Proof   The existence of a solution to (5.2.17) follows from Theorem 5.2.3. Since $\langle Qx_0, x_0 \rangle_H$ is the minimum cost, Q is unique. ⬚

If we keep the term F in (5.2.13), we may consider optimal stationary controls in terms of invariant measures. Precisely, we consider the stochastic control system (5.2.1) together with the class of feedback controls $K(x)$: $H \to U$ with

$$\|K(x) - K(y)\|_U \le c\|x - y\|_H, \quad c > 0, \quad x, y \in H. \tag{5.2.19}$$

We say that the feedback control $u = -K(x)$ is *admissible* if it satisfies (5.2.19) and the Markov process given by (5.2.1) with $u = -K(x)$ has an invariant measure μ_K such that

$$\int_H \|x\|_H^2 \mu_K(dx) < \infty. \tag{5.2.20}$$

Our control problem is to minimize

$$J(u; x_0) = \int_H \left\{ \langle Mx, x \rangle_H + \langle NK(x), K(x) \rangle_U \right\} \mu_K(dx)$$

over all admissible controls. Clearly, the cost is independent of the initial value x_0, thus it is also denoted by $J(u)$.

Theorem 5.2.5   *Suppose that there exist an admissible control $\bar{u} = -\bar{K}(x)$, a number γ and a real-valued function $\Lambda(x)$ on H such that*

(a) $\Lambda(x)$ is twice Fréchet differentiable and $\Lambda(x)$, $\Lambda'(x)$, $\Lambda''(x)$ are continuous,

(b) $|\Lambda(x)| + \|x\|_H\|\Lambda'(x)\|_H + \|x\|_H^2\|\Lambda''(x)\| \le c(1 + \|x\|_H^2)$, $x \in H$, $c > 0$,

(c)

$$\gamma = (\mathbf{L}_{\bar{u}}\Lambda)(x) + \langle Mx, x\rangle_H + \langle N\bar{u}, \bar{u}\rangle_U$$
$$\le (\mathbf{L}_u\Lambda)(x) + \langle Mx, x\rangle_H + \langle Nu, u\rangle_U$$

for any $x \in \mathcal{D}(A)$ and $u \in U$.

Then $\bar{u} = -\bar{K}(x)$ is optimal and $J(\bar{u}) = \gamma$.

Proof   Let $\bar{X}_t(\xi)$ be the mild solution of (5.2.1) with $u = -\bar{K}(x)$ and $x_0 = \xi$. As in Theorem 5.2.1 we can derive

$$E\Lambda\left(\bar{X}_t(\xi)\right) - \Lambda(\xi)$$
$$= \int_0^t \left\{\gamma - E\left[\langle M\bar{X}_r(\xi), \bar{X}_r(\xi)\rangle_H + \langle N\bar{K}(\bar{X}_r(\xi)), \bar{K}(\bar{X}_r(\xi))\rangle_U\right]\right\}dr.$$
$$(5.2.21)$$

Now let $\bar{\mu}$ be an invariant measure of $\bar{X}_t$ and $P(\cdot, t, \cdot)$ the transition function. Then

$$\int_H E\Lambda(\bar{X}_t(\xi))\bar{\mu}(d\xi) = \int_H \left[\int_H \Lambda(\eta)P(\xi, t, d\eta)\right]\bar{\mu}(d\xi) = \int_H \Lambda(\xi)\bar{\mu}(\xi).$$

Taking expectations of (5.2.21) with respect to $\bar{\mu}$ yields that

$$0 = \int_0^t \left\{\gamma - \int_H \left[\langle M\xi, \xi\rangle_H + \langle N\bar{K}(\xi), \bar{K}(\xi)\rangle_U\right]\bar{\mu}(d\xi)\right\}dr.$$

Hence,

$$\gamma = \int_H \left[\langle Mx, x\rangle_H + \langle N\bar{K}(x), \bar{K}(x)\rangle_U\right]\bar{\mu}(dx) = J(\bar{u}).$$

Similarly, for any admissible control $u = -K(x)$ we have

$$\gamma \le J(u).$$

☐

It is natural as we did in Theorem 5.2.3 to seek a function $\Lambda(x)$ of the form

$$\Lambda(x) = \langle Qx, x\rangle_H.$$

Then (c) above yields the Riccati equation (5.2.17), the control law (5.2.18) and

$$\gamma = tr(F^*QFQ_2).$$

In particular, by a similar argument to those in Theorems 5.3.3 and 5.3.4 we can prove:

Theorem 5.2.6   *Suppose that there exists some $K \in \mathcal{L}(H, U)$ such that $(A - BK, C, G)$ is stable. Then the optimal control is the feedback law (5.2.18) and the minimum cost is $J(\bar{u}) = tr(F^*QFQ_2)$ where $Q \geq 0$ is the unique solution of the Riccati equation (5.2.17).*

Example 5.2.1   Consider the scalar stochastic difference differential equation

$$\begin{cases} dy(t) = [a_0 y(t) + a_1 y(t - b) + u(t)]dt + \sigma y(t)dB_t, \\ y(\theta) = y_0(\theta), \quad -b \leq \theta \leq 0, \quad a_0 > 0, \ a_1 > 0, \ b > 0, \end{cases}$$

and the cost functional

$$J(u) = \int_0^\infty [y^2(t) + Nu^2(t)]dt, \quad N > 0,$$

where B_t, $t \geq 0$, is a standard one dimensional Brownian motion. Let us take the feedback control $u = -ky$. It is possible to show that all the conditions in Theorem 5.2.4 are satisfied for k sufficiently large. Then the feedback control $\bar{u}(\cdot) = -N^{-1}[Q_{00}y(t) + \int_{-b}^0 Q_{01}(\theta)y(t + \theta)d\theta]$ is optimal where

$$Q = \begin{pmatrix} Q_{00} & Q_{01} \\ Q_{10} & Q_{11} \end{pmatrix}$$

and kernels $Q_{01}(\theta)$, $Q_{11}(\theta)$ of Q_{01}, Q_{11} and Q_{00} satisfy the following:

$$2Q_{00}a_0 + Q_{10}(0) + Q_{01}(0) + 1 - N^{-1}Q_{00}^2 + \sigma^2 Q_{00} = 0,$$

$$\frac{dQ_{01}(\theta)}{d\theta} = (a_0 - N^{-1}Q_{00})Q_{01}(\theta) + a_0 Q_{11}(\theta, 0), \quad Q_{01}(-b) = a_1 Q_{00},$$

$$\left(\frac{\partial}{\partial\theta} + \frac{\partial}{\partial\eta}\right)Q_{11}(\theta, \eta) - N^{-1}Q_{10}(\theta)Q_{01}(\eta) = 0,$$

$$Q_{11}(-b, \eta) = a_1 Q_{01}(\eta), \quad Q_{11}(\theta, -b) = Q_{10}(\theta)a_1.$$

The reader is also referred to Delfour, McCalla and Mitter [1] for some more details of this formulation.

5.3   Feedback Stabilization of Stochastic Differential Equations

If a control problem is posed as a well-formulated mathematical optimization problem, as in the introduction to Section 5.2, then it is natural at least to attempt to compute the optimizing control. Owing to the difficulty of the computational problem, this is not always possible. In addition, practical control problems are not usually posed as well-formulated mathematical optimization problems.

The goal which the control is designed to accomplish may be phrased somewhat loosely. One may desire a control which will guarantee that a given target set is attained with probability one at some random time. It may also be desired that the control, which accomplishes a given task, does not take "large" values with a high probability. In finite dimensional cases, it is a long story in the use of Lyapunov function methods to design controls which will satisfy such qualitative requirements. See, for example, Geiss [1] for the deterministic and Kushner [1], [2] for the stochastic cases. In this section, based on the perturbation theory of semigroups and some stability results established in Section 3.3 we shall present a theory to design a proper state feedback control law $u(t)$, $t \geq 0$, so that the resulting process has exponential stability uniformly with respect to perturbations.

Let H, U and K be three real separable Hilbert spaces. The question of stability of the semilinear stochastic evolution equation

$$dX_t = (AX_t + F(X_t))dt + G(X_t)dW_t, \qquad X_0 = x_0, \quad t \geq 0,$$

has been considered in Section 3.3, where A generates a strongly continuous semigroup $T(t)$, $t \geq 0$, on H, $F(\cdot)$ and $G(\cdot)$ are nonlinear and satisfy Lipschitz conditions with $F(0) = 0$ and $G(0) = 0$. That is, F and G are bounded perturbations. But in many practical situations, F and G might be unbounded, even uncertain. The problem of stabilization is to design a state feedback control law that assures exponential stability of the zero state uniformly with respect to the perturbations F and G. Precisely, we wish to consider the following controlled stochastic system

$$dX_t = (AX_t + F(X_t))dt + G(X_t)dW_t + Bu(t)dt, \qquad X_0 = x_0, \quad t \geq 0, \quad (5.3.1)$$

on H. Here B is a bounded linear operator from U to H, and $G : H \rightarrow \mathcal{L}(K, H)$ is nonlinear and $F : H \rightarrow H$ is nonlinear unbounded operators. As usual, W_t, $t \geq 0$, is some K-valued process with covariance operator Q such that $trQ < \infty$.

Definition 5.3.1 A linear operator L from a subset $\mathcal{D}(L) \subset H$, domain of L, to H is said to be *dissipative* if

$$\langle x, Lx \rangle_H \leq 0 \qquad \text{for all} \qquad x \in \mathcal{D}(L).$$

The following perturbation result for linear operators is standard but essential in the arguments which follow. The reader is referred to, for instance, Pazy [1], for its proof.

Lemma 5.3.1   *Let L and T be linear operators in H such that $\mathcal{D}(L) \subset \mathcal{D}(T)$. Assume that L is the infinitesimal generator of a C_0-semigroup of contractions. Suppose T is dissipative and satisfies*

$$\|Tx\|_H \le a\|Lx\|_H + b\|x\|_H \quad for \quad x \in \mathcal{D}(L)$$

where $0 \le a < 1$ and $b \ge 0$. Then $L + T$ is the infinitesimal generator of a C_0-semigroup of contractions.

We shall impose the following assumptions on the operators A, $F(\cdot)$ and $G(\cdot)$ of (5.3.1).

A1. The pair $\{A, B\}$ is exponentially stabilizable, i.e., there exists a $D_1 \in \mathcal{L}(H, U)$ such that $\bar{A} := A + BD_1$ generates an exponentially stable contraction semigroup $S(t)$, $t \ge 0$, in H satisfying $\|S(t)\| \le e^{-\mu t}$, $t \ge 0$, for some $\mu > 0$.

A2. $F(\cdot) = F_1(\cdot) + F_2(\cdot)$, where $F_1(\cdot) \in \{F : F$ is a linear operator in H and $\mathcal{D}(A) \subset \mathcal{D}(F) \subset H$, $\langle x, Fx \rangle_H \le k\langle x, Ax \rangle_H$, $k \ge 0$, and there exist constants $0 \le a < 1$, $b \ge 0$ such that $\|F(x)\|_H \le a\|Ax\|_H + b\|x\|_H$ for all $x \in \mathcal{D}(A)\}$, and $F_2(\cdot) \in \{F : F$ maps H into H and satisfies Lipschitz conditions with $F(0) = 0$, i.e., there exists a contant $l_1 > 0$ such that $\|F(x) - F(y)\|_H \le l_1\|x - y\|_H$, $x, y \in H\}$.

A3. The linear operator BB^* is coercive, i.e., $\langle x, BB^*x \rangle_H \ge \theta\|x\|_H^2$ for some $\theta > 0$ and $x \in H$, where B^* is the adjoint operator of B.

A4. The operator $G : H \to \mathcal{L}(K, H)$ is nonlinear and satisfies Lipschitz conditions with $G(0) = 0$, i.e., $\|G(x) - G(y)\| \le l_2\|x - y\|_H$, $x, y \in H$ for some constant $l_2 > 0$.

Theorem 5.3.1   *Suppose the assumptions A1, A2, A3 and A4 hold. Then the mild solution of the system (5.3.1) with a linear feedback control law given by $u(t) = D_2 X_t$, $t \ge 0$, where*

$$D_2 := (1 + k)D_1 - \frac{1}{2\theta}(2l_1 + l_2^2 \cdot trQ)B^* \in \mathcal{L}(H, U), \qquad (5.3.2)$$

is exponentially stable in mean square uniformly with respect to the perturbations $F(\cdot)$ and $G(\cdot)$. That is,

$$E\|X_t(x_0)\|_H^2 \le M\|x_0\|_H^2 e^{-\lambda t} \quad for \ some \quad M \ge 1 \quad and \quad \lambda > 0.$$

Moreover, the null solution is also pathwise exponentially stable in the almost sure sense. That is, there exists a random variable $0 < T(\omega) < \infty$ and constants $M_1 \ge 1$, $\lambda_1 > 0$ such that for all $t \ge T(\omega)$,

$$\|X_t(x_0)\|_H \le M_1\|x_0\|_H e^{-\lambda_1 t} \qquad a.s.$$

Proof   Using the feedback control law (5.3.2), define

$$A_r = (A + BD_1) + (F_1 + kBD_1) = \bar{A} + A_p,$$
$$\bar{F} = F_2 - [(2l_1 + l_2^2 \cdot trQ)/2\theta]BB^*,$$

then the system (5.3.1) deduces to

$$dX_t = [A_r X_t + \bar{F}(X_t)]dt + G(X_t)dW_t, \quad X_0 = x_0 \in H, \quad t \geq 0, \quad (5.3.3)$$

with $\bar{F}(0) = 0$ and $G(0) = 0$.

First of all, we show that A_r generates a C_0-semigroup. By the assumption A1, the semigroup $S(t)$, $t \geq 0$, generated by $\bar{A}$ satisfies $\|S(t)\| \leq e^{-\mu t} \leq 1$ for $t \geq 0$. Therefore, if one can show that $A_r = \bar{A} + A_p$ is dissipative and A_p is relatively bounded with respect to $\bar{A}$ with relative bound $a < 1$, it is easy to deduce by Lemma 5.3.1 that A_r generates a C_0-semigroup of contractions in H. Indeed, by means of A2 and Proposition 2.1.4 we have

$$\langle x, A_r x \rangle_H \leq \langle x, F_1 x \rangle_H + k\langle x, BD_1 x \rangle_H$$
$$\leq k\langle x, (A + BD_1)x \rangle_H$$
$$\leq -k\mu\|x\|_H^2 \leq 0 \quad \text{for} \quad x \in \mathcal{D}(A) \subset \mathcal{D}(A_r).$$

Hence, A_r is dissipative. On the other hand, we have

$$\|A_p x\|_H \leq \|F_1 x\|_H + k\|BD_1 x\|_H$$
$$\leq a\|Ax\|_H + b_1\|x\|_H$$
$$\leq a\|\bar{A}x\|_H + b_2\|x\|_H \quad \text{for} \quad x \in \mathcal{D}(A) = \mathcal{D}(\bar{A}) \subset \mathcal{D}(A_p),$$

where $b_1 = b + k\|BD_1\|$, $b_2 = b_1 + a\|BD_1\|$. Hence, $A_r = \bar{A} + A_p$ generates a C_0-semigroup of contractions in H.

In order to conclude the proof, using the assumptions A1-A4, we obtain

$$2\langle x, (A + BD_2)x + F(x) \rangle_H + tr(G(x)QG^*(x))$$

$$\leq 2\langle x, (A + BD_1)x \rangle_H + 2\langle x, (F_1 + kBD_1)(x) \rangle_H + 2\left\langle x, F_2(x) - \frac{l_1}{\theta}BB^*x \right\rangle_H$$

$$+ trQ\|G(x)G^*(x)\| - \frac{l_2^2 \cdot trQ}{\theta}\langle x, BB^*x \rangle_H$$

$$\leq -2(1+k)\mu\|x\|_H^2 = -d\|x\|_H^2, \quad \text{for} \quad x \in \mathcal{D}(A),$$

where $d = 2(1+k)\mu > 0$. By means of Theorems 3.3.1 and 3.3.2, it is easy to complete the proof. $\qquad\square$

We shall present a stochastic version analogous to Burgers' mathematical models of turbulence to illustrate the theory derived above.

Example 5.3.1   Consider a controlled semilinear stochastic diffusion equation

$$\begin{cases} dz_t(x) - c\Delta z_t(x)dt + f(z_t(x))dt - u(t,x)dt - g(z_t(x))dB_t = 0, \\ \quad t \geq 0, \quad x \in \mathcal{O} = (0,1), \\ z_t(x)|_{\partial\mathcal{O}} = 0, \quad t \geq 0; \quad z_0(x) = h(x) \quad \text{for} \quad x \in (0,1), \end{cases}$$

which describes a large class of physical problems such as stochastic heat transfer, chemical diffusions and turbulences. Here B_t, $t \geq 0$, is a one dimensional standard Brownian motion. In particular, let $c = 1/R$, $R > 0$,

$$f(z) = -\frac{6x}{\sqrt{R}} \cdot \frac{\partial z(x)}{\partial x} - \left(x^2 + \frac{6}{\sqrt{R}}\right)z(x) - R^2 \left(\int_0^1 z^2(x)dx\right)^{1/2}, \quad h(x) = \sin x,$$

and $g(\xi) = \sin \xi$. On this occasion, one may associate with this equation in an obvious way a semilinear stochastic evolution equation of parabolic type

$$\begin{cases} dX_t = (AX_t + F(X_t))dt + Bu(t)dt + G(X_t)dB_t, \\ X_0(x) = \sin x, \end{cases}$$

on the real Hilbert space $H = L^2(0,1)$ where $A = (1/R)(d^2/dx^2)+x^2+(6/\sqrt{R})$ with $\mathcal{D}(A) = \{z \in H : z', z'' \in H, z(0) = z(1) = 0\}$ and $B = I$. It is also clear that for $v, w \in H$,

$$\|G(v) - G(w)\|_H \leq \|v - w\|_H. \tag{5.3.4}$$

If we assume that $R = 2$ and $D_1 = -1$, it is easy to deduce by Proposition 2.1.4 that $\bar{A} = A + BD_1$ generates an exponentially stable contraction semigroup in H. Indeed,

$$\langle z, \bar{A}z \rangle_H \leq \left(-\frac{\pi^2}{2} + x^2 + \frac{6}{\sqrt{2}} - 1\right)\|z\|_H^2 \leq -\frac{1}{2}\|z\|_H^2$$

for $z \in \mathcal{D}(\bar{A}) = \mathcal{D}(A)$. The operator $F(z)$ is given by

$$F(z) = F_1(z) + F_2(z) = \frac{6x}{\sqrt{2}} \cdot \frac{dz(x)}{dx} + 4\|z\|_H.$$

Since

$$\langle z, F_1(z) \rangle_H = \frac{6}{\sqrt{2}} \int_0^1 xz(x)\frac{dz(x)}{dx}dx$$

$$= \frac{3}{\sqrt{2}}xz^2(x)\Big|_0^1 - \frac{3}{\sqrt{2}}\int_0^1 z^2(x)dx,$$

therefore,

$$\langle z, F_1(z) \rangle_H = -\frac{3}{\sqrt{2}}\int_0^1 z^2(x)dx \leq 0,$$

$$z(x) \in \mathcal{D}(F_1) = \{z \in H : z'(x) \in H, \ z(0) = z(1) = 0\}.$$

Thus F_1 is dissipative. Furthermore,

$$\|F_1(z)\|_H^2 = 18 \int_0^1 x^2 \left(\frac{dz(x)}{dx}\right)^2 dx.$$

Hence, it is easy to verify that for any $\varepsilon > 0$,

$$\|F_1(z)\|_H \leq \varepsilon \|Az\|_H + \nu(\varepsilon)\|z\|_H \quad \text{for} \quad z \in \mathcal{D}(A),$$

where $\nu(\varepsilon) \to \infty$ as $\varepsilon \to 0$, i.e., F_1 is bounded with A-bound zero in H. For $F_2(z)$,

$$\|F_2(v) - F_2(w)\|_H = 4|\|v\|_H - \|w\|_H| \leq 4\|v - w\|_H \quad \text{for} \quad v, \ w \in H.$$

By choosing $L_1 = 4$ and $k = 0$ in D_2, it is easy to see, using (5.3.4), that all the conditions of Theorem 5.3.1 are satisfied. Hence, the feedback system is exponentially stable in the mean square and almost sure senses.

5.4 Stochastic Models in Mathematical Physics

This section is devoted to the problem of existence and uniqueness of invariant measures for two classes of important stochastic models, stochastic reaction-diffusion equations and Navier-Stokes equations, under various conditions on the coefficients of equations, motivated by applications. To this end, we use the results derived in Section 3.5 in the spirit of Lyapunov functions.

5.4.1 Stochastic Reaction-Diffusion Equations

Assume $\mathcal{O} \subset \mathbf{R}^n$ is a bounded domain with smooth boundary $\partial\mathcal{O}$. Let $A(x)$ denote a second-order elliptic operator in $\mathcal{O}$ defined as

$$A(x)\psi(x) = \sum_{i,j=1}^n \frac{\partial}{\partial x_i}\left[a_{ij}(x, \nabla\psi)\frac{\partial\psi(x)}{\partial x_j}\right] \quad \text{for any} \quad \psi \in C^2(\mathcal{O}),$$

where $\nabla\psi = (\frac{\partial\psi}{\partial x_1}, \cdots, \frac{\partial\psi}{\partial x_n})$ and a_{ij}'s are arbitrarily given functions on $\mathcal{O} \times \mathbf{R}^n$ such that the matrix $[a_{ij}]_{n \times n}$ is positive-definite. Let $f(x, r, y)$ and $\sigma_i(x, r, y)$, $i = 1, 2, \cdots, m$, be real-valued continuous functions in $(x, r, y) \in \mathcal{O} \times \mathbf{R}^1 \times \mathbf{R}^n$ and $W_t^i(x)$, for $i = 1, \cdots, m$, be independent Wiener processes in $H = L^2(\mathcal{O})$ with norm $\|\cdot\|_H$ so that $W_t = (W_t^1, \cdots, W_t^m)$ is a Q-Wiener process in $K = H^m = H \times \cdots \times H$. The kernel matrix of the covariance operator Q of W_t is denoted by $q(x, y) = [\delta_{ij}q_i(x, y)]_{m \times m}$, where δ_{ij} is the Kronecker delta function and q_i is the covariance kernel of W_t^i. We assume that q_i's are continuous in $\mathcal{O} \times \mathcal{O}$ such that $q_i(x) = q_i(x, x)$ satisfies

$$\sup_{\substack{x \in \mathcal{O} \\ i=1,\cdots,m}} q_i(x) \le q_0.$$

Now consider the parabolic Itô equation of the form:

$$du(t, x) = A(x)u(t, x)dt + f(x, u(t, x), \nabla u)dt + \sum_{i=1}^{m} \sigma_i(x, u(t, x), \nabla u)dW_t^i(x),$$

$$u(t, \cdot)\big|_{\partial \mathcal{O}} = 0, \quad t \ge 0,$$
$$u(0, x) = \xi(x), \quad x \in \mathcal{O},$$

$$\tag{5.4.1}$$

where $\xi(\cdot)$ is supposed to satisfy

$$\|\xi\|_H^2 = \int_{\mathcal{O}} |\xi(x)|^2 dx < \infty.$$

In Equation (5.4.1), the operator $A(x)$ on u is to be interpreted in a variational sense as follows:

$$\langle Au, v \rangle_H = \sum_{i,j=1}^{n} \int_{\mathcal{O}} a_{ij}(x, \nabla u)\left(\frac{\partial u}{\partial x_i}\right)\left(\frac{\partial v}{\partial x_j}\right)dx, \quad \forall v \in C_b^1(\mathcal{O}). \tag{5.4.2}$$

For our purpose, let us impose the following conditions:

(P.1). The functions a_{ij} and f are continuous and there exist constants $\alpha > 0$, $p \ge 2$ and $q \ge 2$ such that

$$\sum_{i,j=1}^{n} a_{ij}(x, y)y_i y_j + |rf(x, r, y)| \le \alpha(1 + \|y\|_{\mathbf{R}^n}^p + |r|^q),$$

$$r \in \mathbf{R}^1, \quad x \in \mathcal{O}, \quad y \in \mathbf{R}^n.$$

Then the operator
$$\tilde{A}(v) = A(\cdot)v + f(\cdot, \cdot, \nabla v) \tag{5.4.3}$$

is well defined as a continuous operator from $V = W_0^{1,p}(\mathcal{O}) \cap L^q(\mathcal{O})$ into its dual V^*. For $w \in K = H^m$, we set

$$B(v)w = \sum_{i=1}^{m} \sigma_i(\cdot, \cdot, \nabla v)w_i.$$

Then $B(\cdot) : V \to \mathcal{L}_2^0(K, H)$ if we assume that:

(P.2). $\sigma_i(x, r, y)$'s are continuous in x and locally Lipschitz continuous in r and y such that

$$\|\sigma(x, r, y)\|_{\mathbf{R}^m}^2 \le c(1 + \|y\|_{\mathbf{R}^n}^2 + |r|^2) \quad \text{for some} \quad c > 0.$$

In addition, suppose that the following two conditions hold:

(P.3). There exist constants $\alpha > 0$ and θ, $\gamma \in \mathbf{R}^1$ such that

$$\sum_{i,j=1}^n a_{ij}(x, y) y_i y_j - r f(x, r, y) - \frac{q_0}{2} \|\sigma(x, r, y)\|_{\mathbf{R}^m}^2 \ge \alpha \|y\|_{\mathbf{R}^n}^p + \theta |r|^2 + \gamma$$

for any $r \in \mathbf{R}^1$, $x \in \mathcal{O}$ and $y \in \mathbf{R}^n$;

(P.4). The operator $\tilde{A} : V \to V^*$ defined by (5.4.3) is monotone in the sense that

$$2\langle u - v, \tilde{A}(u) - \tilde{A}(v) \rangle_{V,V^*} + q_0 \|\sigma(\cdot, \cdot, \nabla u) - \sigma(\cdot, \cdot, \nabla v)\|_K^2 \le -\delta \|u - v\|_V^2,$$

for some $\delta \ge 0$ and arbitrary u, $v \in V$.

Under the conditions (P.1)–(P.4), it is easy to check that all the assumptions of Corollary 3.6.1 are satisfied. Hence, there exists an invariant measure for Equation (5.4.1). If, in Condition (P.4), the constant δ is strictly positive, then the condition (3.6.12) in Corollary 3.6.1 is true so that the invariant measure is unique. Precisely, the following result holds.

Proposition 5.4.1   *If the conditions (P.1)–(P.4) hold, then the parabolic Itô equation (5.4.1) has an invariant measure μ with support in $W_0^{1,p}(\mathcal{O}) \cap L^q(\mathcal{O})$. Moreover, if $\delta > 0$ in (P.4), the invariant measure μ is unique.*

Let $A = A_0$ be a linear strongly elliptic operator in $\mathcal{O} \subset \mathbf{R}^3$ with $a_{ij}(x, y) = \tilde{a}_{ij}(x) \in C_b(\mathcal{O})$ and there exists $\alpha_0 > 0$ such that

$$\sum_{j=1}^3 \tilde{a}_{ij}(x) y_i y_j \ge \alpha_0 \|y\|_{\mathbf{R}^3}^2, \qquad \forall x \in \mathcal{O}, \ y \in \mathbf{R}^3. \tag{5.4.4}$$

Consider the following stochastic reaction-diffusion equation:

$$\begin{cases} du(t, x) = A_0(x) u(t, x) dt + f(u) dt + \sum_{j=1}^3 \sigma_j(x) \frac{\partial u}{\partial x_j} dB_t^j(x) \\ \qquad\qquad + \sigma_4(x) dB_t^4(x), \qquad t \ge 0, \qquad x \in \mathcal{O}, \\ u(t, \cdot)\big|_{\partial \mathcal{O}} = 0, \quad t \ge 0, \\ u(0, x) = \xi(x) \in L^2(\mathcal{O}), \end{cases} \tag{5.4.5}$$

where $f(u) = (\mu u - \nu u^3)$, μ and ν are positive constants, $\sigma_i(x)$'s $\in C_b(\mathcal{O})$ and $B_t^i(x)$'s are independent Wiener random fields with covariance functions q_i's, $i = 1, \cdots, 4$.

Let $H = L^2(\mathcal{O})$, $V = H_0^1(\mathcal{O}) \cap L^4(\mathcal{O})$ and $K = H^4$. It is easy to check that the conditions (P.1)–(P.3) hold for $p = 2$ and $q = 4$. To verify (P.4), we note that, with $\tilde{A} = A_0 + f$ and by (5.4.4) and (5.4.5),

$$
\begin{aligned}
2\langle u - v, \tilde{A}(u) - \tilde{A}(v)\rangle_{V,V^*} &+ q_0\|\sigma(u) - \sigma(v)\|_K^2 \\
&\leq -2\alpha_0\|u - v\|_V^2 + 2\nu\|u - v\|_H^2 + 3q_0\sigma_0^2\|u - v\|_V^2 \\
&\leq -(2\alpha_0 - 2\nu/\lambda_0 - 3q_0\sigma_0^2)\|u - v\|_V^2, \quad \forall u, \ v \in V,
\end{aligned}
$$

where

$$
\lambda_0 = \inf_{\substack{v \in H_0^1 \\ \|v\|_H \neq 0}} \frac{\|v\|_V^2}{\|v\|_H^2} > 0, \qquad \|v\|_V^2 = \int_{\mathcal{O}} \|\nabla v(x)\|_{\mathbf{R}^3}^2 \, dx,
$$

and $\sigma_0 = \sup_{x \in \mathcal{O}} \sigma_i(x)$ for $i = 1, 2, 3, 4$. Hence, (P.4) holds for ν and q_0 so small that

$$
\delta = 2\alpha_0 - 2\nu/\lambda_0 - 3q_0\sigma_0^2 > 0.
$$

In this case, Equation (5.4.5) has a unique invariant measure supported in $H_0^1(\mathcal{O}) \cap L^4(\mathcal{O})$ by Proposition 5.4.1.

Now we may consider a generalization of Equation (5.4.1) where the elliptic operator $A_0(x)$ is replaced by the nonlinear operator

$$
A(x)v(x) = A_0(x)v(x) + A_1v(x), \tag{5.4.6}
$$

where

$$
A_1v(x) = \sum_{j=1}^3 \frac{\partial}{\partial x_j}\left(\left|\frac{\partial v(x)}{\partial x_j}\right|^{p-2}\frac{\partial v(x)}{\partial x_j}\right), \quad p > 2.
$$

Then the corresponding stochastic reaction-diffusion equation is

$$
\begin{aligned}
\frac{\partial u(t,x)}{\partial t} &= \sum_{i,j=1}^3 \frac{\partial}{\partial x_i}\left[\left(\tilde{a}_{ij}(x) + \delta_{ij}\left|\frac{\partial u(t,x)}{\partial x_j}\right|^{p-2}\right)\frac{\partial u(t,x)}{\partial x_j}\right] + f(u(t,x)) \\
&\quad + \sum_{i=1}^3 \sigma_i(x)\frac{\partial u(t,x)}{\partial x_j}\dot{B}_t^j(x) + \sigma_4\dot{B}_t^4(x).
\end{aligned} \tag{5.4.7}
$$

It is well known that the nonlinear elliptic operator is coercive and monotone in $W_0^{1,p}(\mathcal{O})$ (cf. see Lions [1]). Let $V = W_0^{1,p} \cap L^4(\mathcal{O})$. Under the same conditions as in (5.4.4) and (5.4.5), we can apply Proposition 5.4.1 to conclude the existence of a unique invariant measure for Equation (5.4.5) with support in $W_0^{1,p} \cap L^4(\mathcal{O})$.

5.4.2   Stochastic Navier-Stokes Equations

The theory of Navier-Stokes equations occupies a central position in the study of nonlinear partial differential equations, dynamical systems, and modern scientific computation, as well as classical fluid dynamics. The mathematical theory of stochastic Navier-Stokes equations is rather technical and already very large despite being still incomplete. In spite of their intuitive physical background from fluids, these stochastic equations have also been extensively studied without direct reference to the conventional theory of turbulence. Some pioneer work in this respect goes back at least to Bensoussan and Teman [1] and Viot [1] among others. Some recent developments of this expanding topic can be found in the existing literature such as Da Prato and Zabczyk [1], [2] and references therein.

As an application of Lyapunov function approaches, we investigate in this subsection invariant measures of a randomly perturbed Navier-Stokes equation describing a two-dimensional turbulent flow.

Let $\mathcal{O} \subset \mathbf{R}^2$ be a bounded domain with a smooth boundary $\partial\mathcal{O}$. Let $u(t, x) = (u_1, u_2)(t, x)$ be the velocity field and $p(t, x)$ be the pressure field in an incompressible fluid. Then, under a random perturbation by Gaussian white noises, a fluid flow is governed by the stochastic Navier-Stokes equation:

$$du_i(t, x) + \sum_{j=1}^{2} u_j \frac{\partial u_i(t, x)}{\partial x_j} dt = -\frac{1}{\rho} \frac{\partial p(t, x)}{\partial x_i} dt + \nu \sum_{j=1}^{2} \frac{\partial^2 u_i(t, x)}{\partial x_j^2} dt$$

$$+ \sigma_i dB_t^i(x),$$

$$\sum_{j=1}^{2} \frac{\partial u_j(t, x)}{\partial x_j} = 0, \quad x \in \mathcal{O}, \quad i = 1, 2, \quad t \geq 0,$$

where $\rho > 0$ is the constant fluid density, $\nu > 0$ the kinematic viscosity, σ_i's the variance parameters, and $B_t(x) = (B_t^1(x), B_t^2(x))$ is a random force with associated covariance operators Q_i, $trQ_i < \infty$, given by a positive definite kernel

$$q_i(x, y) \in L^2(\mathcal{O} \times \mathcal{O}), \qquad q_i(x, x) \in L^2(\mathcal{O}), \qquad i = 1, 2.$$

In a vectorial notation, the above equation takes a simpler form:

$$\begin{cases} du(t, x) + (u \cdot \nabla)u dt = -\frac{1}{\rho}\nabla p dt + \nu \Delta u dt + \sigma dB_t(x), & t > 0, \quad x \in \mathcal{O}, \\ \nabla \cdot u = 0, \end{cases}$$

$$(5.4.8)$$

where ∇, $\nabla\cdot$ and Δ are the conventional notations for the gradient, divergence and Laplacian operators, respectively, and

$$\sigma = \begin{pmatrix} \sigma_1 & 0 \\ 0 & \sigma_2 \end{pmatrix}.$$

The equation is subject to the initial-boundary conditions:

$$\begin{cases} u(0, x) = \xi(x), \\ u|_{\partial\mathcal{O}} = 0, \end{cases}$$

where $\xi(x)$ is some proper initial velocity field. Let $\mathcal{C}_0^\infty = \{v \in [C_0^\infty(\mathcal{O})]^2 : \nabla \cdot v = 0\}$ and H the closure of $\mathcal{C}_0^\infty$ in $[L^2(\mathcal{O})]^2$, $V = \{v \in [H_0^1(\mathcal{O})]^2 : \nabla \cdot v = 0\}$.

Let V^* denote the dual of V, then it is known that $V \hookrightarrow H \hookrightarrow V^*$ and the embeddings are compact.

Let Π be the orthogonal projector from $[L^2(\mathcal{O})]^2$ unto H, and define for any $v \in \mathcal{C}_0^\infty$,

$$A(v) = \nu\Pi\Delta v - \Pi[(v \cdot \nabla)v].$$

Then $A(\cdot)$ can be extended as a continuous operator from V to V^*. The above equation can be recast as a stochastic differential equation in the form:

$$\begin{cases} du(t) = A(u(t))dt + \sigma dW_t, \\ u(0) = \xi, \quad \xi \in V, \end{cases} \tag{5.4.9}$$

where W_t is a Q-Wiener process in H instead of $[L^2(\mathcal{O})]^2$.

Indeed, let $u \in V$ be a solution of (5.4.8). By the well-known embedding theorem for Sobolev spaces, we obtain

$$E\Big\|\sum_{j=1}^2 u_j \frac{\partial u}{\partial x^j}\Big\|^2_{[L^2(\mathbf{R}_+ \times \mathcal{O})]^2} \leq C \sup_{t\geq 0} E\|u(t)\|^2_{[L^1(\mathcal{O})]^2} \int_0^\infty E\|u(t)\|^2_{[L^2(\mathcal{O})]^2} dt$$

for some $C > 0$. This estimate implies

$$E\Big\|\sum_{j=1}^2 u_j \frac{\partial u}{\partial x^j}\Big\|^2_{[L^2(\mathbf{R}_+ \times \mathcal{O})]^2} \leq ME\|u\|^2_{1,2},$$

where $M > 0$ does not depend on u. Hence, all the terms of the system (5.4.8) containing $u(t, \cdot)$ for any $t \geq 0$ belong to $L^2(\Omega \times \mathcal{O})$, and $\nabla p \in [L^2(\Omega \times \mathbf{R}_+ \times \mathcal{O})]^2$. Therefore, the operator Π can be applied to both sides of (5.4.8). Since

$$\int_{\mathcal{O}} \nabla p \cdot w dx = -\int_{\mathcal{O}} p \operatorname{div} w dx = 0,$$

for any $w \in V$, then $\Pi\nabla p(t, \cdot) = 0$ for almost all $t \geq 0$. Note that by the definition of V, we have

$$\Pi \partial u/\partial t = \partial u/\partial t$$

for almost all $t \geq 0$. Let $-\nu\Pi\Delta H^2 = \{-\nu\Pi\Delta v : v \in H^2(\mathcal{O})\}$ and then we intend to show

$$-\nu\Pi\Delta H^2 = H.$$

But this is immediate. Indeed, it is known (cf. Solonnikov [1]) that for any $f \in H$, there exists $v \in H^2(\mathcal{O})$ such that

$$-\nu\Pi\Delta v = f \quad \text{and} \quad \|v\|_{H^2} \leq C\|f\|_H,$$

where $C > 0$ does not depend on $f \in H$. Hence,

$$-\nu\Pi\Delta H^2 \supset H.$$

Clearly, $-\Pi\Delta H^2 \subset H$. Applying Π to (5.4.8), we see easily from the above that u is a solution of (5.4.8).

On the other hand, by employing a Galerkin approximation technique it can be deduced (cf. Vishik and Fursikov [1], p. 347) that the problem (5.4.9) above has a unique strong solution $\{u^\xi(t); t \geq 0\}$ satisfying:

$$E\|u^\xi(T)\|_H^2 + \nu E \int_0^T \sum_{i=1}^2 \left\|\frac{\partial u^\xi(t)}{\partial x_i}\right\|_H^2 dt \leq E\|\xi\|_H^2 + \frac{T}{2}trQ.$$

Hence, using the fact that $\|u^\xi(t)\|_V$ is equivalent to $\left(\sum_{i=1}^2 \left\|\frac{\partial u^\xi(t)}{\partial x_i}\right\|_H^2\right)^{1/2}$, we get

$$\liminf_{T\to\infty} \frac{1}{T} \int_0^T E\|u^\xi(t)\|_V^2 dt \leq \frac{C}{2\nu}trQ$$

where $C > 0$ is some constant. Using Markov's inequality, we obtain

$$\lim_{R\to\infty} \liminf_{T\to\infty} \frac{1}{T} \int_0^T P\left\{\|u^\xi(t)\|_V > R\right\} dt = 0$$

which, together with Corollary 3.5.2, immediately yields the desired result, namely the existence of an invariant measure for the stochastic Navier-Stokes equation.

The existence of an invariant measure for the Stochastic Navier-Stokes problem was also studied by sophisticated asymptotic analysis in Da Prato and Zabczyk [2] and, for the periodic boundary condition in Albeverio and Cruzerio [1] by the Galerkin approximation and the method of averaging. However, it is worth mentioning that the uniqueness and ergodicity questions cannot be answered by the method shown here and have to be dealt with by different approaches, for instance, that one in Flandoli and Maslowski [1].

5.5 Stochastic Systems Related to Multi-Species Population Dynamics

We shall study a system of two interacting populations in which each population density follows a stochastic partial differential equation. Precisely, consider two populations both living in a bounded domain $\mathcal{O} \subset \mathbf{R}^n$. The evolution of their densities u_i, $i = 1, 2$, defined as number of individuals per unit volume or area, is a result of three competing factors. Firstly, both populations can migrate in $\mathcal{O}$ according to a macroscopic diffusion described by the Laplacian Δ. As the second factor, the interaction between these populations is modelled by the functions

$$f_i(u_1, u_2) = u_i - a_i u_i^2 \pm b_i u_1 u_2, \tag{5.5.1}$$

where $u_i \in \mathbf{R}^1$, a_i, $b_i \geq 0$, $i = 1$, 2. This just implies that in the absence of the other population, the considered one will grow in accordance with the logistic law. The term $\pm b_i u_1 u_2$ describes two typical interactions, namely the *predator-prey* model when the signs in front of b_i are different and the *competition* one when the sign is negative in both terms. As the third factor, it is assumed that each population is randomly distributed by Gaussian noise $\dot{B}_t^i$, $t \geq 0$, defined on a complete probability space $(\Omega, \mathcal{F}, P)$ with a state-dependent noise intensity $\sigma_i(u_i)$, $i = 1$, 2. Thus, the system in which we are interested can be modelled as:

$$\begin{cases} \frac{\partial}{\partial t} u_1(t,x) = \nu_1 \Delta u_1(t,x) + f_1(u_1(t,x), u_2(t,x)) + \sigma_1(u_1(t,x)) \dot{B}_t^1(x), \\ \frac{\partial}{\partial t} u_2(t,x) = \nu_2 \Delta u_2(t,x) + f_2(u_1(t,x), u_2(t,x)) + \sigma_2(u_2(t,x)) \dot{B}_t^2(x), \end{cases}$$
$$(5.5.2)$$

ν_1, $\nu_2 > 0$, $t \in (0,T]$, $x \in \mathcal{O}$, where $T > 0$ is an arbitrary number. In addition, u_i, $i = 1$, 2, satisfy the initial condition

$$u_i(0,x) = \theta_i(x) \geq 0, \quad x \in \bar{\mathcal{O}},$$

where $\theta_i : \Omega \times \bar{\mathcal{O}} \to \mathbf{R}_+$ and here $\bar{\mathcal{O}} = \mathcal{O} \cup \partial\mathcal{O}$ is the closure of $\mathcal{O}$ in $\mathbf{R}^n$. Moreover, u_i, $i = 1$, 2, satisfy homogeneous Dirichlet boundary conditions, i.e.,

$$u_i(t,x) = 0, \quad x \in \partial\mathcal{O}, \quad t \in [0,T],$$

or homogeneous Neumann conditions, i.e.,

$$\frac{\partial}{\partial n} u_i(t,x) = 0, \quad x \in \partial\mathcal{O}, \quad t \in [0,T],$$

where n denotes the normal vector. We require that the boundary $\partial\mathcal{O}$ is smooth enough to guarantee that a Green function to $\frac{\partial}{\partial t} - \Delta$ with the corresponding boundary conditions does exist. The formal problem (5.5.2) is a stochastic counterpart of a logistic population growth model with migration (cf. Murray [1]).

Let B' be a Borel subset of some separable Banach space B and H be a separable Hilbert space. Consider a strong Markov process $(V(t), U(t))_{t \geq 0}$ in the product space $B' \times H$ such that its second component U is governed by the stochastic evolution equation in H

$$U(t) = T(t)u_0 + \int_0^t T(t-s)F(V(s), U(s))ds + \int_0^t T(t-s)G(U(s))dW_s \quad (5.5.3)$$

where $u_0 \in H$, $T(t)$, $t \geq 0$, is some strongly continuous semigroup on H with its infinitesimal generator A (generally unbounded) and W stands for a Wiener process with a trace class incremental covariance operator Q in a separable Hilbert space K. Both $F : B' \times H \to H$ and $G : H \to \mathcal{L}(K, H)$ are Lipschitz continuous. Furthermore, the process $V(t)$, $t \geq 0$, is assumed to be pathwise continuous in B such that

$$\int_0^T E\|V(t)\|_B^p dt < \infty$$

holds for some $p > 2$ and each $T \geq 0$.

Define the operator $\mathbf{L} : C^2(H) \to C(B' \times \mathcal{D}(A))$ by

$$(\mathbf{L}\Lambda)(g, h) = \langle Ah, \Lambda'(h) \rangle_H + \langle F(g, h), \Lambda'(h) \rangle_H + \frac{1}{2} Tr[\Lambda''(h)G(h)QG^*(h)]$$

for any $\Lambda \in C^2(H)$, $(g, h) \in B' \times \mathcal{D}(A)$. Furthermore, let $K(r)$ be a centered open ball in H with radius $r > 0$. In addition to employing a strong solution approximation type of argument, the proofs of the following theorem can be similarly worked out to Theorem 3.3.3.

Theorem 5.5.1   *Assume $F(g, \cdot) = 0$ and $G(0) = 0$ for each $g \in B$. Suppose there exists a mapping $\Lambda \in C^2(H \backslash \{0\})$ satisfying that for each $r > 0$ and $h \in H \backslash K(r)$*

$$|\Lambda(h)| + \|\Lambda'(h)\|_H + \|\Lambda''(h)\| \leq c(r)(1 + \|h\|_H^q), \tag{5.5.4}$$

for some $q > 0$, $c(r) > 0$ and

$$(\mathbf{L}\Lambda)(g, h) \leq -b(\|g\|_B)\Lambda(h), \quad (g, h) \in B' \times (\mathcal{D}(A) \backslash \{0\}), \tag{5.5.5}$$

where $b : \mathbf{R}_+ \to \mathbf{R}_+$ is continuous such that

$$b(u) \leq k(1 + u), \quad u \in \mathbf{R}_+$$

for some $k > 0$ and

$$\liminf_{t \to \infty} b(\|V(t)\|_B) > 0 \quad a.s. \tag{5.5.6}$$

Moreover, let Λ satisfy

$$l_r := \inf_{\|z\|_H > r} \Lambda(z) > 0 \tag{5.5.7}$$

for each $r > 0$, $\Lambda(0) = 0$ and Λ is continuous at zero. Then the solution U of (5.5.3) has the property

$$\lim_{t \to \infty} \|U(t)\|_H = 0$$

almost surely for every initial point $u_0 \in H$.

Next, we shall use Theorem 5.5.1 to derive some sufficient Lyapunov type conditions for stability of the null solution to the system (5.5.2). In particular, we shall consider the predator-prey case, i.e.,

$$f_1(u_1, u_2) = u_1 - a_1 u_1^2 + b_1 u_1 u_2,$$
$$f_2(u_1, u_2) = u_2 - a_2 u_2^2 - b_2 u_1 u_2,$$

where $u_i \in \mathbf{R}_+$, a_i, $b_i \geq 0$, $i = 1, 2$. The corresponding results for the competition can be similarly obtained.

Let $H = L^2(\mathcal{O})$, $A_i = \nu_i\Delta$ with domains equal either to $H^2(\mathcal{O}) \cap H_0^1(\mathcal{O})$ or to $\{\phi \in H^2(\mathcal{O}) : \frac{\partial}{\partial n}\phi(x) = 0,\ x \in \partial\mathcal{O}\}$ in the case of Dirichlet or Neumann boundary conditions, respectively. Furthermore, let

$$(G_i(\phi)\psi)(x) = \sigma_i(\phi(x))\psi(x), \quad \phi,\ \psi \in H, \quad x \in \bar{\mathcal{O}},$$

be the multiplication operators defined by the real-valued mappings σ_i, $i = 1, 2$. Denote by Q_i the covariance operators of W^i, $i = 1, 2$. The basic idea to obtain stability is a comparison of the solution to (5.5.2) with an essentially simpler system which is given by

$$\begin{cases} dU_1(t) = (A_1U_1(t) + U_1(t) + b_1U_1(t)U_2(t))dt + G_1(U_1(t))dW_t^1, \\ dU_2(t) = (A_2U_2(t) + U_2(t))dt + G_2(U_2(t))dW_t^2, \end{cases} \quad (5.5.8)$$

in the space $H \times H$ with initial conditions $U_1(0) = \theta_1$ and $U_2(0) = \theta_2$. For our purpose, we impose the following conditions:

(i) $\theta_i : \Omega \times \bar{\mathcal{O}} \to \mathbf{R}^1$ are measurable, pathwise continuous and there exists a deterministic constant $k_\theta > 0$ such that

$$0 \le \theta_i(\omega, x) \le k_\theta$$

for any $\omega \in \Omega$ and $x \in \bar{\mathcal{O}}$, $i = 1, 2$.
(ii) The mappings $\sigma_i : \mathbf{R}^1 \to \mathbf{R}^1$, $i = 1, 2$, are globally Lipschitz continuous.
(iii) $\sigma_i(0) = 0$, $i = 1, 2$.

The strongly continuous semigroups defined by their infinitesimal generators A_i have the form

$$(T_i(t)\phi)(x) = \int_{\mathcal{O}} G_i(t, x, y)\phi(y)dy, \quad \phi \in H, \quad x \in \mathcal{O}, \quad t \ge 0,$$

where $G_i(t, x, y) = g(\nu_i t, x, y)$, $i = 1, 2$, and g is the Green function to $\frac{\partial}{\partial t} - \Delta$. It can be shown that under the conditions (i)–(iii), the equation (5.5.2) has a pathwise unique nonnegative solution. So does (5.5.8) since the system (5.5.8) represents a particular case of (5.5.2). In what follows we need a comparison theorem for a single stochastic partial differential equation with progressively measurable random coefficients. This is certainly a stochastic version of the standard finite dimensional stochastic comparison principle. The reader is referred to, for instance, Manthey and Zausinger [1] for its proofs.

Consider the following two stochastic equations ($k = 1, 2$)

$$\frac{\partial}{\partial t}v^{(k)}(t, x) = (Av^{(k)})(t, x) + F^{(k)}(\omega, t, x, v^{(k)}(t, x))$$

$$+ G(\omega, t, x, v^{(k)}(t, x))\dot{W}_t(x),$$

where $t \ge 0$, $x \in \mathcal{O}$ with initial data $\theta^{(k)} : \Omega \times \bar{\mathcal{O}} \to \mathbf{R}^1$ and homogeneous Dirichlet or Neumann boundary conditions and $A = \nu \cdot \Delta$, $\nu > 0$.

Theorem 5.5.2 *Assume that $\theta^{(k)}$ are uniformly bounded and let $F^{(k)}$ and G be uniformly Lipschitz continuous and uniformly bounded, $k = 1, 2$. If*

$$\theta^{(1)} \geq \theta^{(2)} \quad a.s.$$

and

$$F^{(1)}(\omega, t, x, v) \geq F^{(2)}(\omega, t, x, v) \quad a.s.$$

where $(t, x) \in [0, \infty) \times \bar{\mathcal{O}}$, $v \in \mathbf{R}^1$, then $v^{(1)} \geq v^{(2)}$ holds almost surely.

By using Theorem 5.5.2, we can carry out a truncation procedure for solutions to show that for the equations (5.5.2) and (5.5.8),

$$\begin{aligned} 0 \leq u_1(t, x) \leq U_1(t, x), \\ 0 \leq u_2(t, x) \leq U_2(t, x), \end{aligned} \tag{5.5.9}$$

almost surely for any $(t, x) \in \mathbf{R}_+ \times \bar{\mathcal{O}}$.

Next we show a stability assertion that requires stability of the null solution for the second component of (5.5.3) in a stronger sense. This result is then used to derive stability for both components of (5.5.2) in H.

Theorem 5.5.3 *Suppose that for some nonnegative initial function $\theta_2 \in C(\bar{\mathcal{O}})$, we have*

$$\lim_{t \to \infty} \|U_2(t)\|_{C(\bar{\mathcal{O}})} = 0 \tag{5.5.10}$$

almost surely, and let one of the following conditions (a) and (b) be satisfied

(a)

$$-\beta_1 + 1 + \frac{1}{2} k_{\sigma_1}^2 Tr Q_1 < 0, \tag{5.5.11}$$

where $-\beta_1$ stands for the first eigenvalue of the operator A_1 and k_{σ_1} for the Lipschitz constant of σ_1;

(b) W^1 is a standard scalar Wiener process,

$$\|\sigma_1(u)\|_H \geq \kappa u, \quad u \geq 0,$$

for some $\kappa \geq 0$ and

$$-\beta_1 + \frac{1}{2} k_{\sigma_1}^2 - \kappa^2 < 0. \tag{5.5.12}$$

Then every solution (u_1, u_2) of (5.5.2) such that $u_1(0, x) = \theta_1(x)$ with an arbitrary nonnegative $\theta_1 \in C(\bar{\mathcal{O}})$ and $u_2(0, x) = \theta_2$ satisfies

$$\lim_{t \to \infty} \|u_1(t, \cdot)\|_H = 0 \quad a.s.$$

and

$$\lim_{t \to \infty} \|u_2(t, \cdot)\|_{C(\bar{\mathcal{O}})} = 0 \quad a.s.$$

Proof   Obviously, according to (5.5.9) it is enough to show $\|U_1(t)\|_H \to 0$ as $t \to \infty$ almost surely. To this end, we use Theorem 5.5.1 with the function $\Lambda(h) = \|h\|_H^{2\gamma}$ where $\gamma > 0$ will be specified later on. The conditions on Λ required in Theorem 5.5.1 are satisfied trivially except the condition (5.5.5), which is to be checked now with $B' = B = C(\bar{\mathcal{O}})$. In the case (a), letting $\gamma = 1$, we get

$$(\mathbf{L}\Lambda)(g, h) = 2(\langle A_1 h, h\rangle_H + \|h\|_H^2 + b_1\langle g \cdot h, h\rangle_H)$$
$$+ \frac{1}{2}Tr[G_1(h)Q_1 G_1^*(h)], \quad (g, h) \in B \times \mathcal{D}(A_1),$$

where $g \cdot h$ means the usual multiplication of functions g and h. It follows that

$$(\mathbf{L}\Lambda)(g, h) \le 2\|h\|_H^2\left(-\beta_1 + 1 + b_1\|g\|_{C(\bar{\mathcal{O}})} + \frac{1}{2}k_{\sigma_1}^2 TrQ_1\right), \quad (g, h) \in B \times \mathcal{D}(A_1).$$

Thus letting

$$b(u) := 2\left(\beta_1 - 1 - \frac{1}{2}k_{\sigma_1}^2 TrQ_1 - b_1 u\right), \quad u \in \mathbf{R}_+,$$

we obtain (5.5.5). Conditions (5.5.10) and (5.5.11) trivially yield (5.5.5) in the present case, so the assumptions of Theorem 5.5.1 are verified and we obtain $\|U_1(t)\| \to 0$, $t \to \infty$ almost surely which proves the case (a).

In the case (b), we have

$$(\mathbf{L}\Lambda)(g, h) \le 2\gamma\|h\|_H^{2\gamma}\left(-\beta_1 + 1 + b_1\|g\|_{C(\bar{\mathcal{O}})} + \frac{1}{2}k_{\sigma_1}^2 + (\gamma - 1)\kappa^2\right),$$

for any $(g, h) \in B \times (\mathcal{D}(A_1)\backslash\{0\})$, so we obtain (5.5.5) with

$$b(u) := 2\gamma\left(\beta_1 - 1 - \frac{1}{2}k_{\sigma_1}^2 + (1 - \gamma)\kappa^2 - b_1 u\right), \quad u \in \mathbf{R}_+.$$

Hence, choosing $\gamma > 0$ small enough and taking into account (5.5.10) and (5.5.12), we see that (5.5.6) in Theorem 5.5.1 is verified and, consequently, $\|U_1(t)\|_H \to 0$, $t \to \infty$ almost surely which proves the case (b).   □

By virtue of Theorem 5.5.3, in order to show stability of the zero solution of (5.5.2) it suffices to prove stability for a single equation for U_2 in (5.5.8) where the "drift" is linear. However, the convergence of U_2 in the $C(\bar{\mathcal{O}})$-norm is needed, which is more difficult to prove than the analogous convergence in H. In the following theorem, we present some results on stability in $C(\bar{\mathcal{O}})$ for the second component in (5.5.8). Together with Theorem 5.5.3, this will give sufficient conditions for the stability of the original system (5.5.2).

Theorem 5.5.4   *Suppose that W^2 is a standard scalar Wiener process.*

(a) *Consider the second equation in (5.5.8) with Neumann boundary conditions and assume*

$$\kappa u \leq \|\sigma_2(u)\|_H, \quad u \geq 0,$$

for some $0 \leq \kappa \leq k_{\sigma_2}$, *and*

$$1 + \frac{1}{2}k_{\sigma_1}^2 - \kappa^2 < 0.$$

Then

$$\lim_{t \to \infty} \|U_2(t)\|_{C(\bar{\mathcal{O}})} = 0 \quad a.s.$$

(b) *If* $\sigma_2(u) = k_{\sigma_2}u$ *and*

$$1 - \frac{1}{2}k_{\sigma_2}^2 < 0,$$

then

$$\lim_{t \to \infty} \|U_2(t)\|_{C(\bar{\mathcal{O}})} = 0 \quad a.s.$$

Proof The second equation of (5.5.8) with Neumann boundary conditions and one-dimensional Wiener process is spatially homogeneous. So if we start from a spatially constant initial function the solution evolves in the space of constant functions. Therefore, taking a constant $\bar{\theta}_2 > 0$ such that $\bar{\theta}_2 \geq \theta_2(x)$, $x \in \bar{\mathcal{O}}$, a comparison theorem argument yields

$$U_2(t, x) \leq v_2(t), \quad x \in \bar{\mathcal{O}},$$

where $v_2(t)$ solves the one-dimensional stochastic differential equation

$$dv_2(t) = v_2(t)dt + \sigma_2(v_2(t))dW_t^2$$

with $v_2(0) = \bar{\theta}_2$. A simple one-dimensional version of the proof of Thorem 5.5.3 yields $v_2(t) \to 0$ as $t \to \infty$ almost surely.

In (b) the Neumann boundary condition is no longer demanded. To prove this assertion note that the Itô formula implies

$$U_2(t, x) = (\tilde{T}_2(t)\theta_2)(x) \cdot \exp\left[(1 - k_{\sigma_2}^2/2)t + k_{\sigma_2}W_t^2\right]$$

for $x \in \bar{\mathcal{O}}$, where $\tilde{T}_2(t)$ is the semigroup generated by A_2 in the space $C(\bar{\mathcal{O}})$. Obviously, $\tilde{T}_2(t)$ is a contraction semigroup and

$$\exp\left\{\left[(1 - k_{\sigma_2}^2/2) + k_{\sigma_2}W_t^2 t^{-1}\right]t\right\} \to 0,$$

as $t \to \infty$ almost surely since $W_t^2/t \to 0$ almost surely. Hence, $\|U_2(t)\|_{C(\bar{\mathcal{O}})} \to 0$ as $t \to \infty$, which completes this proof. □

The interpretation of the above stability results in terms of a predator-prey system is very natural: in case of the prey is extinction, the predator that would have died out if the prey did not exist, dies out in presence of the prey as well.

In the competition case the situation is much easier. We can choose here $b_1 = 0$ in the first equation of (5.5.8) and obtain a system of two decoupled equations. Hence, all results for the second equation summarized in Theorem 5.5.4 are true for both populations.

5.6   Notes and Comments

Semigroup models have been adopted by many researchers to deal with controlled partial differential equations for a long time. For instance, such a formulation for parabolic equations with boundary control has been extensively studied in Balakrishnan [1] and Lasiecka [1] among others. Some important problems such as quadratic control and stabilizability have been investigated by Lasiecka and Triggiani [1], [2]. Following their approaches, a semigroup model (5.1.2) for parabolic equations with boundary and pointwise noise is proposed in Ichikawa [7]. Semigroup models for boundary noise can be also found in Curtain [3] and Zabczyk [3]. Curtain [3] deals with the model (5.1.2) in which $f(y)$ is independent of y and Zabczyk [3] gives a model in which boundary values satisfy stochastic differential equations. The existence and uniqueness of invariant measures for a class of stochastic evolution equations with boundary and pointwise noise has been investigated by Maslowski [2]. The control problem of (5.1.2) can be also considered as in Da Prato and Ichikawa [1].

The existence of an optimal control law is closely connected with the behavior of the Riccati operator equations as shown in Section 5.2.1. The quadratic problem (5.2.13) is discussed without using Theorem 5.2.3 in Ichikawa [1], and the optimal control problem with average cost is also investigated. For discrete-time quadratic control problems, Zabczyk [4] studies the existence of an optimal stationary strategy for a general system described by a linear difference equation. The uniqueness of the stationary measure related to this strategy is also investigated. There is an important problem which is not discussed in this book, that is, the quadratic cost problem for partially observable systems. In this case admissible controls are those dependent only on observations. It is known that the problem can be decomposed into two parts, filtering and control, and this fact is known as the separation principle; see Curtain and Ichikawa [1] and Curtain [2]. After reducing the problems to those with complete observation one can use the results in Section 5.2 and obtain an optimal feedback control law on filters. The filtering part can be solved as a dual problem to the deterministic regulator problem. Hence, quadratic problems with incomplete observation may be solved using the approach in Section 5.2. A different formulation is also presented by Bensoussan and Viot [1] in which they give necessary and sufficient conditions of optimality for linear stochastic distributed parameter systems, with convex differentiable payoffs and partial observation.

In Theorem 5.3.1, the diffusion coefficient G is defined for all $x \in H$, which has the limitation that it does not allow unbounded operators. However, under some circumstances these stability results can be extended to the case as was shown in Li and Ahmed [1], [2]. The existence of an invariant distribution and the stationary solution for the 2-D stochastic Navier-Stokes problem was studied in Vishik and Fursikov [1] and, for periodic boundary conditions, in Albeverio and Cruzerio [1] by using the Galerkin approximation and method of averaging. Flandoli and Gatarek [1] considered rather general white noise and looked for solutions which are martingales or stationary for a class of stochastic Navier-Stokes equations. By the same techniques as in Section 3.9, the stabilization problem for a class of specific models, 2-D stochastic Navier-Stokes equations, was investigated in Caraballo, Langa and Taniguchi [1]. The material in Section 5.5 is mainly borrowed from Manthey and Maslowski [1].

Appendix

A   The Proof of Proposition 4.2.4

Proof   For any positive integer q, define

$$O(q) = \{h_j : h_j = n_1 r_1 + \cdots + n_m r_m, \ n_1 + \cdots + n_m = q\}.$$

An elementary counting argument leads to

$$\#O(q) = \binom{q + m - 1}{m - 1} = \frac{(q + m - 1)!}{q!(m - 1)!}.$$

Let

$$\sum(q) = \sum_{j=1}^{q} \binom{j + m - 1}{m - 1}.$$

Since

$$\binom{j + m - 1}{m - 1} = \frac{j^{m+1}(1 + \frac{1}{j}) \cdots (1 + \frac{m-1}{j})}{(m - 1)!} < \frac{m^m}{(m - 1)!} j^{m-1},$$

we have

$$\sum(q) \leq \frac{m^m}{(m - 1)!} \int_1^{q+1} x^{m-1} dx < \frac{m^m}{m!}(q + 1)^m.$$

We now claim that if $h_\alpha = q r_1$ for a positive integer q, then $\alpha \leq \sum(q)$. Suppose the contrary: there exist at least $\sum(q)$ terms of $\{h_j\}$ which are less than $q r_1$. This implies the existence of an h_j such that

$$q r_1 > h_j = n_1 r_1 + \cdots + n_m r_m, \quad n_1 + \cdots + n_m \geq q.$$

This is impossible since

$$q r_1 > h_j \geq r_1(n_1 + \cdots + n_m) \geq q r_1.$$

Hence,

$$\alpha \leq \sum(q) \leq \frac{m^m}{m!}(q + 1)^m \leq m^m(q + 1)^m.$$

Observe that $\sum(q) < \sum(q+1)$ for all q. We claim that for any $n \in [\sum(q), \sum(q+1)]$, $h_n \geq qr_1$. This is true. Indeed, note that if $h_n < qr_1$, then $h_n < qr_1 = h_\alpha$, where $n < \alpha \leq \sum(q)$ which is impossible. Thus, for this n,

$$\frac{\ln n}{h_n} \leq \ln \sum(q+1) \leq \frac{m \ln m(q+1)}{r_1 q}.$$

On the other hand, for arbitrarily given $\varepsilon > 0$, there exists q_0 such that $(m \ln m(q+2))/(r_1 q)$ is decreasing for $q \geq q_0$ and less then ε. Hence for that q_0 and $n \geq \sum(q_0)$, $\frac{\ln n}{h_n} < \varepsilon$. This completes the proof. $\square$

B   Existence and Uniqueness of Strong Solutions of Stochastic Delay Differential Equations

Theorem   *In addition to (a)–(e) of (4.3.2) to (4.3.4), assume that ϕ and τ satisfy the hypotheses at the beginning of Section 4.3. Then, there exists a unique strong solution to (4.3.1) on $[0, T]$ for all $T \geq 0$.*

Proof   Uniqueness. Suppose that $u(t)$ and $v(t)$ are two strong solutions of (4.3.1) on $[0, T]$. Then, writing $\rho(t) = t - \tau(t)$, $t \geq 0$, it follows that

$$u(t) - v(t) = \int_0^t \Big(A(s, u(s), u(\rho(s))) - A(s, v(s), v(\rho(s))) \Big) ds$$

$$+ \int_0^t \Big(B(s, u(s), u(\rho(s))) - B(s, v(s), v(\rho(s))) \Big) dW_s, \quad \forall t \in [0, T].$$

Now, Itô's formula, the condition (c) in Section 4.3 and the fact $u(\rho(t)) = v(\rho(t))$ as $\rho(t) \leq 0$ yield

$$E\|u(t) - v(t)\|_H^2$$

$$= 2 \int_0^t E\langle u(s) - v(s), A(s, u(s), u(\rho(s))) - A(s, v(s), v(\rho(s)))\rangle_{V, V^*} ds$$

$$+ \int_0^t E\|B(s, u(s), u(\rho(s))) - B(s, v(s), v(\rho(s)))\|_{\mathcal{L}_2^0}^2 ds$$

$$\leq \lambda \Big[\int_0^t E\|u(s) - v(s)\|_H^2 ds + \int_0^t E\|u(\rho(s)) - v(\rho(s))\|_H^2 ds \Big]$$

$$\leq \lambda \Big[\int_0^t E\|u(s) - v(s)\|_H^2 ds + \int_{\rho(0)}^{\rho(t)} E\|u(s) - v(s)\|_H^2 ds \Big]$$

$$\leq \lambda\Big[\int_0^t E\|u(s)-v(s)\|_H^2 ds + \int_0^t E\|u(s)-v(s)\|_H^2 ds\Big], \quad \forall t \in [0,T],$$

and then a Gronwall's lemma type argument yields the required uniqueness.

Existence. First of all, notice that since $\tau'(t) \leq 0$ and $\tau(t) \in [0,r]$ for all $t \geq 0$, there exist only three possible situations:

Case i): $\lim_{t\to\infty} \tau(t) = \delta > 0$.

Case ii): $\lim_{t\to\infty} \tau(t) = 0$ but $\tau(t) > 0$ for all $t \geq 0$.

Case iii): There exists $T^* > 0$ such that $\tau(t) > 0$ for $t \in [0,T^*)$ and $\tau(t) = 0$ for $t \geq T^*$.

Let us analyze each of them separately:

Case i): As $\tau(t) \geq \delta$ for all $t \geq 0$, we get that $\rho(t) \leq t - \delta$ for all $t \geq 0$. So, $\rho(t) \leq t - \delta \leq 0$ for $t \in [0,\delta]$ and therefore the problem on $[0,\delta]$ can be rewritten as

$$u(t) = \phi(0) + \int_0^t A(s,u(s),\phi(\rho(s)))ds + \int_0^t B(s,u(s),\phi(\rho(s)))dW_s, \quad t \in [0,\delta],$$
$$u(t) = \phi(t), \quad t \in [-r,0],$$

which is a nondelay problem. Now, observe that in the case without delays considered by Pardoux [1], the existence of strong solutions is proved under the following similar assumptions to (a)–(e). In fact, consider $A(t,\cdot) : V \to V^*$, a family of nonlinear operators, and $B(t,\cdot) : V \to \mathcal{L}(K,H)$, satisfying

(a)' (Coercivity). There exist $\alpha > 0$, $p > 1$ and $\lambda, \gamma \in \mathbf{R}^1$ such that:

$$2\langle x, A(t,x)\rangle_{V,V^*} + \|B(t,x)\|_{\mathcal{L}_2^0}^2 \leq -\alpha\|x\|_V^p + \lambda\|x\|_H^2 + \gamma, \quad \forall x \in V.$$

(b)' (Growth). There exists $c > 0$ such that

$$\|A(t,x)\|_{V^*} \leq c(1 + \|x\|_V^{p-1}), \quad \forall x \in V.$$

(c)' (Measurability). $t \in (0,T) \to A(t,x) \in V^*$ is Lebesgue-measurable $\forall x \in V$, $\forall T > 0$.

(d)' (Continuity). $\xi \in \mathbf{R}^1 \to \langle v, A(t, x + \xi y)\rangle_{V,V^*} \in \mathbf{R}^1$ is continuous for all x, y, $v \in V$.

(e)' (Monotonicity). For all x, $y \in V$,

$$2\langle x - y, A(t,x) - A(t,y)\rangle_{V,V^*} + \|B(t,x) - B(t,y)\|_{\mathcal{L}_2^0}^2 \leq \lambda\|x - y\|_H^2.$$

(f)' There exists $k > 0$ such that

$$\|B(t,x) - B(t,y)\|_{\mathcal{L}_2^0}^2 \leq k\|x - y\|_V^2, \quad \forall x, y \in V.$$

(g)' $t \in (0,T) \to B(t,x) \in \mathcal{L}(K,H)$ is Lebesgue-measurable $\forall x \in V$, $\forall T > 0$.

However, it is not difficult to check that the proofs in Pardoux [1] are also valid if one assumes some integral versions of the hypotheses (a)', (b)', (e)' and (f)'. In fact, it is sufficient to make the following assumptions instead of (a)', (b)', (e)' and (f)':

(A) There exist $\alpha > 0$, $p > 1$ and λ, $\gamma \in \mathbf{R}^1$ such that for all $w \in L^p(\Omega \times (0,T); V) \cap L^2(\Omega \times (0,T); H)$ and all $t \in [0,T]$,

$$2 \int_0^t E\langle w_s, A(s, w_s)\rangle_{V,V^*} ds + \int_0^t E\|B(s, w_s)\|^2_{\mathcal{L}^0_2} ds$$

$$\leq -\alpha \int_0^t E\|w_s\|^p_V \, ds + \lambda \int_0^t E\|w_s\|^2_H \, ds + \gamma t;$$

(B) There exist positive constants c_1, $c_2 > 0$ such that for all $w \in L^p(\Omega \times (0,T); V) \cap L^2(\Omega \times (0,T); H)$,

$$\int_0^T E\|A(t, w_t)\|^{p/(p-1)}_{V^*} \, dt \leq \int_0^T (c_1 E\|w_t\|^p_V + c_2) \, dt;$$

(E) For all w^1, $w^2 \in L^p(\Omega \times (0,T); V) \cap L^2(\Omega \times (0,T); H)$ and all $t \in [0,T]$,

$$2 \int_0^t E\langle w^1_s - w^2_s, A(s, w^1_s) - A(s, w^2_s)\rangle_{V,V^*} ds + \int_0^t E\|B(s, w^1_s) - B(s, w^2_s)\|^2_{\mathcal{L}^0_2}$$

$$\leq \lambda \int_0^t E\|w^1_s - w^2_s\|^2_H ds.$$

(F) There exists $k > 0$ such that for all w^1, $w^2 \in L^p(\Omega \times (0,T); V) \cap L^2(\Omega \times (0,T); H)$ and all $t \in [0,T]$,

$$\int_0^t E\|B(s, w^1_s) - B(s, w^2_s)\|^2_{\mathcal{L}^0_2} \, ds \leq k \int_0^t E\|w^1_s - w^2_s\|^2_V ds.$$

Let $A_1(t, w_t) = A(t, w(t), \phi(\rho(t)))$ and $B_1(t, w_t) = B(t, w(t), \phi(\rho(t)))$ for $w \in L^p(\Omega \times (0,T); V) \cap L^2(\Omega \times (0,T); H)$ and $t \in [0, \delta]$. Our existence result will hold if we can prove that A_1 and B_1 satisfy (A), (B), (E) and (F). But this follows immediately from the assumptions (a), (b), (c) and (e) in Section 4.3.

Let us prove (A), for instance. Indeed, for $w. \in L^p(\Omega \times (0,\delta); V) \cap L^2(\Omega \times (0,\delta); H)$ and $t \in [0, \delta]$, we obtain

$$2 \int_0^t E\langle w_s, A_1(s, w_s)\rangle_{V,V^*} + \int_0^t E\|B_1(s, w_s)\|^2_{\mathcal{L}^0_2} ds$$

$$\leq -\alpha \int_0^t E\|w_s\|^p_V ds + \lambda \int_0^t E\|w_s\|^2_H ds + \theta \int_0^t E\|\phi(\rho(s))\|^2_H ds + \gamma t$$

$$\leq -\alpha \int_0^t E\|w_s\|^p_V ds + \lambda \int_0^t E\|w_s\|^2_H ds + \left(\beta^2|\theta| \sup_{-r \leq s \leq 0} E\|\phi(s)\|^2_H + \gamma\right) t.$$

Now, since (B), (E) and (F) can be similarly proved, we then obtain the existence of a strong solution on $[0, \delta]$. By induction, the problem can be solved on $[n\delta, (n+1)\delta]$ for all natural numbers $n \geq 0$ and therefore on $[0, \infty)$.

Case ii): In this case, we can choose a strictly increasing sequence $\{\delta_n\}$ such that $\delta_n \to \infty$, $\delta_0 = 0$, $\rho(\delta_{n+1}) = \delta_n$ and $\rho(t) \in [\delta_{n-1}, \delta_n]$ for all $t \in [\delta_n, \delta_{n+1}]$ and for all $n \geq 0$, where we write $\delta_{-1} = -r$. Consequently, our equation can be solved on each $[\delta_n, \delta_{n+1}]$ exactly as in Case i), and further on $[0, \infty)$.

Case iii): Firstly, we can prove the existence of a strong solution on $[0, T^*)$ in the same way as Case ii). Then, it is not difficult to show that the solution $u(t)$ tends to certain $u(T^*) \in L^2(\Omega, \mathcal{F}_{T^*}, P; H)$, as $t \to T^*$. Now, on $[T^*, \infty)$ the problem becomes

$$u(t) = u(T^*) + \int_{T^*}^t A(s, u(s), u(s))ds + \int_{T^*}^t B(s, u(s), u(s))dW_s,$$

which obviously has a unique strong solution and our proof is complete. □

In the deterministic framework, there exists a large literature on the existence of different kinds of solutions to functional differential equations; see, for instance, Ruess [1] for a comprehensive description of recent results. For the stochastic case in a variational setting, Real [1] investigated the existence and uniqueness for a class of linear systems. In a similar spirit to Pardoux [1], Caraballo, Liu and Truman [1] and Caraballo, Garrido-Atienza and Real [1] established conditions to ensure existence and uniqueness of solutions of general stochastic functional differential equations.

References

N. U. Ahmed.

[1] *Semigroups Theory with Applications to Systems and Control.* Longman Scientific and Technical, (1991).

S. Albeverio and A. B. Cruzerio.

[1] Global flows with invariant (Gibbs) measures for Euler and Navier-Stokes two dimensional fluids. *Comm. Math. Phys.* **129**, (1990), 432-444.

L. Arnold.

[1] *Stochastic Differential Equation: Theory and Applications.* Wiley, New York, (1974).
[2] A formula connecting sample and moment stability of linear stochastic systems. *SIAM J. Appl. Math.* **44**, (1984), 793-802.
[3] A new example of an unstable system being stabilized by random parameter noise. *Infor. Comm. Math. Chem.* (1979), 133-140.
[4] Stabilization by noise revisited. *Z. Angew. Math. Mech.* **70**, (1990), 235-246.
[5] *Random Dynamical Systems.* Springer-Verlag, New York, (1998).

L. Arnold, H. Crauel and V. Wihstutz.

[1] Stabilization for linear systems by noise. *SIAM J. Control Optim.* **21**, (1983), 451-461.

L. Arnold, W. Kliemann and E. Oeljeklaus.

[1] Lyapunov exponents of linear stochastic systems. *Lecture Notes in Math.* **1186**, Springer-Verlag, (1984), 85-128.

L. Arnold, E. Oeljeklaus and E. Pardoux.

[1] Almost sure and moment stability for linear Itô equations. *Lecture Notes in Math.* **1186**, Springer-Verlag, (1984), 129-159.

L. Arnold and V. Wihstutz.

[1] Lyapunov exponents: A survey. *Lecture Notes in Math.* **1186**, Springer-Verlag, (1984), 1-26.

A. V. Balakrishnan.

[1] *Applied Functional Analysis.* Springer-Verlag, New York, (1976).

D. R. Bell and S. E. A. Mohammed.

[1] On the solution of stochastic ordinary differential equations via small delays. *Stochastics.* **29**, 293-299, (1989).

R. Bellman.

[1] *Introduction to Matrix Analysis.* McGraw-Hill, New York, (1969).

C. D. Benchimol.

[1] [1] Feedback stabilizability in Hilbert sapces. *Applied Math. Optim.* **4**, (1978), 225-248.

A. Bensoussan, M. Delfour and S. Mitter.

[1] The linear dimensional systems over an infinite horizon; survey and examples. Proceedings of IEEE Conference on: *Decision and Control,* (1976), 746-751.

A. Bensoussan and R. Temam.

[1] Équations aux dérivées partielles stochastiques non linéaires. *Israel J. Math.* **11**, (1972), 95-121.

[2] Equations stochastiques du type Navier-Stokes. *J. Funct. Anal.* **13**, (1973), 195-222.

A. Bensoussan and M. Viot.

[1] Optimal control of stochastic linear distributed parameter systems. *SIAM J. Control.* **13**, (1975), 904-926.

F. Browder.

[1] On the spectral theory of elliptic differential operators. *Ann. of Math.* **142**, (1964), 22-130.

Z. Brzeźniak, M. Capiński and F. Flandoli.

[1] Stochastic partial differential equations and turbulence. *Math. Mods. Meth. Appl. Sci.* **1**, (1991), 41-59.

J. M. Burgers.

[1] *The Nonlinear Diffusion Equation: Asymptotic Solutions and Statistical Problems.* Reidel, (1970).

A. G. Butkovskii.

[1] *Theory of Optimal Control of Distributed Parameter Systems.* Elsevier, New York, (1969).

P. Cannarsa and V. Vespri.

[1] Existence and uniqueness of solutions to a class of stochastic partial differential equations. *Stoch. Anal. Appl.* **3**, (1985), 315-339.

M. Capiński and N. Cutland.

[1] Stochastic Navier-Stokes equations. *Acta Appl. Math.* **25**, (1991), 59-85.

T. Caraballo.

[1] Asymptotic exponential stability of stochastic partial differential equations with delay. *Stochastics.* **33**, (1990), 27-47.

T. Caraballo, M. Garrido-Atienza and J. Real.

[1] Existence and uniqueness of solutions for delay stochastic evolution equations. *Stoch. Anal. Appl.* **20**, (2002), 1225-1256.

[2] Asymptotic stability of nonlinear stochastic evolution equations. *Stoch. Anal. Appl.* **21**, (2003), 301-327.

T. Caraballo and J. Langa.

[1] Comparison of the long-time behavior of linear Itô and Stratonovich partial differential equations. *Stoch. Anal. Appl.* **19**, (2001), 183-195.

T. Caraballo, J. Langa and T. Taniguchi.

[1] The exponential behaviour and stabilizability of stochastic 2D-Navier-Stokes equations. *J. Differential Equas.* **179**, (2002), 714-737.

T. Caraballo and K. Liu.

[1] On exponential stability criteria of stochastic partial differential equations. *Stoch. Proc. Appl.* **83**, (1999), 289-301.

[2] Exponential stability of mild solutions of stochastic partial differential equations with delays. *Stoch. Anal. Appl.* **17**, (1999), 743-764.

[3] Asymptotic exponential stability property for diffusion processes driven by stochastic differential equations in duals of nuclear spaces. *Publ. RIMS, Kyoto Univ.* **37**, (2001), 239-254.

T. Caraballo, K. Liu and X. R. Mao.

[1] Stabilization of partial differential equations by stochastic noise. *Nagoya Math. J.* **161**, (2001), 155-170.

T. Caraballo, K. Liu and A. Truman.

[1] Stochastic functional partial differential equations: existence, uniqueness and asymptotic stability. *Proc. Royal Soc. London* **A**. **456**, (2000), 1775-1802.

M. Chappell.

[1] Bounds for average Lyapunov exponents of gradient stochastic systems. *Lecture Notes in Math.* **1186**, Springer-Verlag, (1984), 292-307.

A. Chojnowska-Michalik and B. Goldys.

[1] Existence, uniqueness and invariant measures for stochastic semilinear equations on Hilbert spaces. *Probab. Theory Relat. Fields.* **102**, (1995), 331-356.

P. L. Chow.

[1] Function-space differential equations associated with a stochastic partial differential equation. *Indiana Univ. Math. J.* **25**, (1976), 609-627.

[2] Stability of nonlinear stochastic evolution equations. *J. Math. Anal. Appl.* **89**, (1982), 400-419.

P. L. Chow and J. L. Jiang.

[1] Almost sure convergence of some approximate solutions for random parabolic equations. Proceeding of the conference on: *Randon partial differential equations*, U. Hornung, P. Kotelenze. al. (eds.) Vol. **102**. (1991), 45-54.

P. L. Chow and R. Z. Khas'minskii.

[1] Stationary solutions of nonlinear stochastic evolution equations. *Stoch. Anal. Appl.* **15(5)**, (1997), 671-699.

I. Chueshov and P. Vuillermot.

[1] Long time behavior of solutions to a class of stochastic parabolic equations with homogeneous white noise: Itô's case. *Stoch. Anal. Appl.* **18**, (2000), 581-615.

[2] Long time behavior of solutions to a class of stochastic parabolic equations with homogeneous white noise: Stratonovitch's case. *Probab. Theory Relat. Fields.* **112**, (1998), 149-202.

M. G. Crandall.

[1] Nonlinear semigroups and evolution governed by accretive operators. *Proc. Symp. Pure Math. Amer. Math. Soc.* **45**, (1976), 305-337.

M. G. Crandall and T. Liggett.

[1] Generation of semi-groups of nonlinear transformations on general Banach spaces. *Amer. J. Math.* **93**, (1971), 265-298.

R. F. Curtain.

[1] Stability of stochastic partial differential equation. *J. Math. Anal. Appl.* **79**, (1981), 352-369.

[2] Estimation and stochastic control for linear infinite dimensional systems. In A. T. Bharucha-Reid (Ed.), *Probabilistic Analysis and Related Topics*, Vol. 1, Academic Press, New York, (1978), 45-86.

[3] Stochastic distributed systems with point observation and boundary control: An abstract theory, *Stochastics*. **3**, (1979), 85-104.

R. F. Curtain and P. L. Falb.

[1] Itô lemma in infinite dimensions. *J. Math. Anal. Appl.* **31**, (1970), 434-448.

R. F. Curtain and A. Ichikawa.

[1] The separation principle for stochastic evolution equations *SIAM. J. Control Optim.* **15**, (1977), 367-383.

R. F. Curtain and A. Pritchad.

[1] *Infinite Dimensional Linear Systems Theory.* Lecture Notes in Control and Information Science, **8**, Springer-Verlag, Berlin, (1978).

R. F. Curtain and H. J. Zwart.

[1] *Introduction to Infinite Dimensional Linear Systems Theory.* (Texts in Applied Math.) Springer-Verlag, (1985).

C. M. Dafermos.

[1] Asymptotic stability in viscoelasticity. *Arch. Rational Mech. Anal.* **37**, (1970), 297-308.

G. Da Prato.

[1] Some results on linear stochastic evolution equations in Hilbert spaces by semigroup method. *Stoch. Anal. Appl.* **1**, (1983), 57-88.

G. Da Prato and D. Gątarek.

[1] Stochastic Burgers equation with correlated noise. *Stochastics*. **52**, (1995), 29-41.

G. Da Prato, D. Gątarek and J. Zabczyk.

[1] Invariant measures for semilinear stochastic equations, *Stoch. Anal. Appl.* **10**, (1992), 387-408.

G. Da Prato, M. Iannelli and L. Tubaro.

[1] Semilinear stochastic differential equations in Hilbert spaces. *Boll. Un. Mat. It.* (**5**), (1979), 168-185.

G. Da Prato and A. Ichikawa.

[1] [1] Stability and quadratic control for linear stochastic equations with unbounded coefficients. *Boll. Un. Mat. It.* **B4**, (1985), 987-1001.

G. Da Prato and J. Zabczyk.

[1] *Stochastic Equations in Infinite Dimensions*. Encyclopedia of Mathematics and its Applications, Cambridge University Press, (1992).

[2] *Ergodicity for Infinite Dimensional Systems*. London Mathematical Society Lecture Note Series. **229**, Cambridge University Press, (1996).

L. Daletskii and G. Krein.

[1] *Stability of solutions of differential equations in Banach spaces*. Trans. Amer. Math. Soc. **43**, Providence, R.I. (1974).

R. Datko.

[1] Extending a theorem of A. Lyapunov to Hilbert space. *J. Math. Anal. Appl.* **32**, (1970), 610-616.

[2] Uniform asymptotic stability of evolutionary processes in a Banach space. *SIAM J. Math. Anal.* **3**, (1973), 428-445.

[3] Representation of solutions and stability of linear differential-difference equations in a Banach space. *J. Differential Equas.* **29**, (1978), 105-166.

[4] Lyapunov functionals for certain linear delay differential equations in a Hilbert space. *J. Math. Anal. Appl.* **76**, (1980), 37-57.

[5] The uniform exponential stability of a class of linear differential-difference equations in a Hilbert space. *Proc. Royal Soc. Edinburgh,* **89A**, (1981), 201-215.

[6] Remarks concerning the asymptotic stability and stabilization of linear delay differential equations. *J. Math. Anal. Appl.* **111**, (1985), 571-584.

[7] An example of an unstable neutral differential equation. *International J. Control.* **20**, (1983), 263-267.

[8] The Laplace transform and the integral stability of certain linear processes. *J. Differential Equas.* **48**, (1983), 386-403.

E. B. Davies.

[1] *Spectral Theory and Differential Operators.* (Cambridge Studies in Advanced Mathematics, Vol. 42), Cambridge University Press, (1996).

D. A. Dawson.

[1] Stochastic evolution equations and related measures processes. *J. Multivariate Anal.* **5**, (1975), 1-52.

M. Delfour, C. McCalla and S. Mitter.

[1] Stability and the infinite-time quadratic cost problem for linear hereditary differential systems. *SIAM J. Control.* **13**, (1975), 48-88.

C. Donati-Martin and E. Pardoux.

[1] White noise driven SPDEs with reflection. *Probab. Theory Relat. Fields.* **95**, (1993), 413-425.

A. Drozdov.

[1] [1] Stability of a class of stochastic integro-differential equations. *Stoch. Anal. Appl.* **13(5)**, (1995), 517-530.

N. Dunford and J. T. Schwartz.

[1] *Linear Operators, Part I.* Interscience Publishers, (1958).

S. N. Ethier and T. G. Kurtz.

[1] *Markov Processes. Characterization and Convergence.* Wiley and Sons, New York, (1986).

F. Flandoli.

[1] Dissipativity and invariant measures for stochastic Navier-Stokes equations. *NoDEA.* **1**, (1994), 403-423.

F. Flandoli and D. Gątarek.

[1] Martingale and stationary solutions for stochastic Navier-Stokes equations. *Probab. Theory Related Fields.* **102**, (1995), 367-406.

F. Flandoli and B. Maslowski.

[1] Ergodicity of the 2-D Navier-Stokes equations under random perturbations. *Comm. Math. Phys.* **171**, (1995), 119-141.

A. Friedman.

[1] *Partial Differential Equations.* Holt, Rinehart and Winston, New York, (1969).

[2] *Stochastic Differential Equations and Applications.* Vols. 1 and 2, Academic Press, New York, (1975).

G. Geiss.

[1] *The Analysis and Design of Nonlinear Control Systems via Lyapunov's Direct Method.* Grumman Aircraft Corp. Rept. RTD-TDR-63-4076, (1964).

I. J. Gihman and A. V. Skorohod.

[1] *Stochastic Differential Equations.* Springer-Verlag, New York, (1972).

E. Govindan.

[1] Existence and stability of solutions of stochastic semilinear functional differential equations. *Stoch. Anal. Appl.* **20**, (2002), 1257-1280.

G. Greiner, J. Voigt and M. Wolff.

[1] On the spectral bound of the generator of semigroups of positive operators. *J. Operator Theory.* **5**, (1981), 245-256.

I. Gyöngy.

[1] On the approximation of stochastic partial differential equations. I. *Stochastics.* **25**, (1988), 59-86.

[2] On the approximation of stochastic partial differential equations. II. *Stochastics.* **26**, (1989), 129-174.

[3] The stability of stochastic partial differential equations and applications. Theorems on supports. *Lecture Notes in Math.* **1390**, G. Da Prato and L. Tubaro (eds), Springer-Verlag, (1989), 91-118.

W. Hahn.

[1] *Stability of Motion.* Springer-Verlag, New York, (1967).

J. Hale.

[1] *Theory of Functional Differential Equations.* Springer-Verlag, New York, (1977).

U. G. Haussmann.

[1] Asymptotic stability of the linear Itô equation in infinite dimensional. *J. Math. Anal. Appl.* **65**, (1978), 219-235.

N. D. Hayes.

[1] Roots of the transcendental equation associated with a certain differential difference equations. *J. London Math. Soc.* **25**, (1950), 226-232.

D. Henry.

[1] *Geometric Theory of Semilinear Parabolic Equations.* Lecture Notes in Math. **840,** Springer-Verlag, Berlin, (1984).

E. Hille and R. S. Phillips.

[1] *Functional Analysis and Semigroups.* American Mathematical Society. Providence, (1957).

A. Ichikawa.

[1] Optimal control of a linear stochastic evolution equation with state and control dependent noise, Proc. IMA Conference, *Recent Theoretical Developments in Control,* (Leicester, England, 1976), Academic Press.
[2] Dynamic programming approach to stochastic evolution equations. *SIAM J. Control Optim.* **17**, (1979), 152-174.
[3] Stability of semilinear stochastic evolution equations. *J. Math. Anal. Appl.* **90**, (1982), 12-44.
[4] Absolute stability of a stochastic evolution equation. *Stochastics.* **11**, (1983), 143-158.

[5] Semilinear stochastic evolution equations: boundedness, stability and invariant measure. *Stochastics.* **12**, (1984), 1-39.

[6] Equivalence of L_p stability and exponential stability for a class of non-linear semilinear semigroups. *Nonlinear Analysis, Theory, Methods and Applications.* **8**, (1984), 805-815.

[7] A semigroup model for parabolic equations with boundary and pointwise noise. In: *Stochastic Space-Time Models and Limit Theorems*, D. Reidel Publishing Company, (1985), 81-94.

[8] Stability of parabolic equations with boundary and pointwise noise. In: *Stochastic Differential Systems*, (Proceedings), Lecture Notes in Control and Information Sci. **69**, Springer-Verlag, Berlin, (1985), 55-66.

A. Ichikawa and A. J. Pritchard.

[1] Existence, uniqueness and stability of nonlinear evolution equations. *J. Math. Anal. Appl.* **68**, (1979), 454-476.

N. Ikeda and S. Watanabe.

[1] *Stochastic Differential Equations and Diffusion Processes.* North-Holland and Kodansha, Tokyo, (1981).

R. Jahanipur.

[1] Stability of stochastic delay evolution equations with monotone nonlinearity. *Stoch. Anal. Appl.* **21**, (2003), 161-181.

G. Kallianpur, I. Mitoma and R. L. Wolpert.

[1] Diffusion equation in duals of nuclear spaces. *Stochastics.* **29**, (1990), 1-45.

G. Kallianpur and V. Perez-Abreu.

[1] Stochastic evolution equations driven by nuclear-space-valued martingale. *Applied Math. Optim.* **17**, (1988), 237-272.

G. Kallianpur and R. L. Wolpert.

[1] Infinite dimensional stochastic differential equation models for spatially distributed neurons. *Applied Math. Optim.* **12**, (1984), 125-172.

G. Kallianpur and J. Xiong.

[1] *Stochastic Differential Equations in Infinite Dimensional Spaces.* Lecture Notes Monograph Series, Vol. 26, Inst. Math. Statis. (1995).

L. V. Kantorovich and G. P. Akilov.

[1] *Functional Analysis.* Second Edition, Pergamon Press, Oxford, (1982).

I. Karatzas and S. E. Shreve.

[1] *Brownian Motion and Stochastic Calculus.* Springer-Verlag, Berlin Heidelberg, New York, 2nd ed., (1991).

N. EI Karoui, S. Peng and M. C. Quenez.

[1] Backward stochastic differential equations in finance. *Math. Finance.* **7**, (1997), 1-71.

V.B. Kolmanovskii and V.R. Nosov.

[1] *Stability of Functional Differential Equations.* Academic Press, New York, (1986).

V.B. Kolmanovskii and A. Myshkis.

[1] *Introduction to the Theory and Applications of Functional Differential Equations.* Kluwer Academic Publishers, (1999).

R. Khas'minskii.

[1] *Stochastic Stability of Differential Equations.* (Sijthoff and Noordfoff), (1980).

[2] On robustness of some concepts in stability of SDE. *Stochastic Modeling and Nonlinear Dynamics*, (W. Kliemann and N.S. Namachchivaya, eds.), CRC Press, (1996), 131-137.

R. Khas'minskii and V. Mandrekar.

[1] On the stability of solutions of stochastic evolution equations. *The Dynkin Festsch. Prog. Probab.* **34**, Birkhäuser, Boston, MA, (1994), 185-197.

F. Kozin.

[1] On almost surely asymptotic sample properties of diffusion processes defined by stochastic differential equations, *J. Math. Kyoto Univ.* **4**, (1965), 515-528.

[2] Stability of the linear stochastic system. *Lecture Notes in Math.* **294**, Sringer-Varlag, (1972).

N. V. Krylov and B. L. Rozovskii.

[1] On Cauchy problem for linear stochastic partial differential equations, *Math. USSR Izvestija*, **11**, (1977), 1267-1284.

[2] Stochastic evolution equations. *J. Sov. Math.* **16**, (1981), 1233-1277.

H. Kunita.

[1] *Stochastic Flows and Stochastic Differential Equations.* Cambridge University Press, (1990).

[2] Stochastic partial differential equations connected with non-linear filtering. *Lecture Notes in Math.* **972**, Springer-Verlag, (1982), 100-169.

H. Kushner.

[1] *Stochastic Stability and Control.* Academic, New York, (1967).

[2] *Introduction to Stochastic Control Theory.* Holt, Rinehart and Winston, New York, (1971).

A. Kwiecińska.

[1] Stabilization of partial differential equations by noise. *Stoch. Proc. Appl.* **79**, (1999), 179-184.

[2] Stabilization of evolution equations by noise. *Proc. Amer. Math. Soc.* **130**, (2001), 3067-3074.

B. Lapeyre.

[1] A priori bound for the superemum of solution of stable stochastic differential equations. *Stochastics.* **28**, (1989), 145-160.

I. Lasiecka.

[1] Unified theory for abstract parabolic boundary problems: A semigroup approach, *Applied Math. Optim.* **6**, (1980), 281-333.

I. Lasiecka and R. Triggiani.

[1] Feedback semigroups and cosine operators for boundary feedback parabolic and hyperbolic equations, *J. Differential Equas.* **47**, (1983), 246-272.

[2] Dirichlet boundary control problem for parabolic equations with quadratic cost: Analyticity and Riccati feedback synthesis, *SIAM J. Control Optim.* **21**, (1983), 41-67.

G. Leha and G. Ritter.

[1] Lyapunov type of conditions for stationary distributions of diffusion processes on Hilbert spaces. *Stochastics.* **48**, (1994), 195-225.

G. Leha, G. Ritter and B. Maslowski.

[1] Stability of solutions to semilinear stochastic evolution equations. *Stoch. Anal. Appl.* **17**, (1999), 1009-1051.

S. M. Lenhart and C .C. Travis.

[1] Stability of functional partial differential equations. *J. Differential Equas.* **58**, (1985), 212-227.

J. J. Levin and J. Nohel.

[1] On a nonlinear delay equation. *J. Math. Anal. Appl.* **8**, (1964), 31-44.

P. Li and N. U. Ahmed.

[1] Feedback stabilization of some nonlinear stochastic systems on Hilbert space. *Nonlinear Analysis, Theory, Methods and Applications.* **17**, (1991), 31-43.

[2] A note on stability of stochastic systems with unbounded perturbations. *Stoch. Anal. Appl.* **7(4)**, (1989), 425-434.

J. L. Lions.

[1] *Équations Differentielles, Opérationelles et Problèmes aux Limites.* Springer-Verlag, Berlin, New York, (1961).

[2] *Optimal Control of Systems Governed by Partial Differential Equations.* Springer-Verlag, Berlin, New York, (1971).

J. L. Lions and E. Magenes.

[1] *Nonhomogeneous Boundary Value Problems and Applications.* (I), (II) and (III). Springer-Verlag, Berlin, New York, (1972).

K. Liu.

[1] On stability for a class of semilinear stochastic evolution equations. *Stoch. Proc. Appl.* **70**, (1997), 219-241.

[2] Carathéodory approximate solutions for a class of semilinear stochastic evolution equations with time delays. *J. Math. Anal. Appl.* **220**, (1998), 349-364.

[3] Lyapunov functionals and asymptotic stability of stochastic delay evolution equations. *Stochastics.* **63**, (1998), 1-26.

[4] Almost sure growth bounds for infinite dimensional stochastic evolution equations. *Quart. J. Math. Oxford (2).* **50**, (1999), 25-35.

[5] Necessary and sufficient conditions for exponential stability and ultimate boundedness of systems governed by stochastic partial differential equations. *J. London Math. Soc.* **62,** (2000), 311-320.

[6] Some remarks on exponential stability of stochastic differential equations. *Stoch. Anal. Appl.* **19(1),** (2001), 59-65.

[7] Uniform stability of autonomous linear stochastic functional differential equations in infinite dimensions. *Stoch. Proc. Appl.* **115**, (2005), 1131-1165.

K. Liu and A. Y. Chen.

[1] Moment decay rates of solutions of stochastic differential equations. *Tohoku Math. J.* **53,** (2001), 81-93.

K. Liu and X. R. Mao.

[1] Exponential stability of non-linear stochastic evolution equations. *Stoch. Proc. Appl.* **78**, (1998), 173-193.

[2] Large time decay behaviour of dynamical equations with random perturbation features. *Stoch. Anal. Appl.* **19(2),** (2001), 295-327.

K. Liu and Y. F. Shi.

[1] Razumikhin-type stability theorems of stochastic functional differential equations in infinite dimensions. *Preprint.*

K. Liu and A. Truman.

[1] Moment and almost sure Lyapunov exponents of mild solutions of stochastic evolution equations with variable delays via approximation approaches. *J. Math. Kyoto Univ.* **41,** (2002), 749-768.

K. Liu and X.W. Xia.

[1] On the exponential stability in mean square of neutral stochastic functional differential equations. *Syst. & Control Letts.* **37**, (1999), 207-215.

R. Liu and V. Mandrekar.

[1] Ultimate boundedness and invariant measures of stochastic evolution equations. *Stochastics.* **56**, (1996), 75-101.

[2] Stochastic semilinear evolution equations: Lyapunov function, stability and ultimate boundedness. *J. Math. Anal. Appl.* **212**, (1997), 537-553.

Z. H. Luo, B. Z. Guo and O. Morgul.

[1] *Stability and Stabilization of Infinite Dimensional with Applications.* Springer Verlag, London, Berlin Heidelberg, (1999).

V. Mandrekar.

[1] On Lyapounov stability theorems for stochastic (deterministic) evolution equations. Proc. of the NATO-ASI School on: *Stochastic analysis and applications in physics*, (L. Streit, Ed.), (1994), 219-237.

A. Manitius.

[1] Optimal control of hereditary systems. In *Control Theory and Topics in Functional Analysis,* Vol. III, International Atomic Energy Agency, Vienna, (1976), 43-178.

R. Manthey and B. Maslowski.

[1] A random continuous model for two interacting populations, *Applied Math. Optim.* **45(2)**, (2002), 213-236.

R. Manthey and K. Mittmann.

[1] On the qualitative behaviour of the solution to a stochastic partial functional differential equation arising in population dynamics, *Stochastics.* **66**, (1999), 153-166.

R. Manthey and T. Zausinger.

[1] Stochastic evolution equations in $L_p^{2\nu}$, *Stochastics.* **66**, (1999), 37-85.

X. R. Mao.

[1] *Stability of Stochastic Differential Equations with respect to Semimartingales.* Longman Scientific and Technical, (1991).

[2] Stochastic stabilization and destabilization. *Syst. & Control Letts.* **23**, (1994), 279-290.

[3] Razumikhin-type theorems on exponential stability of stochastic functional differential equations. *Stoch. Proc. Appl.* **65**, (1996), 233-250.

[4] Razumikhin-type theorems on exponential stability of neutral stochastic functional differential equations, *SIAM J. Math. Anal.* **28(2)**, (1997), 389-401.

G. Maruyama.

[1] Continuous Markov processes and stochastic equations. *Rend. Circ. Mat. Palermo.* **4**, (1955), 48-90.

E. J. Mashane.

[1] *Stochastic Calculus and Stochastic Models.* Academic Press, New York, San Francisco, London, (1974).

B. Maslowski.

[1] Uniqueness and stability of invariant measures for stochastic differential equations in Hilbert spaces. *Stochastics.* **28**, (1989), 85-114.

[2] Stability of semilinear equations with boundary and pointwise noise. *Annali Scuola Normale Superiore di Pisa,* **IV**, (1995), 55-93.

M. Métivier.

[1] *Semimartingales.* Walter de Gruyter, Berlin New York, (1982).

M. Métivier and J. Pellaumail.

[1] *Stochastic Integration.* Academic Press, New York, (1980).

J. Milota.

[1] Stability and saddle-point property for a linear autonomous functional parabolic equations. *Comment, Math. Univ. Carolia.* **27**, (1986), 87-101.

Y. Miyahara.

[1] Ultimate boundedness of the systems governed by stochastic differential equations. *Nagoya Math. J.* **47**, (1972), 111-144.

[2] Invariant measures of ultimately bounded stochastic processes. *Nagoya Math. J.* **49**, (1973), 149-153.

V. Mizel and V. Trutzer.

[1] Stochastic hereditary equations: existence and asymptotic stability. *J. Integral Equations.* **7**, (1984), 1-72.

A. E. Mohammed.

[1] *Stochastic Functional Differential Equations.* Pitman, (1984).

[2] *Stochastic Differential Systems with Memory.* In preparation.

T. Morozan.

[1] Boundedness properties for stochastic systems. *Stability of Stochastic Dynamical Systems,* R. F. Curtain (Ed.) Lecture Notes in Math. **294**, Springer-Verlag, New York, (1972).

J. D. Murray.

[1] *Mathematical Biology I: An Introduction.* Springer-Verlag, New York, (2001).

R. D. Nussbaum.

[1] The radius of the essential spectrum. *Duke Math. J.* **37**, (1970), 473-478.

V. Oseledec.

[1] A multiplicative ergodic theorem Lyapunov characteristic number for dynamical systems. *Trans. Moscow Math. Soc.* **19**, (1969), 197-231.

E. Pardoux.

[1] Équations aux dérivées partielles stochastiques non linéaires monotones. Thesis, Université Paris XI, (1975).

[2] Stochastic partial differential equations and filtering of diffusion processes. *Stochastics.* **3**, (1979), 127-167.

A. Pazy.

[1] *Semigroups of Linear Operators and Applications to Partial Differential Equations.* (Applied Mathematical Sciences, Vol. 44). Springer Verlag, New York, (1983).

[2] On the applicability of Lyapunov's theorem in Hilbert space, *SIAM J. Math. Anal. Appl.* **3**, (1972), 291-294.

[3] Asymptotic behavior of the solution of an abstract evolution equation and some applications, *J. Differential Equas.* **4**, (1968), 493-509.

S. Peszat and J. Zabczyk.

[1] Strong Feller property and irreducibility for diffusions on Hilbert spaces. *Ann. Probab.* **23**, (1995), 157-172.

A. J. Pritchard and J. Zabczyk.

[1] Stability and stabilizability of infinite dimensional systems. *SIAM Review.* **23**, (1981), 25-52.

B. S. Razumikhin.

[1] On stability of systems with a delay. *Prikl. Mat. Meh.* **20**, (1956), 500-512.

[2] Application of Liapunov's methods to problems in stability of systems with a delay. *Automat. i Telemeh.* **21**, (1960), 740-749.

J. Real.

[1] Stochastic partial differential equations with delays. *Stochastics.* **8**, (1982), 81-102.

D. Revuz and M. Yor.

[1] *Continuous Martingales and Brownian Motion.* (3th edition), Springer Verlag, Berlin Heidelberg, (1999).

F. Riesz and B. Sz-Nagy.

[1] *Functional Analysis.* Frederick Ungar, New York, (1955).

B. L. Rozovskii.

[1] *Stochastic Evolution Systems: Linear Theory and Applications to Non-Linear Filtering.* Kluwer Academic Publishers, (1990).

W. Ruess.

[1] Existence of solutions to partial functional differential equations with delays. In: *Theory and Applications of Nonlinear Operators of Accretive and Monotone Type*, Lecture Notes Pure Appl. Math. **178**, Kartsatos, A.G. Ed.; Marcel-Dekker, New York, (1996), 259-288.

M. Scheutzow.

[1] Stabilization and destabilization by noise in the plane. *Stoch. Anal. Appl.* **11**, (1993), 97-113.

T. Shiga.

[1] Ergodic theorems and exponential decay of sample paths for certain interactive diffusion systems. *Osaka J. Math.* **29**, (1992), 789-807.

A. V. Skorohod.

[1] *Asymptotic Methods in the Theory of Stochastic Differential Equations.* Amer. Math. Soc. (1989).

V. A. Solonnikov.

[1] Estimate of Green tensors for certain boundary problems. *Dokl. Akad. Nauk. SSR.* **130**, (1960), 988-991.

O. J. Staffans.

[1] A neutral functional differential equation with stable D-operator is retarded. *J. Differential Equas.* **49**, (1983), 208-217.

T. Taniguchi.

[1] Asymptotic stability theorems of semilinear stochastic evolution equations in Hilbert spaces. *Stochastics.* **53**, (1995), 41-52.

[2] Moment asymptotic behavior and almost sure Lyapunov exponent of stochastic functional differential equations with finite delays via Lyapunov Razumikhin method. *Stochastics.* **58**, (1996), 191-208.

T. Taniguchi, K. Liu and A. Truman.

[1] Existence, uniqueness and asymptotic behavior of mild solutions to stochastic functional differential equations in Hilbert spaces. *J. Differential Equas.* **181**, (2002), 72-91.

R. Temam.

[1] *Infinite Dimensional Dynamical Systems in Mechanics and Physics.* (2th edition), Springer Verlag, New York, (1988).

C. C. Travis and G. F. Webb.

[1] Existence and stability for partial functional differential equations. *Trans. Amer. Math. Soc.* **200**, (1974), 395-418.

[2] Existence, stability and compactness in the α-norm for partial functional differential equations. *Trans. Amer. Math. Soc.* **240**, (1978), 129-143.

L. Tubaro.

[1] An estimate of Burkholder type for stochastic processes defined by the stochastic integral. *Stoch. Anal. Appl.* **2**, (1984), 187-192.

H. Tuckwell.

[1] *Stochastic Processes in the Neurosciences.* Society for Industrial and Applied Mathematics, Philadelphia, (1989).

M. Viot.

[1] *Solutions faibles aux équations aux dérivées partielles stochastiques non linéaires.* Thèse, Université Pierre et Marie Curie, (1976).

M. J. Vishik and A. V. Fursikov.

[1] *Mathematical Problems in Statistical Hydromechanics*. Kluwer Academic Publishers, (1988).

J. B. Walsh.

[1] An introduction to stochastic partial differential equations. École d'eté de Probabilité de Saint Flour XIV, ed. P.L. Hennequin, *Lecture Notes in Math.* **1180**. (1984), 265-439.

[2] A stochastic model of neural response. *Adv. Appl. Probab.* **13**, (1981), 231-281.

P. K. Wang.

[1] On the almost sure stability of linear stochastic distributed parameter dynamical systems. *A.S.M.E. Transactions, J. Applied Mechanics.* (1966), 182-186.

[2] Control of distributed parameter systems. In *Advances in Control Systems*, C. T. Leondes (Ed.), Vol. I, Academic Press, New York, (1964), 75-172.

W. M. Wonham.

[1] Random differential equations in control theory. *Probabilistic Methods in Applied Mathematics.* **2**, A. Bharucha-Reid (Ed.) Academic Press, New York, (1970), 131-212.

[2] Lyapunov criteria for weak stochastic stability, *J. Differential Equas.* **2**, (1966), 195-207.

J. H. Wu.

[1] *Theory and Applications of Partial Functional Differential Equations.* (Applied Mathematical Sciences, Vol. 119), Springer-Verlag, New York, (1996).

Y. Yavin.

[1] On the stochastic stability of a parabolic type system, *Int. J. Syst. Sci.* **5**, (1974), 623-632.

[2] On the modelling and stability of a stochastic distributed parameter system, *Int. J. Syst. Sci.* **6**, (1975), 301-311.

K. Yosida.

[1] *Functional Analysis.* (6th edition), Springer-Verlag, New York, (1980).

J. Zabczyk.

[1] A note on C_0 semigroups, *Bull. Polish Acad. Sci. (Math.)* **162**, (1975), 895-898.

[2] On stability of infinite dimensional stochastic systems. Probab. Theory, Z. Ciesislski (ED), Banach Center Publications, **5**, Warswa, (1979), 273-281.

[3] On decomposition of generators. *SIAM J. Control Optim.* **16**, (1978), 523-534.

[4] On optimal stochastic control of discrete-time systems in Hilbert space. *SIAM J. Control.* **13**, (1975), 1217-1234.

M. Zakai.

[1] On the ultimate boundedness of moments associated with solutions of stochastic differential equations. *SIAM J. Control.* **5**, (1967), 588-593.

[2] A Lyapunov criterion for the existence of stationary probability distributions for systems perturbed by noise. *SIAM J. Control.* **7**, (1969), 390-397.

Index

Milton Keynes UK
Ingram Content Group UK Ltd.
UKHW021619071024
449327UK00020BA/1113